Analysis of Heat Equations on Domains

London Mathematical Society Monograph Series

The London Mathematical Society Monographs Series was established in 1968. Since that time it has published outstanding volumes that have been critically acclaimed by the mathematics community. The aim of this series is to publish authoritative accounts of current research in mathematics and high-quality expository works bringing the reader to the frontiers of research. Of particular interest are topics that have developed rapidly in the last ten years but that have reached a certain level of maturity. Clarity of exposition is important and each book should be accessible to those commencing work in its field.

The original series was founded in 1968 by the Society and Academic Press; the second series was launched by the Society and Oxford University Press in 1983. In January 2003, the Society and Princeton University Press united to expand the number of books published annually and to make the series more international in scope.

Senior Editor: Brian Davies, King's College, London
Honorary Editors: Martin Bridson, Imperial College, London, and Peter Sarnak, Princeton University and Courant Institute, New York
Editorial Advisers: J. H. Coates, University of Cambridge, W. S. Kendall, University of Warwick, and János Kollár, Princeton University

Vol. 31, *Analysis of Heat Equations on Domains,* by El Maati Ouhabaz

ANALYSIS OF HEAT EQUATIONS ON DOMAINS

El Maati Ouhabaz

PRINCETON UNIVERSITY PRESS

PRINCETON AND OXFORD

Copyright 2005 by Princeton University Press

Published by Princeton University Press, 41 William Street,
Princeton, New Jersey 08540

In the United Kingdom: Princeton University Press, 3 Market Place,
Woodstock, Oxfordshire OX20 1SY

Library of Congress Control Number: 2004103591

ISBN: 0-691-12016-1

British Library Cataloging-in-Publication Data is available

This book has been composed in LaTex

The publisher would like to acknowledge the author of this volume for
providing the camera-ready copy from which this book was printed.

Printed on acid-free paper.

pup.princeton.edu

Printed in the United States of America

10 9 8 7 6 5 4 3 2 1

A mes parents,
A Zahra, Nora et Ilias

Contents

Preface

The influence of the theory of linear evolution equations upon developments in other branches of mathematics, as well as physical sciences, would be hard to exaggerate. The theory has a rich interplay with other subjects in functional analysis, stochastic analysis and mathematical physics. Of particular interest are evolution equations associated with second-order elliptic operators in divergence form. Such equations arise in many models of physical phenomena; the classical heat equation is a prototype example. They are also of interest for nonlinear analysis; the proof of existence of local solutions to many nonlinear partial differential equations uses linear theory.

The theory for self-adjoint second-order elliptic operators is well documented, and there is an increasing interest in the non-self-adjoint case. It is one of the aims of the present book to give a systematic study of L^p theory of evolution equations associated with non-self-adjoint operators

$$A = -\sum_{k,j} \frac{\partial}{\partial x_j}\left(a_{kj}\frac{\partial}{\partial x_k}\right) + \sum_k b_k \frac{\partial}{\partial x_k} + \frac{\partial}{\partial x_k}(c_k.) + a_0.$$

We consider operators with bounded measurable coefficients on arbitrary domains of Euclidean space. The sesquilinear form technique provides the right tool to define such operators, and associates them with analytic semigroups on L^2. We are interested in obtaining contractivity properties of these semigroups as well as Gaussian upper bounds on their associated heat kernels. Gaussian upper bounds are then used to prove several results in the L^p-spectral theory.

A special feature of the present book is that several important properties of semigroups are characterized in terms of verifiable inequalities concerning their sesquilinear forms. The operators under consideration are subject to various boundary conditions and do not need to be self-adjoint. We also consider second-order elliptic operators with possibly complex-valued coefficients. Such operators have attracted attention in recent years as their associated heat kernels do not have the same properties as those of their analogues with real-valued coefficients. This book is also motivated by new developments and applications of Gaussian upper bounds to spectral theory. A large number of the results given here have been proved during the last decade.

The approach using sesquilinear form techniques avoids heavy use of so-phisticated results from the theory of partial differential equations or Sobolev embedding properties for which smoothness of the boundary is required. On the other hand, as we consider heat equations on arbitrary domains, we shall not address regularity properties (with respect to the space variable) of their solutions.

This book is for researchers and graduate students who require an in-troductory text to sesquilinear form technique, semigroups generated by second-order elliptic operators in divergence form, heat kernel bounds, and their applications. It should also be of value for mathematical physicists. We tried to keep the text self-contained and most of the material needed is introduced here. A few standard results are stated without proofs, but we provide the reader with several references.

We now give an outline of the content of each chapter. Chapter 1 is de-voted to sesquilinear forms and their associated operators and semigroups. It provides the necessary background from functional analysis and evolu-tion equations. Most of the material on sesquilinear forms is known, but our presentation differs from that in other books on this topic. We give a sys-tematic account on the interplay between forms, operators, and semigroups. Chapter 2 is devoted to contractivity properties of semigroups associated with sesquilinear forms. We give criteria in terms of forms for positivity, irreducibility, L^∞-contractivity, and domination of semigroups. These cri-teria are obtained as simple consequences of a result on invariance of closed convex sets under the action of the semigroup (see Theorems 2.2 and 2.3). We also include a section on semigroups acting on vector-valued functions. All the results in this chapter are in the spirit of the famous Beurling-Deny criteria for sub-Markovian semigroups. Chapter 3 contains Kato type in-equalities for generators of sub-Markovian semigroups. For symmetric sub-Markovian semigroups, a partial description of the domain of the corre-sponding generator in L^p is given. Chapter 4 is devoted to uniformly elliptic operators of type A as above. We discuss some examples of boundary con-ditions and apply the criteria of Chapter 2 to describe precisely, in terms of the boundary conditions and the coefficients, when the semigroup generated by $-A$ is positive, irreducible, or L^p-contractive. Chapter 2 also gives the right tools to compare (in the pointwise sense) semigroups associated with two different divergence form operators. Some results are extended in Chap-ter 5 to the case of degenerate-elliptic operators. Gaussian upper bounds for heat kernels of uniformly elliptic operators are proved in Chapter 6. We prove sharp bounds for operators with real-valued symmetric principal co-efficients a_{kj}. Gaussian upper bounds are derived from L^p-contractivity re-sults together with a well-known perturbation argument due to E.B. Davies.

We also derive bounds for the time derivatives as well as weighted gradient estimates for heat kernels. In Chapter 7, we use Gaussian upper bounds to prove several spectral properties. This includes L^p-analyticity of the semigroup, p-independence of the spectrum, L^p-estimates for Schrödinger and wave type equations. Although the book is devoted to uniformly elliptic operators on domains of Euclidean space, this chapter is written in a general setting of abstract operators on domains of metric spaces. The framework includes uniformly elliptic operators on domains of Euclidean space or more general Riemannian manifolds, sub-Laplacians on Lie groups, or Laplacians on fractals. In the last chapter we review the Kato square root problem for uniformly elliptic operators. We include at the end of each chapter a section of notes where the reader can find references to the literature and supplementary information.

Acknowledgments: I wish to express my hearty thanks to the many colleagues and friends who have contributed to my understanding of the subject of this book. I want to thank Wolfgang Arendt, Pascal Auscher, Sonke Blunck, Thierry Coulhon, Brian Davies, Xuan Thinh Duong, Alan McIntosh, and Rainer Nagel for their help and encouragement. I'm grateful to Philippe Depouilly for his unstinting help with the many tasks involved in typing the manuscript.

Notation

$C_c(\Omega)$: The space of continuous functions with compact support in Ω.

$C_c^\infty(\Omega)$: The space of C^∞-functions with compact support in Ω.

$(C_c^\infty(\Omega))'$: The space of distributions on Ω.

$\mathrm{supp}(u)$: The support of the function u.

$\Sigma(\psi) := \{z \in \mathbb{C}, z \neq 0, |\arg z| < \psi\}$, $\mathbb{C}^+ := \Sigma(\frac{\pi}{2})$.

$u^+ := \sup(u, 0)$ the positive part of u, $u^- := \sup(-u, 0)$ the negative part.

$f \wedge g := \inf(f, g)$, $f \vee g := \sup(f, g)$.

$$\mathrm{sign}\, u(x) = \begin{cases} \frac{u(x)}{|u(x)|} & \text{if } u(x) \neq 0, \\ 0 & \text{if } u(x) = 0. \end{cases}$$

$\Re$: Real part, $\Im$: Imaginary part.

χ_Ω: Characteristic function of Ω.

$L^p(X, \mu, \mathbb{K})$: The classical Lebesgue spaces of functions with values in $\mathbb{K}$.

$\|.\|_p$: The norm of $L^p(X, \mu, \mathbb{K})$.

dx: Lebesgue measure.

$W^{s,p}$: Sobolev spaces.

$H^1(\Omega) := W^{1,2}(\Omega)$, $H_0^1(\Omega)$ is the closure of $C_c^\infty(\Omega)$ in $H^1(\Omega)$.

$D_i := \frac{\partial}{\partial x_i}$ and $\Delta = \frac{\partial^2}{\partial x_1^2} + \cdots + \frac{\partial^2}{\partial x_d^2}$ is the Laplacian.

$\mathcal{L}(E, F)$: The space of bounded linear operators from E into F. $\mathcal{L}(E) := \mathcal{L}(E, E)$.

$\|T\|_{\mathcal{L}(E,F)}$: The operator norm of T in $\mathcal{L}(E, F)$.

$\rho(A)$: Resolvent set of the operator A. $\sigma(A)$: Spectrum of A.

Chapter One

SESQUILINEAR FORMS, ASSOCIATED OPERATORS, AND SEMIGROUPS

1.1 BOUNDED SESQUILINEAR FORMS

Let H be a Hilbert space over $\mathbb{K} = \mathbb{C}$ or $\mathbb{R}$. We denote by $(.;.)$ the inner product of H and by $\|.\|$ the corresponding norm. Let $\mathfrak{a}$ be a sesquilinear form on H, i.e., $\mathfrak{a}$ is an application from $H \times H$ into $\mathbb{K}$ such that for every $\alpha \in \mathbb{K}$ and $u, v, h \in H$:

$$\mathfrak{a}(\alpha u + v, h) = \alpha \mathfrak{a}(u, h) + \mathfrak{a}(v, h) \text{ and } \mathfrak{a}(u, \alpha v + h) = \overline{\alpha}\mathfrak{a}(u, v) + \mathfrak{a}(u, h). \tag{1.1}$$

Here $\overline{\alpha}$ denotes the conjugate number of α. Of course, $\overline{\alpha} = \alpha$ if $\mathbb{K} = \mathbb{R}$ and in this case the form $\mathfrak{a}$ is then bilinear. For simplicity, we will not distinguish the two cases $\mathbb{K} = \mathbb{R}$ and $\mathbb{K} = \mathbb{C}$. We will use the sesquilinear term in both cases and also write conjugate, real part, imaginary part, and so forth of elements in $\mathbb{K}$ as if we had $\mathbb{K} = \mathbb{C}$. These quantities have their obvious meaning if $\mathbb{K} = \mathbb{R}$.

DEFINITION 1.1 *A sesquilinear form* $\mathfrak{a} : H \times H \to \mathbb{K}$ *is continuous if there exists a constant M such that*

$$|\mathfrak{a}(u, v)| \leq M\|u\|\|v\| \text{ for all } u, v \in H.$$

Every continuous form can be represented by a unique bounded linear operator. More precisely,

PROPOSITION 1.2 *Assume that* $\mathfrak{a} : H \times H \to \mathbb{K}$ *is a continuous sesquilinear form. There exists a unique bounded linear operator T acting on H such that*

$$\mathfrak{a}(u, v) = (Tu; v) \text{ for all } u, v \in H.$$

Proof. Fix $u \in H$ and consider the linear continuous functional

$$\phi(v) := \overline{\mathfrak{a}(u, v)}, \; v \in H.$$

By the Riesz representation theorem, there exists a unique vector $Tu \in H$, such that

$$\phi(v) = (v; Tu) \text{ for all } v \in H.$$

The fact that T is a linear and continuous operator on H follows easily from the linearity and continuity of the form $\mathfrak{a}$. The uniqueness of T is obvious. $\square$

The bounded operator T is the operator associated with the form $\mathfrak{a}$. One can study the invertibility of T (or its adjoint T^*) using the form. More precisely, the following basic result holds.

LEMMA 1.3 *(Lax-Milgram) Let $\mathfrak{a}$ be a continuous sesquilinear form on H. Assume that $\mathfrak{a}$ is coercive, that is, there exists a constant $\delta > 0$ such that*

$$\Re\mathfrak{a}(u, u) \geq \delta \|u\|^2 \text{ for all } u \in H.$$

Let ϕ be a continuous linear functional on H. Then there exists a unique $v \in H$ such that

$$\phi(u) = \mathfrak{a}(u, v) \quad \text{for all } u \in H.$$

Proof. It suffices to prove that the adjoint operator T^* is invertible on H. Indeed, by the Riesz representation theorem, there exists a unique $g \in H$ such that

$$\phi(u) = (u; g) \quad \text{for all } u \in H,$$

and hence by writing $g = T^*v$ for some $v \in H$, it follows that

$$\phi(u) = (u; T^*v) = (Tu; v) = \mathfrak{a}(u, v) \text{ for all } u \in H.$$

Now we prove that T^* is invertible. Let $v \in H$ be such that $T^*v = 0$. Thus,

$$0 = (v; T^*v) = (Tv; v) = \Re\mathfrak{a}(v, v) \geq \delta \|v\|^2.$$

Hence $v = 0$ and so T^* is injective.

It remains to show that T^* has range $R(T^*) = H$. We first prove that $R(T^*)$ is dense. If $u \in H$ is such that

$$(u; T^*v) = 0 \text{ for all } v \in H,$$

then by taking $v = u$ and using again the coercivity assumption, we obtain $u = 0$. Finally, we prove that $R(T^*)$ is closed. For this, let $v_k = T^*u_k$ be a sequence which converges to v in H. We have

$$\begin{aligned}
\delta \|u_k - u_j\|^2 &\leq \Re\mathfrak{a}(u_k - u_j, u_k - u_j) \\
&\leq |(u_k - u_j; T^*u_k - T^*u_j)| \\
&\leq \|u_k - u_j\| \|v_k - v_j\|.
\end{aligned}$$

From this, it follows that $(u_k)_k$ is a Cauchy sequence and hence it converges in H. If u denotes the limit, then $v = T^*u$ by continuity of T^*. This proves that $R(T^*)$ is closed.   □

1.2 UNBOUNDED SESQUILINEAR FORMS AND THEIR ASSOCIATED OPERATORS

1.2.1 Closed and closable forms

In this section, we consider sesquilinear forms which do not act on the whole space H, but only on subspaces of H. These forms are unbounded sesquilinear forms. They play an important role in the study of elliptic or parabolic equations (cf. Chapters 4 and 5). We will say, for simplicity, sesquilinear forms rather than "unbounded sesquilinear forms."

Let H be as in the previous section and consider a sesquilinear form $\mathfrak{a}$ defined on a linear subspace $D(\mathfrak{a})$ of H, called the domain of $\mathfrak{a}$. That is,

$$\mathfrak{a} : D(\mathfrak{a}) \times D(\mathfrak{a}) \to \mathbb{K}$$

is a map which satisfies (1.1) for $u, v, h \in D(\mathfrak{a})$.

DEFINITION 1.4 *Let* $\mathfrak{a} : D(\mathfrak{a}) \times D(\mathfrak{a}) \to \mathbb{K}$ *be a sesquilinear form. We say that:*
1) $\mathfrak{a}$ *is densely defined if*

$$D(\mathfrak{a}) \text{ is dense in } H. \tag{1.2}$$

2) $\mathfrak{a}$ *is accretive if*

$$\Re\mathfrak{a}(u, u) \geq 0 \text{ for all } u \in D(\mathfrak{a}). \tag{1.3}$$

3) $\mathfrak{a}$ *is continuous if there exists a non-negative constant M such that*

$$|\mathfrak{a}(u, v)| \leq M\|u\|_{\mathfrak{a}}\|v\|_{\mathfrak{a}} \text{ for all } u, v \in D(\mathfrak{a}) \tag{1.4}$$

where $\|u\|_{\mathfrak{a}} := \sqrt{\Re\mathfrak{a}(u, u) + \|u\|^2}$.
4) $\mathfrak{a}$ *is closed if*

$$(D(\mathfrak{a}), \|.\|_{\mathfrak{a}}) \text{ is a complete space.} \tag{1.5}$$

If $\mathfrak{a}$ satisfies (1.2)−(1.5), one checks easily that $\|.\|_{\mathfrak{a}}$ is a norm on $D(\mathfrak{a})$. It is called the norm associated with the form $\mathfrak{a}$.

DEFINITION 1.5 *Let* $\mathfrak{a}$ *be a sesquilinear form on H. The adjoint form of* $\mathfrak{a}$ *is the sesquilinear form* $\mathfrak{a}^*$ *defined by:*

$$\mathfrak{a}^*(u, v) := \overline{\mathfrak{a}(v, u)} \text{ with domain } D(\mathfrak{a}^*) = D(\mathfrak{a}).$$

The symmetric part of $\mathfrak{a}$ *is defined by*

$$\mathfrak{b} := \frac{1}{2}(\mathfrak{a} + \mathfrak{a}^*), \ D(\mathfrak{b}) = D(\mathfrak{a}).$$

We say that $\mathfrak{a}$ *is a symmetric form if* $\mathfrak{a}^* = \mathfrak{a}$, *that is,*

$$\mathfrak{a}(u, v) = \overline{\mathfrak{a}(v, u)} \ for \ all \ u, v \in D(\mathfrak{a}).$$

Let $\mathfrak{a}$ be a sesquilinear form which satisfies (1.2)–(1.5). Then $D(\mathfrak{a})$ is a Hilbert space. The inner product is given by

$$(u; v)_{\mathfrak{a}} := \frac{1}{2}[\mathfrak{a}(u, v) + \mathfrak{a}^*(u, v)] + (u; v) \text{ for all } u, v \in D(\mathfrak{a}).$$

The norm $\|.\|_{\mathfrak{a}}$ is the same as $\|.\|_{\mathfrak{b}}$, where $\mathfrak{b}$ is the symmetric part of $\mathfrak{a}$.

On a complex Hilbert space H, every sesquilinear form $\mathfrak{a}$ can be written in terms of symmetric forms $\mathfrak{b}$ and $\mathfrak{c}$ as follows:

$$\mathfrak{a} = \mathfrak{b} + i\mathfrak{c}, \ D(\mathfrak{a}) = D(\mathfrak{b}) = D(\mathfrak{c}). \tag{1.6}$$

It suffices indeed to take $\mathfrak{b} := \frac{1}{2}(\mathfrak{a} + \mathfrak{a}^*)$ and $\mathfrak{c} := \frac{1}{2i}(\mathfrak{a} - \mathfrak{a}^*)$. In this way, the symmetric part $\mathfrak{b}$ is seen as the real part of the form $\mathfrak{a}$ and $\mathfrak{c}$ as the imaginary part.

In the present chapter we will consider only accretive forms (i.e., forms that satisfy (1.3)). We could instead consider forms that are merely bounded from below, that is,

$$\Re\mathfrak{a}(u, u) \geq -\gamma(u; u) \text{ for all } u \in D(\mathfrak{a})$$

for some positive constant γ. The general theory of such forms does not differ much from that of accretive ones. A simple perturbation argument (which consists of considering the form $\mathfrak{a} + \gamma$, defined by $(\mathfrak{a} + \gamma)(u, v) := \mathfrak{a}(u, v) + \gamma(u; v)$ for $u, v \in D(\mathfrak{a})$) allows us to consider only accretive forms. According to Section 1.2.3 below, if B denotes the operator associated with the accretive form $\mathfrak{a} + \gamma$, then $A = B - \gamma I$ is the operator associated with $\mathfrak{a}$. Here I denotes the identity operator on H.

If $\mathfrak{a}$ is a symmetric form, the accretivity property (1.3) means that $\mathfrak{a}$ is non-negative, that is,

$$\mathfrak{a}(u, u) \geq 0 \text{ for all } u \in D(\mathfrak{a}).$$

Thus, for symmetric forms, we use both terms non-negative or accretive to refer to the property (1.3).

The condition (1.4) means that the sesquilinear form $\mathfrak{a}$ is continuous on the space $(D(\mathfrak{a}), \|.\|_{\mathfrak{a}})$. The smallest possible constant M for which (1.4) holds is of some interest (see, e.g., Theorem 1.52).

PROPOSITION 1.6 *Let* $\mathfrak{a} : H \times H \to \mathbb{K}$ *be a closed accretive sesquilinear form. Then the norms* $\|.\|$ *and* $\|.\|_{\mathfrak{a}}$ *are equivalent on* H.

Proof. We have for every $u \in H$

$$\|u\| \leq \|u\|_{\mathfrak{a}} = [\|u\|^2 + \Re\mathfrak{a}(u,u)]^{1/2}.$$

In other words, the identity operator $I : (H, \|.\|_{\mathfrak{a}}) \to H$ is continuous. Since I is bijective, its inverse $I^{-1} = I$ is continuous by the closed graph theorem. Hence, there exists a non-negative constant C such that

$$\|u\|_{\mathfrak{a}} \leq C\|u\| \text{ for all } u \in H.$$

This shows that the two norms are equivalent. □

A stronger assumption than continuity is sectoriality, which we introduce in the following definition.

DEFINITION 1.7 *A sesquilinear form* $\mathfrak{a} : D(\mathfrak{a}) \times D(\mathfrak{a}) \to \mathbb{C}$, *acting on a complex Hilbert space* H, *is called sectorial if there exists a non-negative constant* C, *such that*

$$|\Im\mathfrak{a}(u,u)| \leq C\Re\mathfrak{a}(u,u) \text{ for all } u \in D(\mathfrak{a}). \tag{1.7}$$

The numerical range of $\mathfrak{a}$ *is the set*

$$\mathcal{N}(\mathfrak{a}) := \{\mathfrak{a}(u,u), \ u \in D(\mathfrak{a}) \text{ with } \|u\| = 1\}. \tag{1.8}$$

Clearly, $\mathfrak{a}$ satisfies (1.7) if and only if the numerical range $\mathcal{N}(\mathfrak{a})$ is contained in the closed sector $\{z \in \mathbb{C}, |\arg z| \leq \arctan C\}$.

PROPOSITION 1.8 *Every sectorial form acting on a complex Hilbert space* H *is continuous. More precisely, if*

$$|\Im\mathfrak{a}(u,u)| \leq C\Re\mathfrak{a}(u,u) \text{ for all } u \in D(\mathfrak{a}),$$

where $C \geq 0$ *is a constant, then*

$$|\mathfrak{a}(u,v)| \leq (1+C)(\Re\mathfrak{a}(u,u))^{1/2}(\Re\mathfrak{a}(v,v))^{1/2} \text{ for all } u, v \in D(\mathfrak{a}).$$

Proof. By (1.6) we have $\mathfrak{a} = \mathfrak{b} + i\mathfrak{c}$, where $\mathfrak{b}$ and $\mathfrak{c}$ are symmetric forms and $\mathfrak{b}$ is non-negative. By the Cauchy-Schwarz inequality,

$$|\mathfrak{b}(u,v)| \leq \mathfrak{b}(u,u)^{1/2}\mathfrak{b}(v,v)^{1/2}.$$

It remains to estimate $|\mathfrak{c}(u,v)|$. Changing v into $e^{i\psi}v$ for some ψ, we may assume without loss of generality that $\mathfrak{c}(u,v)$ is real. In this case, we have

$$\mathfrak{c}(u,v) = \frac{1}{4}[\mathfrak{c}(u+v,u+v) - \mathfrak{c}(u-v,u-v)].$$

The sectoriality assumption gives

$$|\mathfrak{c}(u,v)| \le \frac{C}{4}[\mathfrak{b}(u+v,u+v) + \mathfrak{b}(u-v,u-v)]$$
$$= \frac{C}{2}[\mathfrak{b}(u,u) + \mathfrak{b}(v,v)].$$

Replacing u by $\sqrt{\varepsilon}u$ in the last estimate gives

$$|\mathfrak{c}(u,v)| \le \frac{C}{2}\left[\sqrt{\varepsilon}\mathfrak{b}(u,u) + \frac{1}{\sqrt{\varepsilon}}\mathfrak{b}(v,v)\right].$$

If $\mathfrak{b}(u,u) \ne 0$, we choose $\varepsilon = \frac{\mathfrak{b}(v,v)}{\mathfrak{b}(u,u)}$ and obtain

$$|\mathfrak{c}(u,v)| \le C\mathfrak{b}(u,u)^{1/2}\mathfrak{b}(v,v)^{1/2} = C(\Re\mathfrak{a}(u,u))^{1/2}(\Re\mathfrak{a}(v,v))^{1/2}.$$

If $\mathfrak{b}(u,u) = 0$, then $\mathfrak{c}(u,v) \le \frac{C}{2}\mathfrak{b}(v,v)$. Replacing v by λv for $\lambda > 0$ and letting $\lambda \to 0$, one obtains $\mathfrak{c}(u,v) = 0$, which gives again the desired conclusion.   □

A converse to Proposition 1.8 is given by the following simple lemma.

LEMMA 1.9 *If $\mathfrak{a}$ is an accretive and continuous sesquilinear form on a complex Hilbert space H, then $1 + \mathfrak{a}$ is sectorial. More precisely, if $\mathfrak{a}$ satisfies (1.4) with some constant M, then*

$$|\Im[(u;u) + \mathfrak{a}(u,u)]| \le M\Re[(u;u) + \mathfrak{a}(u,u)] \text{ for all } u \in D(\mathfrak{a})$$

Proof. The lemma follows immediately from

$$|\Im[(u;u) + \mathfrak{a}(u,u)]| = |\Im\mathfrak{a}(u,u)| \le |\mathfrak{a}(u,u)|$$

and the continuity assumption (1.4).   □

Note that the continuity assumption of the form $\mathfrak{a}$ is sometimes written in the following way:

$$|\mathfrak{a}(u,v)| \le M'[\Re\mathfrak{a}(u,u) + w\|u\|^2]^{1/2}[\Re\mathfrak{a}(v,v) + w\|v\|^2]^{1/2}$$

for some constants w and M'. It is clear that the norms $[\Re\mathfrak{a}(u,u)+w\|u\|^2]^{1/2}$ and $[\Re\mathfrak{a}(u,u) + \|u\|^2]^{1/2}$ are equivalent. For this reason, we have chosen to write (1.4) and $\|.\|_{\mathfrak{a}}$ without the extra constant w.

It may happen in some problems that the starting form $\mathfrak{a}$ satisfies the properties (1.2)–(1.4) but not (1.5). In this case, one tries to find an extension of $\mathfrak{a}$ which is a closed form and acts on a subspace of H.

DEFINITION 1.10 *A densely defined accretive sesquilinear form $\mathfrak{a}$ is called closable if there exists a closed accretive form $\mathfrak{c}$, acting on a subspace $D(\mathfrak{c})$ of H, such that $D(\mathfrak{a}) \subseteq D(\mathfrak{c})$ and $\mathfrak{a}(u,v) = \mathfrak{c}(u,v)$ for all $u,v \in D(\mathfrak{a})$.*

A closed extension, when it exists, is not unique in general.[1] Nevertheless, in that case, one can define the smallest closed extension $\bar{\mathfrak{a}}$. It is tempting to define $\bar{\mathfrak{a}}$ as follows:

$$D(\bar{\mathfrak{a}}) := \{u \in H \text{ s.t. } \exists u_n \in D(\mathfrak{a}) : u_n \to u \text{ (in } H) \text{ and}$$
$$\mathfrak{a}(u_n - u_m, u_n - u_m) \to 0 \text{ as } n, m \to \infty\}$$

and

$$\bar{\mathfrak{a}}(u, v) := \lim_{n \to \infty} \mathfrak{a}(u_n, v_n) \tag{1.9}$$

for $u, v \in D(\bar{\mathfrak{a}})$, where $(u_n)_n$ and $(v_n)_n$ are any sequences of elements of $D(\mathfrak{a})$ which converge respectively to u and v (with respect to the norm of H) and satisfy $\mathfrak{a}(u_n - u_m, u_n - u_m) \to 0$ and $\mathfrak{a}(v_n - v_m, v_n - v_m) \to 0$ as $n, m \to \infty$.

PROPOSITION 1.11 *Let $\mathfrak{a}$ be a densely defined, accretive, and continuous sesquilinear form. If $\mathfrak{a}$ is closable, then $\bar{\mathfrak{a}}$ is well defined and satisfies (1.2)–(1.5). In addition, every closed extension of $\mathfrak{a}$ is also an extension of $\bar{\mathfrak{a}}$.*

Proof. Fix two sequences $(u_n)_n$ and $(v_n)_n$ of elements of $D(\mathfrak{a})$ which converge for the norm of H and satisfy $\mathfrak{a}(u_n - u_m, u_n - u_m) \to 0$ and $\mathfrak{a}(v_n - v_m, v_n - v_m) \to 0$ as $n, m \to \infty$. In order to see that $\lim_{n \to \infty} \mathfrak{a}(u_n, v_n)$ exists, we write using the continuity assumption (1.4):

$$|\mathfrak{a}(u_n, v_n) - \mathfrak{a}(u_m, v_m)| = |\mathfrak{a}(u_n - u_m, v_n) + \mathfrak{a}(u_m, v_n - v_m)|$$
$$\leq M \|u_n - u_m\|_{\mathfrak{a}} \|v_n\|_{\mathfrak{a}} + M \|v_n - v_m\|_{\mathfrak{a}} \|u_m\|_{\mathfrak{a}}.$$

Since $\|u_n - u_m\|_{\mathfrak{a}}$ and $\|v_n - v_m\|_{\mathfrak{a}} \to 0$ as $n, m \to \infty$, the sequences $(\|u_n\|_{\mathfrak{a}})_n$ and $(\|v_n\|_{\mathfrak{a}})_n$ are bounded. It follows from the previous inequality that $\mathfrak{a}(u_n, v_n)$ is a Cauchy sequence, thus it is convergent.

The fact that $\lim_{n \to \infty} \mathfrak{a}(u_n, v_n)$ is independent of the chosen sequences $(u_n)_n$ and $(v_n)_n$ follows by a similar argument. Indeed, if $(u'_n)_n$ and $(v'_n)_n$ satisfy the same properties as $(u_n)_n$ and $(v_n)_n$, then

$$|\mathfrak{a}(u_n, v_n) - \mathfrak{a}(u'_n, v'_n)| = |\mathfrak{a}(u_n - u'_n, v_n) + \mathfrak{a}(u'_n, v_n - v'_n)|$$
$$\leq M \|u_n - u'_n\|_{\mathfrak{a}} \|v_n\|_{\mathfrak{a}} + M \|v_n - v'_n\|_{\mathfrak{a}} \|u'_n\|_{\mathfrak{a}}.$$

Now, if $\mathfrak{a}_1$ is a closed extension of $\mathfrak{a}$ then $\|u_n - u'_n\|_{\mathfrak{a}} = \|u_n - u'_n\|_{\mathfrak{a}_1} \to 0$ as $n \to \infty$, since $(u_n)_n$ and $(u'_n)_n$ converge to the same limit in the Hilbert

---

[1] A simple example is given by the form $\mathfrak{a}(u, v) = \int_{(0,1)} \frac{d}{dx} u \frac{d}{dx} v \, dx$, $D(\mathfrak{a}) = C_c^\infty(0, 1)$. The same expression with domains the Sobolev spaces $H_0^1(0, 1)$ and $H^1(0, 1)$ gives two different closed extensions. See Chapter 4 for more examples.

space $(D(\mathfrak{a}_1), \|.\|_{\mathfrak{a}_1})$. Applying the same argument to $\|v_n - v'_n\|_\mathfrak{a}$, we obtain $|\mathfrak{a}(u_n, v_n) - \mathfrak{a}(u'_n, v'_n)| \to 0$ as $n \to \infty$.

By construction, $D(\mathfrak{a})$ is dense in $(D(\overline{\mathfrak{a}}), \|.\|_{\overline{\mathfrak{a}}})$ and hence (1.2)–(1.4) hold for the form $\overline{\mathfrak{a}}$. This density shows also that every closed extension of $\mathfrak{a}$ is an extension of $\overline{\mathfrak{a}}$.

Finally, we show that $\overline{\mathfrak{a}}$ is closed. Let $(u_n)_n \in D(\mathfrak{a})$ be a Cauchy sequence for the norm of $\overline{\mathfrak{a}}$. It converges with respect to the norm of H to some $u \in H$. It follows from the definition of $\overline{\mathfrak{a}}$ that $u \in D(\overline{\mathfrak{a}})$. In addition,

$$\overline{\mathfrak{a}}(u_n - u, u_n - u) = \lim_m \mathfrak{a}(u_n - u_m, u_n - u_m).$$

Thus,

$$\lim_n \overline{\mathfrak{a}}(u_n - u, u_n - u) = 0,$$

which means that the sequence $(u_n)_n$ is convergent in $(D(\overline{\mathfrak{a}}), \|.\|_{\overline{\mathfrak{a}}})$. This together with the density of $D(\mathfrak{a})$ in $(D(\overline{\mathfrak{a}}), \|.\|_{\overline{\mathfrak{a}}})$ show that $\overline{\mathfrak{a}}$ is a closed form.   □

DEFINITION 1.12 *If the form* $\mathfrak{a}$ *is closable, then* $\overline{\mathfrak{a}}$ *defined by (1.9) with domain* $D(\overline{\mathfrak{a}})$ *is called the closure of the form* $\mathfrak{a}$.

Remark. 1) The proof of Proposition 1.11 shows that if $\mathfrak{a}$ is any sesquilinear form satisfying (1.2)–(1.4) and $(u_n)_n, (v_n)_n$ are convergent sequences in H, such that $\mathfrak{a}(u_n - u_m, u_n - u_m)$ and $\mathfrak{a}(v_n - v_m, v_n - v_m) \to 0$ as $n, m \to \infty$, then the limit in the right-hand side of (1.9) exists. In addition, if $\mathfrak{a}$ is closed then this limit is $\mathfrak{a}(u, v)$, where u and v are the limits in H of $(u_n)_n$ and $(v_n)_n$, respectively.

2) It follows also from the same proof that if $\mathfrak{a}$ is a sesquilinear form satisfying (1.2)–(1.4), then the form $\overline{\mathfrak{a}}$ is closed whenever it is well defined (i.e., the limit in the right-hand side of (1.9) does not depend on the chosen sequences $(u_n)_n$ and $(v_n)_n$).

PROPOSITION 1.13 *Let* $\mathfrak{a}$ *be a densely defined, accretive, and continuous sesquilinear form. Then* $\mathfrak{a}$ *is closable if and only if it satisfies the following property:*
If $(u_n)_n \in D(\mathfrak{a})$, $u_n \to 0$ *in* H *and* $\mathfrak{a}(u_n - u_m, u_n - u_m) \to 0$ *(as* $n, m \to \infty$*), then* $\mathfrak{a}(u_n, u_n) \to 0$ *as* $n \to \infty$.

Proof. Assume that $\mathfrak{a}$ is closable and let $\mathfrak{a}_1$ be a closed extension. If $u_n \to 0$ in H and $\mathfrak{a}(u_n - u_m, u_n - u_m) \to 0$, then $(u_n)_n$ converges to 0 in $(D(\mathfrak{a}_1), \|.\|_{\mathfrak{a}_1})$. The above proposition (or the above Remark 1)) implies that $\mathfrak{a}(u_n, u_n) = \mathfrak{a}_1(u_n, u_n) \to 0$.

Now we prove the converse. We construct a closed extension by taking the completion of $D(\mathfrak{a})$ with respect to the norm $\|.\|_\mathfrak{a}$. That is, we prove that the form $\bar{\mathfrak{a}}$ given by (1.9) with domain $D(\bar{\mathfrak{a}})$ is well defined (by Remark 2) above, $\bar{\mathfrak{a}}$ will be a closed extension of $\mathfrak{a}$). As mentioned in Remark 1) above, the limit in the right-hand side of (1.9) exists. It remains to prove that the limit is independent of the chosen sequences $(u_n)_n$ and $(v_n)_n$. Let $(u_n')_n$ and $(v_n')_n$ be two other sequences satisfying $u_n' \to u, v_n' \to v$ in H and $\|u_n' - u_m'\|_\mathfrak{a}, \|v_n' - v_m'\|_\mathfrak{a} \to 0$ as $n, m \to \infty$. We write again

$$|\mathfrak{a}(u_n, v_n) - \mathfrak{a}(u_n', v_n')| = |\mathfrak{a}(u_n - u_n', v_n) + \mathfrak{a}(u_n', v_n - v_n')|$$
$$\leq M\|u_n - u_n'\|_\mathfrak{a}\|v_n\|_\mathfrak{a} + M\|v_n - v_n'\|_\mathfrak{a}\|u_n'\|_\mathfrak{a}.$$

The sequence $w_n := u_n - u_n'$ satisfies $w_n \to 0$ in H and

$$\|w_n - w_m\|_\mathfrak{a} \leq \|u_n - u_m\|_\mathfrak{a} + \|u_n' - u_m'\|_\mathfrak{a} \to 0 \text{ as } n, m \to \infty.$$

By assumption, this implies $\mathfrak{a}(w_n, w_n) \to 0$ as $n \to \infty$. The same argument applies to $v_n - v_n'$. Hence, $|\mathfrak{a}(u_n, v_n) - \mathfrak{a}(u_n', v_n')| \to 0$ as $n \to \infty$. $\square$

Lemma 1.29 below guarantees the closability for a class of sesquilinear forms. There are several examples of sesquilinear forms which are not closable.

Example 1.2.1 *Consider on $L^2(\mathbb{R})$ (endowed with the Lebesgue measure dx) the symmetric form*

$$\mathfrak{a}(u, v) = u(0)\overline{v(0)}, \ D(\mathfrak{a}) = C_c(\mathbb{R}), \tag{1.10}$$

where $C_c(\mathbb{R})$ is the space of continuous and compactly supported functions on $\mathbb{R}$. Then $\mathfrak{a}$ is densely defined, symmetric, and non-negative but not closable. Indeed, choose a sequence $(u_n)_n \in C_c(\mathbb{R})$ such that $u_n(0) = 1$ for all n and such that $u_n \to 0$ in $L^2(\mathbb{R})$. Thus, $\mathfrak{a}(u_n - u_m, u_n - u_m) = 0$ and $u_n \to 0$ in $L^2(\mathbb{R})$ but $\mathfrak{a}(u_n, u_n) = 1$ for all n. Proposition 1.13 shows that $\mathfrak{a}$ is not closable.

Example 1.2.2 *Consider now on the real space $L^2(\mathbb{R})$ (endowed again with the Lebesgue measure dx) the form*

$$\mathfrak{a}(u, v) = \int_\mathbb{R} \frac{du}{dx}(x)v(x)dx, \ D(\mathfrak{a}) = H^1(\mathbb{R}). \tag{1.11}$$

The form $\mathfrak{a}$ is not closable. Otherwise, the fact that $\mathfrak{a}(u, u) = 0$ for all $u \in H^1(\mathbb{R})$ implies that $D(\bar{\mathfrak{a}}) = L^2(\mathbb{R})$ and one deduces from Proposition 1.2 that there exists a bounded linear operator T on $L^2(\mathbb{R})$ such that $\mathfrak{a}(u, v) = (Tu; v)$ for all $u, v \in H^1(\mathbb{R})$. This is not possible since $Tu = \frac{du}{dx}$ for $u \in H^1(\mathbb{R})$ and T cannot be extended to a bounded operator on $L^2(\mathbb{R})$.

Example 1.2.2 shows also that the conclusion of Proposition 1.13 cannot hold if the form is not continuous.

DEFINITION 1.14 *Let* $\mathfrak{a}$ *be a densely defined accretive sesquilinear form on* H. *A linear subspace* D *of* $D(\mathfrak{a})$ *is called a core of* $\mathfrak{a}$ *if* D *is dense in* $D(\mathfrak{a})$ *endowed with the norm* $\|\cdot\|_{\mathfrak{a}}$.

Let D be a linear subspace of $D(\mathfrak{a})$. The restriction of $\mathfrak{a}$ to D is the form $\mathfrak{a}_{|D}$, defined by

$$\mathfrak{a}_{|D}(u,v) = \mathfrak{a}(u,v), \; D(\mathfrak{a}_{|D}) = D.$$

A relationship between closability and the notion of core is given by the following.

PROPOSITION 1.15 *Let* $\mathfrak{a}$ *be a densely defined, accretive, continuous, and closed sesquilinear form. Denote by* D *a linear subspace of* $D(\mathfrak{a})$. *Then* D *is a core of* $\mathfrak{a}$ *if and only if the closure of* $\mathfrak{a}_{|D}$ *is* $\mathfrak{a}$, *i.e.,* $\overline{\mathfrak{a}_{|D}} = \mathfrak{a}$.

Proof. The form $\mathfrak{a}$ is a closed extension of $\mathfrak{a}_{|D}$, hence it is an extension of $\overline{\mathfrak{a}_{|D}}$.

Assume that D is a core of $\mathfrak{a}$ and let $u \in D(\mathfrak{a})$. There exists a sequence $(u_n) \in D$ such that $\|u_n - u\|_{\mathfrak{a}} \to 0$ as $n \to \infty$. Hence, (u_n) converges to u in H and $\mathfrak{a}(u_n - u_m, u_n - u_m) \to 0$ as $n, m \to \infty$. This shows that $u \in D(\overline{\mathfrak{a}_{|D}})$. Therefore $\overline{\mathfrak{a}_{|D}} = \mathfrak{a}$.

Conversely, assume that $\overline{\mathfrak{a}_{|D}} = \mathfrak{a}$ and let $u \in D(\mathfrak{a}) = D(\overline{\mathfrak{a}_{|D}})$. It follows from the definition of the closure that there exists a sequence (u_n) in D which converges in H to u and such that $\mathfrak{a}_{|D}(u_n - u_m, u_n - u_m) \to 0$ as $n, m \to \infty$. This means that (u_n) converges to u with respect to the norm $\|\cdot\|_{\mathfrak{a}}$, which shows that D is a core of $\mathfrak{a}$. $\qquad\square$

1.2.2 Perturbation of sesquilinear forms

In this section, we study perturbations of forms. The main questions concern closability and continuity of the sum of two sesquilinear forms.

The sum $\mathfrak{a} + \mathfrak{b}$ of two sesquilinear forms $\mathfrak{a}$ and $\mathfrak{b}$ on H is defined by

$$[\mathfrak{a} + \mathfrak{b}](u,v) := \mathfrak{a}(u,v) + \mathfrak{b}(u,v), \; D(\mathfrak{a} + \mathfrak{b}) = D(\mathfrak{a}) \cap D(\mathfrak{b}).$$

THEOREM 1.16 *Let* $\mathfrak{a}$ *and* $\mathfrak{b}$ *be two accretive sesquilinear forms on* H. *Then the sum* $\mathfrak{a} + \mathfrak{b}$ *is accretive. In addition,*
1) If $\mathfrak{a}$ *and* $\mathfrak{b}$ *are continuous, then so is* $\mathfrak{a} + \mathfrak{b}$.
2) If $\mathfrak{a}$ *and* $\mathfrak{b}$ *are closed, then so is* $\mathfrak{a} + \mathfrak{b}$.
3) If $\mathfrak{a}$ *and* $\mathfrak{b}$ *are closable, then so is* $\mathfrak{a} + \mathfrak{b}$.

Proof. The accretivity of the sum as well as assertion 1) are obvious. Assume that $\mathfrak{a}$ and $\mathfrak{b}$ are closed. Let $(u_n)_n \in D(\mathfrak{a}) \cap D(\mathfrak{b})$ be a Cauchy sequence for the norm $\|.\|_{\mathfrak{a}+\mathfrak{b}}$. The inequalities $\|.\|_{\mathfrak{a}} \leq \|.\|_{\mathfrak{a}+\mathfrak{b}}$ and $\|.\|_{\mathfrak{b}} \leq \|.\|_{\mathfrak{a}+\mathfrak{b}}$ imply that $(u_n)_n$ is a Cauchy sequence for the norms $\|.\|_{\mathfrak{a}}$ and for $\|.\|_{\mathfrak{b}}$. It follows that $(u_n)_n$ converges both in $(D(\mathfrak{a}), \|.\|_{\mathfrak{a}})$ and $(D(\mathfrak{b}), \|.\|_{\mathfrak{b}})$. The limit in both spaces is the same since the convergence in each space implies the convergence in H. The limit belongs then to $D(\mathfrak{a}) \cap D(\mathfrak{b})$. The inequality $\|.\|_{\mathfrak{a}+\mathfrak{b}} \leq \|.\|_{\mathfrak{a}} + \|.\|_{\mathfrak{b}}$ implies that $(u_n)_n$ converges with respect to the norm $\|.\|_{\mathfrak{a}+\mathfrak{b}}$. Hence, $\mathfrak{a} + \mathfrak{b}$ is closed.

If both forms $\mathfrak{a}$ and $\mathfrak{b}$ are closable, then the sum of their closures $\overline{\mathfrak{a}} + \overline{\mathfrak{b}}$ is a closed form by assertion 2). Thus, $\overline{\mathfrak{a}} + \overline{\mathfrak{b}}$ is a closed extension of $\mathfrak{a} + \mathfrak{b}$. The latter is then a closable form, its closure $\overline{\mathfrak{a} + \mathfrak{b}}$ is a restriction of $\overline{\mathfrak{a}} + \overline{\mathfrak{b}}$. $\square$

DEFINITION 1.17 *Let $\mathfrak{a}$ be a densely defined, accretive, continuous, and closed sesquilinear form on H. A sesquilinear form $\mathfrak{a}'$ with domain $D(\mathfrak{a}')$ is $\mathfrak{a}$-form bounded if $D(\mathfrak{a}) \subseteq D(\mathfrak{a}')$ and there exist non-negative constants α and β such that*

$$|\mathfrak{a}'(u, u)| \leq \alpha|\mathfrak{a}(u, u)| + \beta\|u\|^2 \ for \ all \ u \in D(\mathfrak{a}). \tag{1.12}$$

The infimum of all possible constants α for which the inequality holds is called the $\mathfrak{a}$-bound of $\mathfrak{a}'$.

Under closability assumptions on the forms, $\mathfrak{a}'$ is $\mathfrak{a}$-bounded as soon as $D(\mathfrak{a}) \subseteq D(\mathfrak{a}')$. More precisely,

PROPOSITION 1.18 *Let $\mathfrak{a}$ and $\mathfrak{a}'$ be accretive and continuous forms. Assume that $\mathfrak{a}$ is closed, $\mathfrak{a}'$ is closable, and $D(\mathfrak{a}) \subseteq D(\mathfrak{a}')$. Then $\mathfrak{a}'$ is $\mathfrak{a}$-bounded.*

Proof. Since the form $\mathfrak{a}'$ is closable, its restriction to $D(\mathfrak{a})$, $\mathfrak{a}'_{|D(\mathfrak{a})} : D(\mathfrak{a}) \times D(\mathfrak{a}) \to \mathbb{K}$ is also closable. Thus, $\mathfrak{a}'_{|D(\mathfrak{a})}$ is a closable form acting on the Hilbert space $(D(\mathfrak{a}), \|.\|_{\mathfrak{a}})$. Its closure (as a form on $(D(\mathfrak{a}), \|.\|_{\mathfrak{a}})$) is itself. By Proposition 1.6, there exists a non-negative constant M such that for all $u \in D(\mathfrak{a})$

$$|\mathfrak{a}'(u, u)| \leq M\|u\|_{\mathfrak{a}}^2 = M[\|u\|^2 + \Re\mathfrak{a}(u, u)].$$

This proves that $\mathfrak{a}'$ is $\mathfrak{a}$-bounded. $\square$

Assume that $\mathfrak{a}$ is an accretive and continuous sesquilinear form. Denote by $\mathfrak{b} := \frac{1}{2}[\mathfrak{a} + \mathfrak{a}^*]$ the symmetric part of $\mathfrak{a}$. Recall that $\mathfrak{b}(u, u) = \Re\mathfrak{a}(u, u)$ for all $u \in D(\mathfrak{a})$. Using the continuity of $\mathfrak{a}$, we have for all $u \in D(\mathfrak{a})$

$$\mathfrak{b}(u, u) \leq |\mathfrak{a}(u, u)| \leq M\|u\|_{\mathfrak{a}}^2 = M[\|u\|^2 + \mathfrak{b}(u, u)].$$

It follows from this that a form $\mathfrak{a}'$ is $\mathfrak{a}$-bounded if and only if it is $\mathfrak{b}$-bounded.

The next theorem shows that continuity and closability properties carry over from a form $\mathfrak{a}$ to $\mathfrak{a} + \mathfrak{a}'$ provided the form $\mathfrak{a}'$ is $\mathfrak{a}$-bounded with small bound.

THEOREM 1.19 *Let $\mathfrak{a}$ be an accretive and continuous sesquilinear form on a complex Hilbert space H. Assume that $\mathfrak{a}'$ is a sesquilinear form such that $D(\mathfrak{a}) \subseteq D(\mathfrak{a}')$ and*

$$|\mathfrak{a}'(u, u)| \le \alpha \Re\mathfrak{a}(u, u) + \beta \|u\|^2 \text{ for all } u \in D(\mathfrak{a}), \qquad (1.13)$$

where α, β are non-negative constants with $\alpha < 1$. Then the form sum $\mathfrak{t} := \mathfrak{a} + \mathfrak{a}' + \beta$ with domain $D(\mathfrak{t}) = D(\mathfrak{a})$ is accretive and continuous. Moreover,
1) $\mathfrak{t}$ is closed if and only if $\mathfrak{a}$ is closed.
2) $\mathfrak{t}$ is closable if and only if $\mathfrak{a}$ is closable.

Proof. The domain of $\mathfrak{t}$ is $D(\mathfrak{a})$, since $D(\mathfrak{a}) \subseteq D(\mathfrak{a}')$. By (1.13) and the fact that $\alpha < 1$, we have for $u \in D(\mathfrak{a})$,

$$\Re\mathfrak{t}(u, u) = \Re\mathfrak{a}(u, u) + \Re\mathfrak{a}'(u, u) + \beta \|u\|^2 \ge (1 - \alpha)\Re\mathfrak{a}(u, u) \ge 0.$$

Thus, $\mathfrak{t}$ is accretive.

Using the continuity of $\mathfrak{a}$, we obtain by Lemma 1.9 and (1.13)

$$\begin{aligned}
|\Im\mathfrak{t}(u, u)| &\le |\Im\mathfrak{a}(u, u)| + |\Im\mathfrak{a}'(u, u)| \\
&\le \alpha \Re\mathfrak{a}(u, u) + \beta \|u\|^2 + M[\Re\mathfrak{a}(u, u) + \|u\|^2] \\
&\le \frac{M + \alpha}{1 - \alpha} \Re\mathfrak{t}(u, u) + (M + \beta)\|u\|^2 \\
&\le C\Re[\mathfrak{t}(u, u) + \|u\|^2]
\end{aligned}$$

for some non-negative constant C. Hence, the form $\mathfrak{t} + 1$ is sectorial. Proposition 1.8 implies that $\mathfrak{t}$ is continuous.

The inequalities

$$\Re\mathfrak{t}(u, u) \ge (1 - \alpha)\Re\mathfrak{a}(u, u) \text{ and } \Re\mathfrak{t}(u, u) \le (1 + \alpha)\Re\mathfrak{a}(u, u) + 2\beta\|u\|^2$$

show that the norms $\|.\|_{\mathfrak{a}}$ and $\|.\|_{\mathfrak{t}}$ are equivalent on $D(\mathfrak{a})$. From this, assertion 1) follows immediately. To prove assertion 2), let us assume that $\mathfrak{a}$ is closable and let $(u_n) \in D(\mathfrak{a})$ such that $u_n \to 0$ in H and $\mathfrak{t}(u_n - u_m, u_n - u_m) \to 0$ as $m, n \to \infty$. Then, $\Re\mathfrak{a}(u_n - u_m, u_n - u_m) \to 0$ as $m, n \to \infty$, since the norms $\|.\|_{\mathfrak{a}}$ and $\|.\|_{\mathfrak{t}}$ are equivalent. By continuity of $\mathfrak{a}$, it follows that $\mathfrak{a}(u_n - u_m, u_n - u_m) \to 0$. Proposition 1.13 asserts that $\mathfrak{a}(u_n, u_n) \to 0$ as $n \to \infty$. As previously, we obtain from this and continuity of $\mathfrak{t}$, that $\mathfrak{t}(u_n, u_n) \to 0$ as $n \to \infty$ and we conclude by Proposition 1.13 that $\mathfrak{t}$ is closable. The converse holds for the same reasons. $\square$

The following proposition is extracted from the previous proof.

PROPOSITION 1.20 *Let $\mathfrak{a}$ and $\mathfrak{a}'$ be two accretive and continuous sesquilinear forms on a Hilbert space H. Assume that $D(\mathfrak{a}) = D(\mathfrak{a}')$ and the norms $\|.\|_\mathfrak{a}$ and $\|.\|_{\mathfrak{a}'}$ are equivalent. Then*
1) $\mathfrak{a}$ is closed if and only if $\mathfrak{a}'$ is closed.
2) $\mathfrak{a}$ is closable if and only if $\mathfrak{a}'$ is closable.

1.2.3 Associated operator

Let $\mathfrak{a}$ be a densely defined, accretive, continuous, and closed sesquilinear form on H. One can define in terms of $\mathfrak{a}$ an unbounded operator A, defined on a linear subspace $D(A)$ of H as follows:

$u \in D(\mathfrak{a})$ *is in the domain $D(A)$ of A, if and only if there exists $v \in H$ such that the equality $\mathfrak{a}(u, \phi) = (v; \phi)$ holds for all $\phi \in D(\mathfrak{a})$. We then set $Au := v$.*

We rewrite this as

$$D(A) = \{u \in H \text{ s.t. } \exists v \in H : \mathfrak{a}(u, \phi) = (v; \phi) \, \forall \phi \in D(\mathfrak{a})\}, \ Au := v.$$

Observe also that $D(A)$ is the set of vectors $u \in D(\mathfrak{a})$ for which the mapping $\phi \mapsto \mathfrak{a}(u, \phi)$ is continuous on $D(\mathfrak{a})$ with respect to the norm of H.

DEFINITION 1.21 *The linear operator A, defined above, is called the operator associated with the form $\mathfrak{a}$.*

There are several important properties of operators which are associated with sesquilinear forms. We start with the following result.

PROPOSITION 1.22 *Denote by A the operator associated with a densely defined, accretive, continuous, and closed sesquilinear form $\mathfrak{a}$. Then A is densely defined and for every $\lambda > 0$, the operator $\lambda I + A$ is invertible (from $D(A)$ into H) and its inverse $(\lambda I + A)^{-1}$ is a bounded operator on H (here I is the identity operator). In addition,*

$$\|\lambda(\lambda I + A)^{-1}f\| \leq \|f\| \text{ for all } \lambda > 0, f \in H.$$

Proof. Fix $\lambda > 0$ and put

$$\|u\|_\lambda := \sqrt{\Re\mathfrak{a}(u, u) + \lambda\|u\|^2}, \ u \in D(\mathfrak{a}).$$

The norm $\|.\|_\lambda$ is equivalent to the norm $\|.\|_\mathfrak{a}$ and hence $V := (D(\mathfrak{a}), \|.\|_\lambda)$ is a Hilbert space. It follows from (1.4) that the form $\lambda + \mathfrak{a}^*$ (defined by $(\lambda + \mathfrak{a}^*)(u, v) = \lambda(u; v) + \mathfrak{a}^*(u, v)$) is bounded on V. It is in addition coercive on V.

Let $f \in H$ and define

$$\phi(v) := (v; f), \ v \in V.$$

Clearly, ϕ is a linear continuous functional on V. Thus by Lemma 1.3, there exists a unique $u \in V$ such that

$$\phi(v) = \mathfrak{a}^*(v, u) + \lambda(v; u) = \overline{\mathfrak{a}(u, v)} + \lambda(v; u) \text{ for all } v \in V.$$

It follows from this and the definition of A that $u \in D(A)$ and $(\lambda I + A)u = f$. Thus, $\lambda I + A$ has range $R(\lambda I + A) = H$. The accretivity assumption (1.3) implies easily that $\lambda I + A$ is injective and hence invertible.

Let now $f \in H$ and let $u \in D(A)$ be such that $(\lambda I + A)u = f$. Taking the inner product with u, and using

$$\Re(Au; u) = \Re\mathfrak{a}(u, u) \geq 0,$$

it follows that

$$\Re(f; u) \geq \lambda \|u\|^2.$$

This implies that $\lambda \|u\| \leq \|f\|$. That is,

$$\|\lambda(\lambda I + A)^{-1}f\| \leq \|f\|.$$

Finally, we show that $D(A)$ is dense in H. Let $v \in H$ be such that

$$(v; u) = 0 \ \text{ for all } u \in D(A).$$

Since $I + A$ is invertible, there exists $\psi \in D(A)$ such that $v = (I + A)\psi$. Applying the above equality with $u = \psi$, we obtain

$$0 = (v; \psi) = ((I + A)\psi; \psi) = \|\psi\|^2 + (A\psi; \psi).$$

This together with the fact that $\Re(A\psi; \psi) = \Re\mathfrak{a}(\psi, \psi) \geq 0$ implies that $\psi = 0$ and hence $v = 0$. $\qquad\square$

Note that if the sesquilinear form $\mathfrak{a}$ satisfies (1.2)–(1.5), then the adjoint form $\mathfrak{a}^*$ satisfies the same conditions. One then associates an operator with $\mathfrak{a}^*$. It turns out that this operator is the adjoint A^* of A. Let us recall the definition of the adjoint for unbounded operators.

DEFINITION 1.23 *Let B be a densely defined operator acting in H. The adjoint of B is the operator B^* defined by*

$$D(B^*) = \{u \in H \text{ s.t. } \exists v \in H : (B\phi; u) = (\phi; v) \text{ for all } \phi \in D(B)\},$$
$$B^*u := v.$$

A symmetric operator B *is an operator such that* $D(B) \subseteq D(B^*)$ *and* $Bu = B^*u$ *for all* $u \in D(B)$.

The operator B is self-adjoint if $B^* = B$. This means that $D(B) = D(B^*)$ and $Bu = B^*u$ for all $u \in D(B)$.

PROPOSITION 1.24 *The operator associated with* $\mathfrak{a}^*$ *is* A^*. *In particular, if* $\mathfrak{a}$ *is symmetric then* A *is self-adjoint.*

Proof. Denote by B the operator associated with $\mathfrak{a}^*$ and let $u \in D(B)$. By definition,

$$\mathfrak{a}^*(u, \phi) = (Bu; \phi) \text{ for all } \phi \in D(\mathfrak{a}^*) = D(\mathfrak{a}).$$

Hence

$$(Bu; \phi) = \mathfrak{a}^*(u, \phi) = \overline{\mathfrak{a}(\phi, u)} = \overline{(A\phi; u)} \text{ for all } \phi \in D(A).$$

This shows that $u \in D(A^*)$ and $A^*u = Bu$. It remains to prove that $D(A^*) \subseteq D(B)$. For this, fix $v \in D(A^*)$. By Proposition 1.22, there exists $\psi \in D(B)$ such that $(I+A^*)v = (I+B)\psi$. Hence $(I+A^*)v = (I+A^*)\psi$. Thus,

$$(v - \psi; (I + A)u) = ((I + A^*)(v - \psi); u) = 0 \text{ for all } u \in D(A).$$

Since $I + A$ is invertible, this implies that $v = \psi \in D(B)$. $\square$

We have seen in Proposition 1.22 that the operator A associated with $\mathfrak{a}$ is densely defined in H. It is also densely defined in $D(\mathfrak{a})$, endowed with the norm $\|.\|_{\mathfrak{a}}$. This is formulated in the following lemma whose proof is postponed to Section 1.4.2.

LEMMA 1.25 *Let* $\mathfrak{a}$ *be a densely defined, accretive, continuous, and closed sesquilinear form and denote by* A *its associated operator. Then* $D(A)$ *is a core of* $\mathfrak{a}$.

Using this lemma and Proposition 1.15 one concludes that the form $\mathfrak{a}$ coincides with the closure of the restriction of $\mathfrak{a}$ to $D(A)$, that is, $\mathfrak{a} = \overline{\mathfrak{a}_{|D(A)}}$.

DEFINITION 1.26 *1) An operator* $B : D(B) \subseteq H \to H$ *is called sectorial if there exists a non-negative constant* C, *such that*

$$|\Im(Bu; u)| \leq C\Re(Bu; u) \text{ for all } u \in D(B). \tag{1.14}$$

2) The numerical range of an operator B *on* H *is the set*

$$\mathcal{N}(B) := \{(Bu; u), u \in D(B) \text{ with } \|u\| = 1\}.$$

Clearly, B satisfies (1.14) if and only if its numerical range $\mathcal{N}(B)$ is contained in the sector $\{z \in \mathbb{C}, |\arg z| \leq \arctan C\}$.

It is also clear that the operator associated with a sectorial form is a sectorial operator. The converse is also true. We formulate this in the following proposition.

PROPOSITION 1.27 *Let $\mathfrak{a}$ be a densely defined, accretive, continuous, and closed sesquilinear form acting on a complex Hilbert space H. Denote by A the operator associated with $\mathfrak{a}$. The following assertions are equivalent:*
1) $\mathfrak{a}$ is a sectorial form.
2) A is a sectorial operator.

The proposition is an immediate consequence of the following lemma.

LEMMA 1.28 *Let $\mathfrak{a}$ be a densely defined, accretive, continuous, and closed sesquilinear form acting on a complex Hilbert space H and denote by A its associated operator. Then, the numerical range $\mathcal{N}(A)$ of A is dense in the numerical range $\mathcal{N}(\mathfrak{a})$ of $\mathfrak{a}$.*

Proof. Apply Lemma 1.25.   □

LEMMA 1.29 *Let A be a densely defined operator on a Hilbert space H such that $\Re(Au; u) \geq 0$ for all $u \in D(A)$. Assume that either:*
1) H is complex and there exists a constant $\alpha \geq 0$, such that the operator $\alpha I + A$ is sectorial,
or
2) A is a symmetric operator (here H may be real).
Then the form defined by

$$\mathfrak{a}(u, v) := (Au; v) \text{ with domain } D(\mathfrak{a}) = D(A)$$

is closable.

Proof. Assume that 1) is satisfied. Write

$$\mathfrak{a}(u, v) = ((\alpha I + A)u; v) - \alpha(u; v).$$

By Proposition 1.8, the sectorial form $(u, v) \to ((\alpha I + A)u; v)$ is continuous and hence $\mathfrak{a}$ is continuous, too.

If 2) is satisfied, then $\mathfrak{a}$ is continuous. This follows from the Cauchy-Schwarz inequality.

In order to prove that $\mathfrak{a}$ is closable, we apply Proposition 1.13. Assume that $(u_n) \in D(A)$ is such that $u_n \to 0$ in H and $\mathfrak{a}(u_n - u_m, u_n - u_m) \to 0$ (as $n, m \to \infty$). By continuity of the form, we have

$$|\mathfrak{a}(u_n, u_n)| \leq |\mathfrak{a}(u_n - u_m, u_n)| + |\mathfrak{a}(u_m, u_n)|$$
$$\leq M\|u_n - u_m\|_{\mathfrak{a}}\|u_n\|_{\mathfrak{a}} + |(Au_m; u_n)|.$$

By assumption, $\|u_n - u_m\|_\mathfrak{a} \to 0$ as $m, n \to \infty$ and thus $\|u_n\|_\mathfrak{a}$ is a bounded sequence. In addition, for each m, $|(Au_m; u_n)| \to 0$ as $n \to \infty$. These properties imply that $\mathfrak{a}(u_n, u_n) \to 0$ as $n \to \infty$. This proves that $\mathfrak{a}$ is closable. □

The role of assumptions 1) and 2) in Lemma 1.29 is to guarantee the continuity of the form

$$\mathfrak{a}(u, v) := (Au; v), \quad D(\mathfrak{a}) = D(A). \tag{1.15}$$

The proof shows that this form is closable whenever it is continuous. The assumption of continuity cannot be removed. Also, on a real Hilbert space, the single assumption that $(Au, u) \geq 0$ for all $u \in D(A)$, is not enough to guarantee the closability of the form $\mathfrak{a}$. All this can be seen from Example 1.2.2.

When the form defined by (1.15) is closable, the operator associated with its closure is clearly an extension of A.

DEFINITION 1.30 *The operator associated with the closure $\bar{\mathfrak{a}}$ of the form $\mathfrak{a}$ defined by (1.15) is called the Friedrichs extension of A.*

PROPOSITION 1.31 *Let $B : D(B) \subseteq H \to H$ be a closed operator (see Definition 1.33 below) with dense domain $D(B)$. Then B^*B defined by*

$$D(B^*B) = \{u \in D(B), Bu \in D(B^*)\}, \ B^*Bu = B^*(Bu)$$

is a densely defined self-adjoint operator.

Proof. Define the symmetric form

$$\mathfrak{a}(u, v) = (Bu; Bv), \quad D(\mathfrak{a}) = D(B).$$

Since B is a closed operator, $\mathfrak{a}$ is a closed form. Thus, there exists a self-adjoint (and densely defined) operator A associated with $\mathfrak{a}$. By definition,

$$D(A) = \{u \in D(\mathfrak{a}), \exists v \in H : \mathfrak{a}(u, \phi) = (v; \phi) \ \forall \phi \in D(\mathfrak{a})\}, \ Au = v.$$

Thus,

$$\begin{aligned} D(A) &= \{u \in D(B), \exists v \in H : (Bu; B\phi) = (v; \phi) \ \forall \phi \in D(B)\} \\ &= \{u \in D(B), Bu \in D(B^*)\}, \end{aligned}$$

and $Au = B^*(Bu)$. This shows that $A = B^*B$ and proves the proposition. □

We finish this section with the following lemma, which is related to the results of the previous section. Its proof requires certain results of the present section.

LEMMA 1.32 *Let* $\mathfrak{a}$ *be a densely defined, accretive, continuous, and closed form on a Hilbert space* H. *Assume that* $(u_n)_n$ *is a bounded sequence in* $(D(\mathfrak{a}), \|.\|_{\mathfrak{a}})$ *which converges in* H *to* u. *Then* $u \in D(\mathfrak{a})$ *and we have* $\Re\mathfrak{a}(u, u) \leq \liminf_n \Re\mathfrak{a}(u_n, u_n)$.

Proof. The sequence $(u_n)_n$ is bounded in the Hilbert space $(D(\mathfrak{a}), \|.\|_{\mathfrak{a}})$, thus it has a weakly convergent subsequence. Let (u_{n_k}) be this subsequence and $\phi \in D(\mathfrak{a})$ be its weak limit. For every $v \in D(\mathfrak{a})$,

$$(u_{n_k}; v) + \mathfrak{b}(u_{n_k}, v) \to (\phi; v) + \mathfrak{b}(\phi, v) \text{ as } n_k \to \infty, \qquad (1.16)$$

where $\mathfrak{b}$ denotes the symmetric part of $\mathfrak{a}$. Denote by B the self-adjoint operator associated with $\mathfrak{b}$. The above convergence holds for $v \in D(B)$ and hence

$$(u_{n_k}; (I + B)v) \to (\phi; (I + B)v).$$

Now the fact that u_{n_k} converges to u in H implies that

$$(u; (I + B)v) = (\phi; (I + B)v) \text{ for all } v \in D(B).$$

By Proposition 1.22, $I + B$ is invertible and hence $u = \phi \in D(\mathfrak{a})$.

Taking $v = u$ in (1.16), yields $\mathfrak{b}(u, u) = \lim_k \mathfrak{b}(u_{n_k}, u)$. This and the Cauchy-Schwarz inequality imply $\mathfrak{b}(u, u) \leq \liminf \mathfrak{b}(u_{n_k}, u_{n_k})$. It follows that $\mathfrak{b}(u, u) \leq \liminf \mathfrak{b}(u_n, u_n)$ since we can replace $(u_n)_n$ in the above arguments by any subsequence. □

Remark. We have used in the proof only that some subsequence of (u_n) converges weakly to u. Therefore, the conclusion of the lemma holds under the weaker assumption that (u_n) is a bounded sequence in $(D(\mathfrak{a}), \|.\|_{\mathfrak{a}})$ which converges weakly in H to u.

1.3 SEMIGROUPS AND UNBOUNDED OPERATORS

1.3.1 Closed and closable operators

Throughout this section, E denotes a Banach space (over $\mathbb{K} = \mathbb{R}$ or $\mathbb{C}$) with norm $\|.\|$. By $\mathcal{L}(E)$, we denote the space of all bounded linear operators on E.

DEFINITION 1.33 *An operator* $B : D(B) \subseteq E \to E$ *is called a closed operator if the graph*

$$G(B) := \{(u; Bu), u \in D(B)\}$$

is closed in $E \times E$.

This definition can be rephrased as follows:

If $(x_n)_n \in D(B)$ is such that $x_n \to x$ and $Bx_n \to y$ in E (as $n \to \infty$), then $x \in D(B)$ and $y = Bx$.

Note also that B is a closed operator if and only if $D(B)$ endowed with the graph norm $\|.\| + \|B.\|$ is a complete space.

DEFINITION 1.34 *Let $B : D(B) \subseteq E \to E$ be an operator on E. A scalar $\lambda \in \mathbb{K}$ is in the resolvent set of B if $\lambda I - B$ is invertible (from $D(B)$ into E) and its inverse $(\lambda I - B)^{-1}$ is a bounded operator on E. For such λ, the operator $(\lambda I - B)^{-1}$ is the resolvent of B at λ.*
The set

$$\rho(B) := \{\lambda \in \mathbb{K}, \lambda I - B \text{ is invertible and } (\lambda I - B)^{-1} \in \mathcal{L}(E)\}$$

is called the resolvent set of B.
    The complement of $\rho(B)$ in $\mathbb{K}$

$$\sigma(B) := \mathbb{K} \setminus \rho(B)$$

is called the spectrum of B.

PROPOSITION 1.35 *1) Assume that B is a closed operator on a Banach space E. Then a scalar λ is in $\rho(B)$ if and only if $\lambda I - B$ is invertible (from $D(B)$ into E).*
2) If the resolvent set $\rho(B)$ is not empty, then B is a closed operator.

Proof. In order to prove the first assertion we have to prove that $(\lambda I - B)^{-1}$ is a continuous operator for every λ such that $\lambda I - B$ is invertible. Let $(y_n)_n$ be a sequence in E which converges to y and such that $(\lambda I - B)^{-1} y_n$ converges to z. Set $x_n := (\lambda I - B)^{-1} y_n$. We have $x_n \in D(B)$ for each n and $(\lambda I - B) x_n$ converges to y. Since B is a closed operator, it follows that $z \in D(B)$ and $y = (\lambda I - B)z$, that is, $z = (\lambda I - B)^{-1} y$. We conclude now by the closed graph theorem that $(\lambda I - B)^{-1}$ is continuous on E.

Assume that $\lambda \in \rho(B)$ for some λ. Let (x_n) be a sequence in $D(B)$ such that $x_n \to x$ and $Bx_n \to y$ in E. Thus, $(\lambda I - B)x_n \to \lambda x - y$ and by continuity of $(\lambda I - B)^{-1}$, we have $x_n \to (\lambda I - B)^{-1}(\lambda x - y)$. Thus, $x = (\lambda I - B)^{-1}(\lambda x - y)$. This implies that $x \in D(B)$ and $Bx = y$. This shows that B is a closed operator. $\square$

DEFINITION 1.36 *An operator B on a Banach space E is closable if there exists a closed operator $C : D(C) \subseteq E \to E$ such that $D(B) \subseteq D(C)$ and $Bu = Cu$ for all $u \in D(B)$. In other words, B has a closed extension C.*

Assume that B is a closable operator on a Banach space E. One can define the smallest closed extension $\overline{B}$ of B as follows:

$$D(\overline{B}) = \{u \in E \ s.t. \ \exists u_n \in D(B) : \lim_n u_n = u, \ \lim_{n,m}[Bu_n - Bu_m] = 0\},$$
$$(1.17)$$

and if u and $(u_n)_n$ are as in (1.17) we set

$$\overline{B}u := \lim_n Bu_n, \qquad\qquad (1.18)$$

where the limits are taken with respect to the norm of E.

One shows easily that $\overline{B}$ is a closed operator and every closed extension of B is also an extension of $\overline{B}$.

If B is an operator such that $\overline{B}$, defined by (1.17) and (1.18), is well defined (i.e., $\overline{B}u = \lim_n Bu_n$ does not depend on the choice of the sequence (u_n)), then $\overline{B}$ is a closed extension of B. Consequently, B is closable if and only if $\overline{B}$ is a well defined operator.

Let now $u \in D(\overline{B})$ and let $u_n \in D(B), v_n \in D(B)$ be two sequences which converge to u and such that $Bu_n - Bu_m \to 0$ and $Bv_n - Bv_m \to 0$ as $n, m \to \infty$. Thus, Bu_n and Bv_n converge to some w and w' in E. Now, $\overline{B}$ is well defined if and only if $w = w'$. Thus, we have proved the following characterization of closable operators.

PROPOSITION 1.37 *A linear operator B on E is closable if and only if it satisfies the following property:*
if $(u_n) \in D(B)$ is any sequence such that $u_n \to 0$ and $Bu_n \to v$ (in E), then $v = 0$.

DEFINITION 1.38 *Assume that B is a closable operator on a Banach space E. The operator $\overline{B}$ defined by (1.17) and (1.18) is called the closure of B.*

DEFINITION 1.39 *Let B be an operator with domain $D(B)$ on a Banach space E. A linear subspace of $D(B)$ is called a core of B if it is dense in $D(B)$, endowed with the graph norm $\|.\| + \|B.\|$.*

Let B act on a Banach space E and D a linear subspace of $D(B)$. The restriction of B to D is the operator

$$B_{|D}u := Bu \text{ for } u \in D = D(B_{|D}).$$

The next result follows easily from the previous definitions.

PROPOSITION 1.40 *Let B be a closed operator on a Banach space E and D a linear subspace of $D(B)$. Then, D is a core of B if and only if the closure of $B_{|D}$ is B, i.e., $\overline{B_{|D}} = B$.*

1.3.2 A rapid course on semigroup theory

In this subsection, we give some definitions and recall several important results and properties of semigroups. Semigroup theory is a well documented subject and we shall not give a detailed study. For more details and proofs of the classical results given below see, e.g., Arendt et al. [ABHN01], Davies [Dav80], Goldstein [Gol85], Engel and Nagel [EnNa99], Kato [Kat80], Nagel et al. [Nag86], Pazy [Paz83], Yosida [Yos65].

DEFINITION 1.41 *1) A semigroup on a Banach space E is a family of bounded linear operators $(T(t))_{t\geq 0}$ acting on E such that*

$$T(0) = I \text{ and } T(t+s) = T(t)T(s) \text{ for all } t, s \geq 0.$$

2) A semigroup $(T(t))_{t\geq 0}$ is called a contraction semigroup (or a contractive semigroup) if $T(t)$ is a contraction operator on E for each $t \geq 0$.
3) We say that a semigroup $(T(t))_{t\geq 0}$ is strongly continuous if for every $u \in E$, we have

$$\lim_{t\downarrow 0} T(t)u = u.$$

Note that the property in 3) is precisely the strong continuity at $t_0 = 0$. From this and the semigroup property it follows that $(T(t))_{t\geq 0}$ is strongly continuous at each $t_0 \in [0, \infty)$.

DEFINITION 1.42 *Let $(T(t))_{t\geq 0}$ be a strongly continuous semigroup on E. The generator of $(T(t))_{t\geq 0}$ is the operator B defined by*

$$D(B) := \{u \in E, \lim_{t\downarrow 0} \frac{1}{t}(T(t)u - u) \text{ exists}\},$$

$$Bu := \lim_{t\downarrow 0} \frac{1}{t}(T(t)u - u) \text{ for all } u \in D(B).$$

The theory of strongly continuous semigroups was developed in order to study existence and uniqueness of solutions to the evolution equations (or the Cauchy problem)

$$(CP) \begin{cases} \frac{d}{dt}u(t) = & Bu(t), \ t \geq 0, \\ u(0) = & f, \end{cases}$$

where $u : [0, \infty) \to E$ satisfies $u(t) \in D(B)$ for all $t > 0$ is the searched for solution.

If B is the generator of the strongly continuous semigroup $(T(t))_{t\geq 0}$, then for every $f \in D(B)$, (CP) has a unique solution. The latter is given by $u(t) = T(t)f$.

The following is a central theorem in semigroup theory.

THEOREM 1.43 *(Hille-Yosida) Let B be a densely defined operator on E. The following assertions are equivalent:*
i) B is the generator of a strongly continuous semigroup.
ii) There exists a constant w such that $(w, \infty) \subseteq \rho(B)$ and

$$\sup_{\lambda > w, n \in \mathbb{N}} \|(\lambda - w)^n (\lambda I - B)^{-n}\|_{\mathcal{L}(E)} < \infty.$$

If the operator B is bounded on E, then it generates a strongly continuous semigroup. In addition, this semigroup is given by

$$e^{tB} = \sum_{k \geq 0} \frac{t^k B^k}{k!}.$$

By analogy to this case, we will denote by $(e^{tB})_{t \geq 0}$ the strongly continuous semigroup generated by the operator B, even when B is not bounded.[2]

The resolvent of the generator B coincides with the Laplace transform of the semigroup, that is,

$$(\lambda I - B)^{-1} = \int_0^\infty e^{-\lambda t} e^{tB} dt \text{ for all } \lambda > w.$$

Conversely, the semigroup can be written in terms of the resolvent. This is given by the exponential formula

$$e^{tB} u = \lim_n (I - \frac{t}{n} B)^{-n} u \text{ for all } u \in E.$$

DEFINITION 1.44 *Let $\psi \in (0, \frac{\pi}{2}]$ and denote by $\Sigma(\psi)$ the open sector*

$$\Sigma(\psi) := \{z \in \mathbb{C}, z \neq 0 \text{ and} | \arg z| < \psi\}.$$

A strongly continuous semigroup $(T(t))_{t \geq 0}$ acting on E is called a bounded holomorphic semigroup on the sector $\Sigma(\psi)$ if $(T(t))_{t \geq 0}$ admits a holomorphic extension $(T(z))_{z \in \Sigma(\psi)}$ such that for each $\theta \in (0, \psi)$, $(T(z))_{z \in \Sigma(\theta)}$ is uniformly bounded and strongly continuous at 0.

If the boundedness assumption on smaller sectors is not required, we say that $(T(t))_{t \geq 0}$ is a holomorphic semigroup on $\Sigma(\psi)$. Finally, by a holomorphic semigroup we mean a semigroup that is holomorphic on some sector of angle > 0.

Note that a holomorphic semigroup on the sector $\Sigma(\psi)$ satisfies

$$T(z + z') = T(z)T(z') \text{ for all } z, z' \in \Sigma(\psi).$$

---

[2]This notation makes sense for self-adjoint operators by the functional calculus.

This is a consequence of holomorphy and the corresponding property for z and $z' \geq 0$.

Holomorphic semigroups play an important role in the theory of evolution equations and functional calculi. In particular, if the semigroup generated by B is holomorphic, then

$$(CP) \begin{cases} \frac{d}{dt}u(t) = & Bu(t), \; t > 0, \\ u(0) = & f \end{cases}$$

have a unique solution for every initial data $f \in E$.

The following theorem characterizes generators of bounded holomorphic semigroups (for a proof see, e.g., [ABHN01], Theorem 3.7.11, [Nag86], Theorem 1.12 Chap. A II, or [EnNa99], Theorem 4.5 Chap. II).

THEOREM 1.45 *Let B be a densely defined operator on a complex Banach space E. Then B generates a semigroup which is bounded holomorphic on $\Sigma(\psi)$ if and only if $\Sigma(\psi + \frac{\pi}{2}) \subseteq \rho(B)$ and for every $\theta \in (0, \psi)$, one has*

$$\sup_{\lambda \in \Sigma(\theta + \frac{\pi}{2})} \|\lambda(\lambda I - B)^{-1}\|_{\mathcal{L}(E)} < \infty.$$

Assume that B generates a semigroup $(T(t))_{t \geq 0}$ which is bounded holomorphic on $\Sigma(\psi)$ for some $\psi > 0$. An application of the Cauchy formula shows that there exists a constant M such that

$$\|BT(t)u\| \leq \frac{M}{t}\|u\| \text{ for all } u \in E \text{ and } t > 0. \tag{1.19}$$

The holomorphy of the semigroup $(T(t))_{t \geq 0}$ on the sector $\Sigma(\psi)$ also implies that for every $\theta \in (-\psi, \psi)$, $(T(e^{i\theta}t))_{t \geq 0}$ is a strongly continuous semigroup on E whose generator is $e^{i\theta}B$. These results and more information on holomorphic semigroups can be found in the books mentioned at the beginning of this section.

1.3.3 Accretive operators on Hilbert spaces

Denote again by H a Hilbert space with inner product $(.;.)$ and norm $\|.\|$. Let A be an operator on H, with domain $D(A)$.

DEFINITION 1.46 *We say that A is an accretive operator if*

$$\Re(Au; u) \geq 0 \text{ for all } u \in D(A).$$

An operator A is m-accretive (or maximal accretive) if it is accretive and $1 \in \rho(-A)$.

It is clear that the operator associated with an accretive form is an accretive operator. If the sesquilinear form is densely defined, accretive, continuous, and closed, then its associated operator is m-accretive (see Proposition 1.22).

Note also that if A is an accretive operator that satisfies the assumptions of Lemma 1.29, its Friedrichs extension is an m-accretive operator. In particular, every densely defined symmetric accretive operator has an m-accretive symmetric extension. Therefore, every densely defined symmetric accretive operator has a self-adjoint extension.

LEMMA 1.47 *Let A be a densely defined accretive operator on H. Then A is closable, its closure $\overline{A}$ is accretive, and for every $\lambda \in \mathbb{K}$, the range $R(\lambda I + A)$ is dense in $R(\lambda I + \overline{A})$.*

Proof. Let $(u_n)_n \in D(A)$ be a sequence such that u_n converges to 0 and Au_n converges to v in H. Take $w \in D(A)$ and apply the accretivity assumption to obtain

$$0 \leq \Re(A(u_n + w); u_n + w)$$
$$= \Re(Au_n; u_n) + \Re(Au_n; w) + \Re(Aw; u_n) + \Re(Aw; w).$$

Letting $n \to \infty$, we obtain $\Re(v; w) + \Re(Aw; w) \geq 0$. We apply this with λw in place of w for $\lambda > 0$ and let $\lambda \to 0$ to obtain $\Re(v; w) \geq 0$. The same inequality applied to $-w$ allows us to conclude that $\Re(v; w) = 0$ and hence $(v; w) = 0$. Since this holds for all $w \in D(A)$, which is dense in H, it follows that $v = 0$. We conclude by Proposition 1.37 that A is closable.

The accretivity of $\overline{A}$ as well as the density of $R(\lambda + A)$ in $R(\lambda + \overline{A})$ follows easily from the definition of $\overline{A}$ and simple approximation arguments. $\square$

LEMMA 1.48 *Let A be a densely defined operator on H.*
1) Assume that A is closed and accretive. Then $I + A$ is injective and has closed range. In particular, A is m-accretive if and only if $I + A$ has dense range.
2) If A is m-accretive, then $(0, \infty) \subseteq \rho(-A)$ and $\lambda(\lambda I + A)^{-1}$ is a contraction operator on H for every $\lambda > 0$.
3) Assume that A is accretive and denote by $\overline{A}$ its closure (cf. Lemma 1.47). Then $\overline{A}$ is m-accretive if and only if there exists $\lambda > 0$ such that $\lambda I + A$ has dense range.

Proof. 1) Let $u \in D(A)$ be such that $u + Au = 0$. The accretivity of A implies that

$$(u; u) \leq \Re(u + Au; u) = 0$$

and hence $u = 0$. This shows that $I + A$ is injective.

In order to show that $I + A$ has closed range, we let $(u_n)_n \in D(A)$ be such that the sequence $u_n + Au_n$ converges to $v \in H$. Since $(u_n; u_n) \le \Re(u_n + Au_n; u_n)$ it follows that u_n is a bounded sequence. Writing

$$(u_n - u_m; u_n - u_m) \le \Re(u_n - u_m; u_n - u_m + Au_n - Au_m)$$
$$\le \|u_n - u_m\| \|u_n + Au_n - u_m - Au_m\|,$$

we see that $(u_n)_n$ is a Cauchy sequence. If u denotes the limit of $(u_n)_n$, then Au_n converges to $v - u$. The fact that A is closed implies that $u \in D(A)$ and $v = u + Au \in R(I + A)$.

By Proposition 1.35, $1 \in \rho(-A)$ if and only if $I + A$ is invertible. Hence, A is m-accretive if and only if $I + A$ has dense range.

2) Assume that A is m-accretive and let $\lambda > 0$. By Proposition 1.35, A is a closed operator. Thus, by the same proposition, $\lambda \in \rho(-A)$ if and only if $\lambda I + A$ is invertible. Applying assertion 1) to the accretive operator $\lambda^{-1}A$, we see that it is enough to prove that $\lambda I + A$ has dense range. Let $f \in H$ be such that

$$(f; \lambda u + Au) = 0 \text{ for all } u \in D(A).$$

Since A is m-accretive we can find $v \in D(A)$ such that $f = v + Av$. Applying the previous equality with $u = v$, gives $v = 0$ and hence $f = 0$. Thus, $R(\lambda I + A)$ is dense.

Now fix $f \in H$ and let $u \in D(A)$ be such that $f = \lambda u + Au$. We write

$$\|f\|^2 = \Re(\lambda u + Au; \lambda u + Au)$$
$$\ge \lambda^2 \|u\|^2 + 2\lambda \Re(Au; u)$$
$$\ge \lambda^2 \|u\|^2.$$

This implies that $\lambda(\lambda I + A)^{-1}$ is a contraction operator on H.

3) By assertion 2), if $\overline{A}$ is m-accretive then $\lambda I + \overline{A}$ is invertible for $\lambda > 0$. By Lemma 1.47, the range $R(\lambda + A)$ is dense in $R(\lambda I + \overline{A}) = H$.

Conversely, assume that $\lambda I + A$ has dense range for some $\lambda > 0$. This implies that $I + \lambda^{-1}A$ has dense range. Thus, by 1), $\lambda^{-1}\overline{A}$ is m-accretive. Assertion 2) implies now that $\alpha I + \lambda^{-1}\overline{A}$ is invertible for all $\alpha > 0$. This implies in particular that $I + \overline{A}$ is invertible and hence $\overline{A}$ is m-accretive. $\square$

The following result is a particular case of the well-known Lumer-Phillips theorem for generators of contraction semigroups.

THEOREM 1.49 *Let A be a densely defined operator on H. The following assertions are equivalent:*
1) The operator A is closable and $-\overline{A}$ is the generator of a strongly continuous contraction semigroup on H.

2) $\overline{A}$ is m-accretive.

3) A is accretive and there exists $\lambda > 0$ such that $\lambda I + A$ has dense range.

Proof. The fact that 2) and 3) are equivalent is already stated in Lemma 1.48.

Assume now that 2) holds. By Lemma 1.48, $(0, \infty) \subseteq \rho(-\overline{A})$ and $\lambda(\lambda I + \overline{A})^{-1}$ is a contraction operator on H. The Hille-Yosida theorem implies that $-\overline{A}$ generates a strongly continuous semigroup $(e^{-t\overline{A}})_{t \geq 0}$ on H. Moreover, for every $u \in D(\overline{A})$, we have

$$\frac{d}{dt}\|e^{-t\overline{A}}u\|^2 = -2\Re(\overline{A}e^{-t\overline{A}}u; e^{-t\overline{A}}u) \leq 0.$$

Hence, $\|e^{-t\overline{A}}u\|^2 \leq \|u\|^2$ for all $t \geq 0$. From this and the density of $D(\overline{A})$, it follows that $e^{-t\overline{A}}$ is a contraction operator on H for every $t \geq 0$. This shows assertion 1).

Conversely, assume that A is closable and $-\overline{A}$ generates a strongly continuous contraction semigroup $(e^{-t\overline{A}})_{t \geq 0}$. Hence, for $t \geq 0$

$$\Re(u - e^{-t\overline{A}}u; u) \geq 0 \text{ for all } u \in H.$$

Applying this to $u \in D(\overline{A})$ yields

$$\Re(\overline{A}u; u) = \lim_{t \downarrow 0} \frac{1}{t}\Re(u - e^{-t\overline{A}}u; u) \geq 0,$$

which shows that $\overline{A}$ is accretive. Since $-\overline{A}$ is the generator of a strongly continuous contraction semigroup, $(0, \infty) \subseteq \rho(-A)$. Thus $\overline{A}$ is m-accretive. $\square$

Let A be a densely defined accretive operator on H and denote by $\overline{A}$ its closure. The operator $\overline{A}$ is accretive, too. The next theorem gives a sufficient condition under which the operator $\overline{A}$ is m-accretive.

THEOREM 1.50 *Let A be an accretive operator on H. Assume that S is an m-accretive operator satisfying the following two conditions:*

1) $D(S) \subseteq D(A)$.

2) There exists a constant $a \in \mathbb{R}$ such that

$$\Re(Au; Su) \geq -a(u; Su) \text{ for all } u \in D(S).$$

Then the closure $\overline{A}$ of A is m-accretive and $D(S)$ is a core of $\overline{A}$.

Proof. Of course, we can assume $a \geq 0$. Considering now $A + aI$ instead of A and applying assertion 3) of Lemma 1.48, we see that we can assume $a = 0$.

Let $B_n := A(I + \frac{1}{n}S)^{-1}$ for $n \geq 1$. For each n, the operator B_n is bounded on H (apply assumption 1) and the closed graph theorem). In addition,

$$\Re\left(Au; \left(I + \frac{1}{n}S\right)u\right) = \Re(Au; u) + \frac{1}{n}\Re(Au; Su) \geq 0 \text{ for all } u \in D(S).$$

This implies that B_n is accretive. On the other hand, the resolvent set of the bounded operator B_n is not empty; thus by assertion 3) of Lemma 1.48 we can conclude that B_n is m-accretive. From the formula

$$I + A + \frac{1}{n}S = (I + B_n)\left(I + \frac{1}{n}S\right)$$

it follows that the operator $I + A + \frac{1}{n}S$ with domain $D(S)$ is invertible. Hence for every $f \in H$ and every n, there exists $u_n \in D(S)$ such that

$$u_n + Au_n + \frac{1}{n}Su_n = f. \tag{1.20}$$

We claim that

$$\|u_n\| \leq \|f\| \quad \text{and} \quad \left\|\frac{1}{n}Su_n\right\| \leq 2\|f\|. \tag{1.21}$$

The first inequality follows form the accretivity of $A + \frac{1}{n}S$, since

$$(u_n; u_n) \leq \Re\left(u_n + Au_n + \frac{1}{n}Su_n; u_n\right)$$
$$= \Re(f; u_n)$$
$$\leq \|f\|\|u_n\|.$$

The second inequality follows from the first one and the following estimates

$$\left\|\frac{1}{n}Su_n\right\|^2 \leq \Re\left(Au_n + \frac{1}{n}Su_n; \frac{1}{n}Su_n\right)$$
$$= \Re\left(f - u_n; \frac{1}{n}Su_n\right)$$
$$\leq (\|f\| + \|u_n\|)\left\|\frac{1}{n}Su_n\right\|.$$

Now we prove that $\overline{A}$ is m-accretive. By Lemma 1.48, it suffices to prove that $I + \overline{A}$ has dense range. Actually, we will show that the operator $I + A$ with domain $D(S)$ has dense range. Let $f \in H$ be such that

$$(f; u + Au) = 0 \text{ for all } u \in D(S). \tag{1.22}$$

Let $(u_n)_n \in D(S)$ be a sequence satisfying (1.20). By (1.22) we have

$$(f;f) = \left(u_n + Au_n + \frac{1}{n}Su_n; f\right) = \left(\frac{1}{n}Su_n; f\right). \qquad (1.23)$$

On the other hand, since S is m-accretive, it follows that the adjoint operator S^* is densely defined.[3] Consider a sequence $(f_k)_k \in D(S^*)$ which converges to f in H. We have

$$\left|\left(\frac{1}{n}Su_n; f\right) - \left(\frac{1}{n}Su_n; f_k\right)\right| \leq \left\|\frac{1}{n}Su_n\right\| \|f - f_k\|$$
$$\leq 2\|f\|\|f - f_k\|.$$

It follows that $|(\frac{1}{n}Su_n; f) - (\frac{1}{n}Su_n; f_k)|$ converges to 0 as $k \to \infty$, uniformly with respect to n. Thus, using the fact that

$$\left|\left(\frac{1}{n}Su_n; f_k\right)\right| = \frac{1}{n}|(u_n; S^*f_k)| \leq \frac{1}{n}\|f\|\|S^*f_k\|$$

(which follows from (1.21)), we obtain

$$\left(\frac{1}{n}Su_n; f\right) \to 0 \text{ as } n \to \infty.$$

We conclude from (1.23) that $f = 0$. Thus, we have proved that the closure B of the restriction of A to $D(S)$ is m-accretive. Since $I + \overline{A}$ is an extension of $I + B$, it follows that $(I + \overline{A})D(\overline{A}) = H$. Thus, $\overline{A}$ is m-accretive.

Finally, it remains to prove the equality $B = \overline{A}$ and conclude that $D(S)$ is a core of $\overline{A}$. If $u \in D(\overline{A})$, there exists $v \in D(B)$ such that

$$u + \overline{A}u = v + Bv = v + \overline{A}v.$$

It follows from the fact that $I + \overline{A}$ is injective that $u = v \in D(B)$. $\qquad \square$

Remark. Let S and A be two operators acting in a Banach space X, with norm $\|.\|$. We assume that S is closed, A is closable and $D(S) \subseteq D(A)$. Then there exist two constants a and b such that

$$\|Au\| \leq a\|Su\| + b\|u\| \text{ for all } u \in D(S).$$

In order to prove this, we first observe that $D(S)$, endowed with the graph norm $\|.\| + \|S.\|$, is a Banach space and the restriction A_S of A to $D(S)$

---

[3]It is easily seen that $I + S^*$ is invertible with inverse $(I + S^*)^{-1} = ((I + S)^{-1})^*$. Now, if $u \in H$ is such that $(u;v) = 0$ for all $v \in D(S^*)$, we write $u = (I + S)\phi$ for some $\phi \in D(S)$ and obtain $(\phi; (I + S^*)v) = 0$. Since this is true for all $v \in D(S^*)$ and $R(I + S^*) = H$, we obtain $\phi = 0$ and then $u = 0$.

can be seen as an operator defined from $D(S)$ into X. Let $(x_n, A_S x_n)$ be a sequence in the graph $G(A_S)$ of A_S. Assume that $(x_n, A_S x_n)$ converges in the Banach space $D(S) \times X$ to (x, y) (here $D(S)$ is endowed with its graph norm). If $\overline{A}$ denotes the closure of A, then $x \in D(\overline{A})$ and $y = \overline{A}x$. But $x \in D(S)$ and hence $y = A_S x$. Consequently, $G(A_S)$ is closed in $D(S) \times X$. We conclude by the closed graph theorem that A_S is a continuous operator from $(D(S), \|.\| + \|S.\|)$ into X, and this implies the desired inequality.

On the basis of this remark, we can add to the conclusions of Theorem 1.50 that every core of S is a core of $\overline{A}$.

1.4 SEMIGROUPS ASSOCIATED WITH SESQUILINEAR FORMS

1.4.1 The semigroup on the Hilbert space H.

In this section we use the same notation as in Section 1.2. Let $\mathfrak{a}$ be a densely defined, accretive, continuous, and closed sesquilinear form on a Hilbert space H (see (1.2)–(1.5)). Denote by A the operator associated with $\mathfrak{a}$. Clearly, A is an accretive operator since $\mathfrak{a}$ is an accretive form. As a consequence of Proposition 1.22 and Theorem 1.49, we have

PROPOSITION 1.51 *The operator* $-A$ *is the generator of a strongly continuous contraction semigroup on* H.

In the next result, we show that the semigroup generated by $-A$ is holomorphic. More precisely,

THEOREM 1.52 *Suppose that H is a complex Hilbert space. Denote by* $(e^{-tA})_{t \geq 0}$ *the semigroup generated by* $-A$ *on H. Then* $(e^{-tA})_{t \geq 0}$ *is a holomorphic semigroup on the sector* $\Sigma(\frac{\pi}{2} - \arctan M)$ *where M is the constant in the continuity assumption (1.4). In addition, for every $\varepsilon \in (0, 1]$,* $e^{-\varepsilon z} e^{-zA}$ *is a contraction operator on H for all* $z \in \Sigma(\frac{\pi}{2} - \arctan \frac{M}{\varepsilon}) = \Sigma(\arctan \frac{\varepsilon}{M})$.

Proof. The continuity assumption (1.4) implies that for every $\varepsilon \in (0, 1]$

$$|\Im(Au; u)| \leq M[\Re(Au; u) + (u; u)]$$
$$\leq \frac{M}{\varepsilon}[\Re(Au; u) + \varepsilon(u; u)] \text{ for all } u \in D(A).$$

Thus, if we set $B := \varepsilon I + A$, the above inequality shows that B is sectorial and

$$|\Im(Bu; u)| \leq \frac{M}{\varepsilon} \Re(Bu; u) \text{ for all } u \in D(B).$$

By Theorem 1.53 or Theorem 1.54 below, we conclude that $-B$ is the generator of a semigroup which is holomorphic on the sector $\Sigma(\frac{\pi}{2} - \arctan \frac{M}{\varepsilon})$. In addition, e^{-zB} is a contraction operator for $z \in \Sigma(\frac{\pi}{2} - \arctan \frac{M}{\varepsilon})$. Since $e^{-tB} = e^{-\varepsilon t}e^{-tA}$ for $t > 0$, we obtain the theorem. $\square$

Observe that if one could write (1.4) as

$$|\mathfrak{a}(u, v)| \leq M'[\mathfrak{Ra}(u, u) + w\|u\|^2]^{1/2}[\mathfrak{Ra}(v, v) + w\|v\|^2]^{1/2}$$

with a constant $M' < M$ (at the cost of enlarging the constant w), then it follows that $(e^{-tA})_{t\geq 0}$ is holomorphic on the larger sector $\Sigma(\frac{\pi}{2} - \arctan M')$. In addition, $e^{-wz}e^{-zA}$ is a contraction operator on H for every z in that sector.

Recall that every densely defined accretive operator B is closable (see Lemma 1.47). We denote by $\overline{B}$ its closure.

THEOREM 1.53 *Let B be a densely defined operator on a complex Hilbert space H. Assume that both B and B^* are sectorial, that is, there exists a non-negative constant C such that for every $u \in D(B)$ and $v \in D(B^*)$,*

$$|\Im(Bu; u)| \leq C\Re(Bu; u) \text{ and } |\Im(B^*v; v)| \leq C\Re(B^*v; v). \qquad (1.24)$$

Then $-\overline{B}$ generates a strongly continuous semigroup $(e^{-t\overline{B}})_{t\geq 0}$ on H. This semigroup is holomorphic on the sector $\Sigma(\frac{\pi}{2} - \arctan C)$ and $e^{-z\overline{B}}$ is a contraction operator on H for every $z \in \Sigma(\frac{\pi}{2} - \arctan C)$.

Proof. Using the definition of $\overline{B}$, we see that the first inequality in (1.24) extends to all $u \in D(\overline{B})$. On the other hand, it follows from the definition of the adjoint operator that B^* is an extension of $(\overline{B})^*$. Thus, (1.24) holds for all $v \in D((\overline{B})^*)$.

Now let $u \in D(\overline{B})$ be such that $\|u\| = 1$ and let $\lambda \in \mathbb{C}$. Denote by dist the usual distance in $\mathbb{C}$. We have

$$\begin{aligned}\|(\lambda I - \overline{B})u\| &\geq |(\lambda u - \overline{B}u; u)| \\ &= |\lambda - (\overline{B}u; u)| \\ &\geq \text{dist}(\lambda, \Sigma(\arctan C)).\end{aligned}$$

Hence, we have

$$\|(\lambda I - \overline{B})u\| \geq \text{dist}(\lambda, \Sigma(\arctan C))\|u\| \text{ for all } u \in D(\overline{B}). \qquad (1.25)$$

It follows that for $\lambda \notin \overline{\Sigma(\arctan C)}$ (where the latter denotes the closed sector), the operator $\lambda I - \overline{B}$ is injective. Moreover, $\lambda I - \overline{B}$ has closed range $R(\lambda I - \overline{B})$ for all $\lambda \notin \overline{\Sigma(\arctan C)}$. To see this, let $v_k = \lambda u_k - \overline{B}u_k$ be a convergent sequence with limit $v \in H$ and apply (1.25) to obtain that $(u_k)_k$

is a Cauchy sequence. Let u be the limit of $(u_k)_k$. The fact that the operator $\overline{B}$ is closed implies that $u \in D(\overline{B})$ and $v = (\lambda I - \overline{B})u \in R(\lambda I - \overline{B})$.

Let us show that $R(\lambda I - \overline{B})$ is dense for $\lambda \notin \overline{\Sigma(\arctan C)}$. If $g \in H$ is such that

$$(g; \lambda u - \overline{B}u) = 0 \text{ for all } u \in D(\overline{B}),$$

then $g \in D((\overline{B})^*)$ and $\overline{\lambda}g - (\overline{B})^*g = 0$. But $\overline{\lambda}$ and the adjoint $(\overline{B})^*$ satisfy the same properties as λ and $\overline{B}$. In particular, $\overline{\lambda}I - (\overline{B})^*$ is injective. This implies that $g = 0$ and thus $R(\lambda I - \overline{B})$ is dense. It follows that $(\lambda I - \overline{B})$ is invertible for all $\lambda \notin \overline{\Sigma(\arctan C)}$. In addition, (1.25) gives

$$\|(\lambda I - \overline{B})^{-1}u\| \leq \frac{1}{\text{dist}(\lambda, \Sigma(\arctan C))}\|u\| \text{ for all } u \in H. \qquad (1.26)$$

Fix now $\theta \in (\arctan C, \pi)$. We have for every $\lambda \notin \Sigma(\theta)$,

$$\sin(\theta - \arctan C) \leq \frac{\text{dist}(\lambda, \Sigma(\arctan C))}{|\lambda|}.$$

It follows from this and (1.26) that for all $u \in H$ and $\lambda \notin \Sigma(\theta)$,

$$\|\lambda(\lambda I - \overline{B})^{-1}u\| \leq \frac{1}{\sin(\theta - \arctan C)}\|u\|.$$

Theorem 1.45 allows us to conclude that $-\overline{B}$ generates a bounded holomorphic semigroup on the sector $\Sigma(\frac{\pi}{2} - \arctan C)$.

It remains to show that $e^{-z\overline{B}}$ is a contraction operator for every $z \in \Sigma(\frac{\pi}{2} - \arctan C)$. Fix $\theta \in (0, \frac{\pi}{2} - \arctan C)$ and consider the semigroup $(e^{-te^{i\theta}\overline{B}})_{t\geq 0}$. For every $u \in H$ and $t > 0$

$$\frac{d}{dt}\|e^{-te^{i\theta}\overline{B}}u\|^2$$

$$= -2\Re(e^{i\theta}\overline{B}e^{-te^{i\theta}\overline{B}}u; e^{-te^{i\theta}\overline{B}}u)$$

$$= -[\Re(\overline{B}e^{-te^{i\theta}\overline{B}}u; e^{-te^{i\theta}\overline{B}}u)\cos\theta - \Im(\overline{B}e^{-te^{i\theta}\overline{B}}u; e^{-te^{i\theta}\overline{B}}u)\sin\theta]$$

$$\leq -\left(\frac{\cos\theta}{C} - \sin\theta\right)|\Im(\overline{B}e^{-te^{i\theta}\overline{B}}u; e^{-te^{i\theta}\overline{B}}u)|.$$

Since $\theta \in (0, \frac{\pi}{2} - \arctan C)$, it follows that $\frac{\cos\theta}{C} - \sin\theta \geq 0$. This shows that $\|e^{-te^{i\theta}B}u\|^2$ is non-increasing (as a function of t). Thus,

$$\|e^{-te^{i\theta}\overline{B}}u\| \leq \|u\| \text{ for all } t > 0$$

and this finishes the proof. $\qquad\qquad\qquad\qquad\qquad\qquad\qquad\qquad\qquad\square$

The sectoriality assumption of the adjoint operator B^* was only used to prove that $\lambda I - \overline{B}$ has dense range. If we assume that there exists $\lambda_0 \in \rho(\overline{B})$ with $\text{dist}(\lambda_0, \Sigma(\arctan C)) > 0$, then the theorem holds without assuming sectoriality of B^*. In order to prove this, we only have to show that $\lambda \in \rho(\overline{B})$ for all λ such that $\text{dist}(\lambda, \Sigma(\arctan C)) > 0$ and argue as in the previous proof. Fix now λ such that $\text{dist}(\lambda, \Sigma(\arctan C)) > 0$ and write

$$\lambda I - \overline{B} = \lambda_0 I - \overline{B} + \lambda I - \lambda_0 I$$
$$= (\lambda_0 I - \overline{B})[I + (\lambda - \lambda_0)(\lambda_0 I - \overline{B})^{-1}].$$

Using (1.26) for λ_0, we see that $\lambda I - \overline{B}$ is invertible for all λ such that $|\lambda - \lambda_0| < \text{dist}(\lambda_0, \Sigma(\arctan C))$. Repeating this procedure, we obtain $\lambda \in \rho(\overline{B})$ for all λ such that $\text{dist}(\lambda, \Sigma(\arctan C)) > 0$. The estimate (1.26) holds for such λ since it is only based on the sectoriality of B. Finally, recall that B is a closed operator whenever its resolvent set $\rho(B)$ is not empty (see Proposition 1.35). Thus, we have proved the following

THEOREM 1.54 *Let B be a densely defined operator on a complex Hilbert space H. Assume that B is sectorial, that is,*

$$|\Im(Bu; u)| \leq C\Re(Bu; u) \ for \ all \ u \in D(B), \tag{1.27}$$

where $C \geq 0$ is a constant. Assume also that there exists $\lambda_0 \in \rho(B)$ with $\text{dist}(\lambda_0, \Sigma(\arctan C)) > 0$. Then $-B$ generates a strongly continuous semigroup which is holomorphic on the sector $\Sigma(\frac{\pi}{2} - \arctan C)$ and such that e^{-zB} is a contraction operator on H for every $z \in \Sigma(\frac{\pi}{2} - \arctan C)$.

Remark. The study of the holomorphy of the semigroup associated with a form $\mathfrak{a}$ requires that the Hilbert space H is complex. In the case where H is real, one uses the following complexification procedure.

Let $H_{\mathbb{C}} := H + iH$ and define the form

$$\tilde{\mathfrak{a}}(u + iv, g + ih) := \mathfrak{a}(u, g) + \mathfrak{a}(v, h) + i[\mathfrak{a}(v, g) - \mathfrak{a}(u, h)] \tag{1.28}$$

for all $u, v, g, h \in D(\mathfrak{a})$. The domain of the form $\tilde{\mathfrak{a}}$ is given by $D(\mathfrak{a}) + iD(\mathfrak{a})$.

One checks easily that the assumptions (1.2)–(1.5) carry over from $\mathfrak{a}$ to $\tilde{\mathfrak{a}}$. The semigroup associated with $\tilde{\mathfrak{a}}$ is given by

$$T(t)(u + iv) := e^{-tA}u + ie^{-tA}v.$$

This is the complexification of the semigroup $(e^{-tA})_{t\geq 0}$. The semigroup $(T(t))_{t\geq 0}$ is holomorphic on $H_{\mathbb{C}}$. From this, one obtains several interesting consequences for the semigroup $(e^{-tA})_{t\geq 0}$ on H. In particular, $e^{-tA}H \subseteq D(A) \subseteq D(\mathfrak{a})$ for all $t > 0$.

1.4.2 Extrapolation to the (anti-) dual space $D(\mathfrak{a})'$

As previously, we denote by $\mathfrak{a}$ a densely defined, accretive, continuous, and closed sesquilinear form on a Hilbert space H and by A the operator associated with $\mathfrak{a}$. The semigroup $(e^{-tA})_{t\geq 0}$ is defined on H and enjoys several interesting properties. In this section, we extend this semigroup to a larger space and prove similar properties for the extension. Our interest in doing this lies in the fact that it is sometimes more flexible to work with the semigroup on the larger space. For example, the function $t \to e^{-tA}u$ has a derivative in H at $t = 0$ only when $u \in D(A)$, whereas in the larger space the derivative may exist for $u \notin D(A)$. This gives a new point of view on the semigroup $(e^{-tA})_{t\geq 0}$ itself, and will allow us to prove other properties of this semigroup and its generator.

Denote by $D(\mathfrak{a})'$ the anti-dual space of $D(\mathfrak{a})$, that is, the space of continuous functionals ϕ such that

$$\phi(u + v) = \phi(u) + \phi(v), \quad \phi(\alpha u) = \overline{\alpha}\phi(u) \text{ for all } \alpha \in \mathbb{K}, u, v \in D(\mathfrak{a}).$$

When H is real, $D(\mathfrak{a})'$ is of course the dual space of $D(\mathfrak{a})$.

Identifying H' with H yields

$$D(\mathfrak{a}) \subset H \subset D(\mathfrak{a})' \tag{1.29}$$

with continuous and dense imbedding. The dualization between $D(\mathfrak{a})'$ and $D(\mathfrak{a})$ is denoted by $< ., . >$ (i.e., $< \phi, u >$ denotes the value of ϕ at u for $u \in D(\mathfrak{a})$ and $\phi \in D(\mathfrak{a})')$. We note that if $\phi \in H$ and $u \in D(\mathfrak{a})$, then $< \phi, u >= (\phi; u)$, the inner product in H.

Fix $u \in D(\mathfrak{a})$ and consider the functional

$$\phi(v) := \mathfrak{a}(u, v), \quad v \in D(\mathfrak{a}).$$

It follows from the continuity assumption (1.4) that ϕ is continuous on $D(\mathfrak{a})$, and hence $\phi \in D(\mathfrak{a})'$. Thus, it can be represented as $\phi(v) = < \mathcal{A}u, v >$, where $\mathcal{A}u \in D(\mathfrak{a})'$ depends on u. Using the fact that $\mathfrak{a}$ is sesquilinear, we see that $\mathcal{A}$ is a linear operator which maps $D(\mathfrak{a})$ into $D(\mathfrak{a})'$. In addition, using again the continuity assumption (1.4), we have

$$\begin{aligned}
\|\mathcal{A}u\|_{D(\mathfrak{a})'} &= \sup_{\|v\|_{\mathfrak{a}}\leq 1} | < \mathcal{A}u, v > | \\
&= \sup_{\|v\|_{\mathfrak{a}}\leq 1} |\mathfrak{a}(u, v)| \\
&\leq M\|u\|_{\mathfrak{a}}.
\end{aligned}$$

Thus, $\mathcal{A}$ is a continuous operator from $D(\mathfrak{a})$ (endowed with the norm $\|.\|_{\mathfrak{a}}$) into $D(\mathfrak{a})'$. The operator $\mathcal{A}$ can also be seen as an unbounded operator on

$D(\mathfrak{a})'$, with domain $D(\mathcal{A}) = D(\mathfrak{a})$, and such that

$$\mathfrak{a}(u, v) = < \mathcal{A}u, v > \quad \text{for all } u, v \in D(\mathfrak{a}). \qquad (1.30)$$

Now let A be the operator associated with $\mathfrak{a}$ (defined in Section 1.2.3). Using the fact that $D(\mathfrak{a})$ is dense in H and the definition of A, we see that A is precisely the part of $\mathcal{A}$ in H. That is,

$$D(A) = \{u \in D(\mathcal{A}); \mathcal{A}u \in H\} \text{ and } Au = \mathcal{A}u \text{ for } u \in D(A).$$

The following result shows that $(e^{-tA})_{t \geq 0}$ extends from H to the larger space $D(\mathfrak{a})'$.

THEOREM 1.55 *Let $\mathfrak{a}$ be a densely defined, accretive, continuous, and closed sesquilinear form on a Hilbert space H. Then the operator $-\mathcal{A}$, with domain $D(\mathcal{A}) = D(\mathfrak{a})$, generates a strongly continuous semigroup $(e^{-t\mathcal{A}})_{t \geq 0}$ on $D(\mathfrak{a})'$. Moreover,*

$$e^{-t\mathcal{A}}f = e^{-tA}f \quad \text{for every } f \in H \text{ and } t \geq 0. \qquad (1.31)$$

If H is complex, the semigroup $(e^{-t\mathcal{A}})_{t \geq 0}$ is holomorphic (on $D(\mathfrak{a})'$) on the sector $\Sigma(\frac{\pi}{2} - \arctan M)$, where M is the constant in (1.4).

Proof. We first assume that the Hilbert space H is complex. Given $u \in D(\mathfrak{a})$, let $\phi = (\lambda + 1 + \mathcal{A})u$. Clearly,

$$< \phi, u > = \lambda(u; u) + (u; u) + < \mathcal{A}u, u >$$
$$= \lambda(u; u) + (u; u) + \mathfrak{a}(u, u).$$

Hence,

$$\|u\|_{\mathfrak{a}}^2 \leq \|\phi\|_{D(\mathfrak{a})'} \|u\|_{\mathfrak{a}} + |\lambda| \|u\|^2. \qquad (1.32)$$

On the other hand, by Lemma 1.9 we have for every $u \in D(\mathfrak{a})$ with $u \neq 0$,

$$\mathfrak{a}\left(\frac{u}{\|u\|}, \frac{u}{\|u\|}\right) + \left(\frac{u}{\|u\|}; \frac{u}{\|u\|}\right) \in \Sigma(\arctan M).$$

Thus,

$$\|u\|_{\mathfrak{a}} \|\phi\|_{D(\mathfrak{a})'} = \|u\|_{\mathfrak{a}} \|(\lambda + 1 + \mathcal{A})u\|_{D(\mathfrak{a})'}$$
$$\geq | < (I + \lambda I + \mathcal{A})u, u > |$$
$$= \left|\lambda + \mathfrak{a}\left(\frac{u}{\|u\|}, \frac{u}{\|u\|}\right) + \left(\frac{u}{\|u\|}; \frac{u}{\|u\|}\right)\right| \|u\|^2$$
$$\geq \text{dist}(\lambda, -\Sigma(\arctan M)) \|u\|^2.$$

We have proved that

$$\|u\|_{\mathfrak{a}}\|\phi\|_{D(\mathfrak{a})'} \geq \mathrm{dist}(\lambda, -\Sigma(\arctan M))\|u\|^2. \qquad (1.33)$$

Fix $\theta \in (\arctan M, \pi)$. We have, as in the proof of Theorem 1.53,

$$\mathrm{dist}(\lambda, -\Sigma(\arctan M)) \geq |\lambda| \sin(\theta - \arctan M)$$

for all $\lambda \notin -\Sigma(\theta)$. Inserting this in (1.33) gives

$$\|u\|_{\mathfrak{a}}\|\phi\|_{D(\mathfrak{a})'} \geq c|\lambda|\|u\|^2 \text{ for all } \lambda \notin -\Sigma(\theta), \qquad (1.34)$$

where c is a positive constant. The last estimate together with (1.32) gives
for all $\lambda \notin -\Sigma(\theta)$

$$\|u\|_{\mathfrak{a}} \leq \left(1 + \frac{1}{c}\right)\|\phi\|_{D(\mathfrak{a})'} = C\|(\lambda I + I + \mathcal{A})u\|_{D(\mathfrak{a})'}. \qquad (1.35)$$

The estimate (1.35) shows that $\lambda I + I + \mathcal{A}$ is invertible on $D(\mathfrak{a})'$ for all
$\lambda \notin -\Sigma(\theta)$. Indeed, it is clear that $\lambda I + I + \mathcal{A}$ is injective. It has dense
range because H is dense in $D(\mathfrak{a})'$ and

$$(\lambda I + I + \mathcal{A})D(\mathfrak{a}) \supseteq (\lambda I + I + A)D(A) = H,$$

where the last equality follows from the fact that $\lambda \in \rho(-A)$ (see Theorems
1.52 and 1.45). Finally, if $(\lambda I + I + \mathcal{A})u_n$ is a convergent sequence in
$D(\mathfrak{a})'$, then we obtain from (1.35) that (u_n) is a Cauchy sequence in $D(\mathfrak{a})$.
It is then convergent in $D(\mathfrak{a})$. From the continuity of $\mathcal{A}$, as an operator from
$D(\mathfrak{a})$ into $D(\mathfrak{a})'$, we obtain that that $\lambda I + I + \mathcal{A}$ has closed range. Thus
$\lambda I + I + \mathcal{A}$ is invertible in $D(\mathfrak{a})'$ for $\lambda \notin -\Sigma(\theta)$.

Now let $v \in D(\mathfrak{a})$. We have

$$|\lambda| | < u, v > | = | < \phi, v > -(u; v) - \mathfrak{a}(u, v)|$$
$$\leq \|\phi\|_{D(\mathfrak{a})'}\|v\|_{\mathfrak{a}} + (M+1)\|u\|_{\mathfrak{a}}\|v\|_{\mathfrak{a}}.$$

Taking the supremum over $\|v\|_{\mathfrak{a}} \leq 1$ and using (1.35), we obtain

$$|\lambda|\|u\|_{D(\mathfrak{a})'} \leq \|\phi\|_{D(\mathfrak{a})'} + (M+1)\|u\|_{\mathfrak{a}}$$
$$\leq \|\phi\|_{D(\mathfrak{a})'} + (M+1)C\|\phi\|_{D(\mathfrak{a})'}$$
$$= C'\|(\lambda I + I + \mathcal{A})u\|_{D(\mathfrak{a})'}.$$

We have proved that $\lambda I + I + \mathcal{A}$ is invertible on $D(\mathfrak{a})'$ for $\lambda \notin -\Sigma(\theta)$ and

$$\sup_{\lambda \notin -\Sigma(\theta)} \|\lambda(\lambda I + I + \mathcal{A})^{-1}\|_{\mathcal{L}(D(\mathfrak{a})')} < \infty.$$

Theorem 1.45 ensures that $-(\mathcal{A} + I)$ generates on $D(\mathfrak{a})'$ a semigroup which is bounded holomorphic on the sector $\Sigma(\frac{\pi}{2} - \arctan M)$.

Now we prove (1.31). Let $f \in H$ and $u(t) := e^{-t\mathcal{A}}f - e^{-tA}f$. From the embedding $H \subset D(\mathfrak{a})'$, it follows that $\frac{d}{dt}e^{-t\mathcal{A}}f$ exists (at each $t > 0$) in $D(\mathfrak{a})'$ and equals $-\mathcal{A}e^{-t\mathcal{A}}f = -\mathcal{A}e^{-t\mathcal{A}}f$. This gives

$$\frac{d}{dt}u(t) = -\mathcal{A}u(t), \ t > 0.$$

Thus, $u(t) = e^{-t\mathcal{A}}u(0) = 0$ and the desired equality holds.

We have proved the theorem in the case where H is complex. Now, if H is real, we use the complexification argument described in the Remark at the end of the previous section. One obtains then a holomorphic semigroup $(T(t))_{t \geq 0}$ on $(D(\mathfrak{a}) + iD(\mathfrak{a}))'$ whose generator is (minus) the operator associated with the form defined in (1.28)). If $\phi \in D(\mathfrak{a})'$, then it can be written as the limit (in $D(\mathfrak{a})'$) of a sequence $(u_n) \in H$ and hence $T(t)\phi$ is the limit of $T(t)u_n = e^{-t\mathcal{A}}u_n \in H$. Consequently, $T(t)\phi \in D(\mathfrak{a})'$ for every $t \geq 0$. Therefore, $T(t)D(\mathfrak{a})' \subseteq D(\mathfrak{a})'$ for all $t \geq 0$. Thus, the restriction of $(T(t))_{t \geq 0}$ to $D(\mathfrak{a})'$ is then a strongly continuous semigroup whose generator is $-\mathcal{A}$. $\square$

It is shown in this proof that if H is complex, the semigroup satisfies

$$\sup_{z \in \Sigma(\psi)} \|e^{-z}e^{-z\mathcal{A}}\|_{\mathcal{L}(D(\mathfrak{a})')} < \infty$$

for all $0 \leq \psi < \frac{\pi}{2} - \arctan M$. For the same reasons as in Theorem 1.52, we have

$$\sup_{z \in \Sigma(\psi)} \|e^{-\varepsilon z}e^{-z\mathcal{A}}\|_{\mathcal{L}(D(\mathfrak{a})')} < \infty$$

for all $0 \leq \psi < \frac{\pi}{2} - \arctan \frac{M}{\varepsilon}$ and all $\varepsilon \in (0, 1]$.

If the form $\mathfrak{a}$ is sectorial, i.e.,

$$|\Im\mathfrak{a}(u, u)\| \leq M\Re\mathfrak{a}(u, u) \text{ for all } u \in D(\mathfrak{a}), \qquad (1.36)$$

then

$$\sup_{z \in \Sigma(\psi)} \|e^{-z\mathcal{A}}\|_{\mathcal{L}(D(\mathfrak{a})')} < \infty$$

for all $0 \leq \psi < \frac{\pi}{2} - \arctan M$. The proof is the same as the previous one, replacing (1.4) by (1.36). That is, we can replace $I + \mathcal{A}$ by $\mathcal{A}$ in the previous proof.

Note also that these estimates hold in $(D(\mathfrak{a}), \|.\|_\mathfrak{a})$. More precisely, for every $\varepsilon \in (0,1]$,

$$\sup_{z \in \Sigma(\psi)} \|e^{-\varepsilon z} e^{-zA}\|_{\mathcal{L}(D(\mathfrak{a}))} < \infty \tag{1.37}$$

for all $0 \leq \psi < \frac{\pi}{2} - \arctan \frac{M}{\varepsilon}$. Indeed, let $u \in D(\mathfrak{a})$ and write

$$\|e^{-zA}u\|_\mathfrak{a}^2 = \Re < e^{-zA}\mathcal{A}u, e^{-zA}u > + \|e^{-zA}u\|^2$$
$$\leq \|e^{-zA}\mathcal{A}u\|_{D(\mathfrak{a})'}\|e^{-zA}u\|_\mathfrak{a} + \|e^{-zA}u\|^2.$$

Hence,

$$\|e^{-zA}u\|_\mathfrak{a}^2 \leq \|e^{-zA}\mathcal{A}u\|_{D(\mathfrak{a})'}^2 + 2\|e^{-zA}u\|^2. \tag{1.38}$$

It follows from Theorem 1.52 and the above observations that both terms $\|e^{-z\varepsilon}e^{-zA}\|_{D(\mathfrak{a})'}$ and $\|e^{-z\varepsilon}e^{-zA}\|_{\mathcal{L}(H)}$ are uniformly bounded on $\Sigma(\psi)$ for $0 \leq \psi < \frac{\pi}{2} - \arctan \frac{M}{\varepsilon}$. Using this and the fact that $\mathcal{A}$ is a bounded operator from $D(\mathfrak{a})$ into $D(\mathfrak{a})'$, we see that (1.37) follows from (1.38).

Again, if we assume the stronger condition (1.36), we obtain

$$\sup_{z \in \Sigma(\psi)} \|e^{-zA}\|_{\mathcal{L}(D(\mathfrak{a}))} < \infty. \tag{1.39}$$

Proof of Lemma 1.25. First, $e^{-tA}H \subseteq D(A)$ for all $t > 0$. Indeed, if H is complex the semigroup generated by $-A$ is holomorphic on H (cf. Theorem 1.52) and this implies trivially the above inclusion. Now if H is real, we use again the complexification argument to obtain a holomorphic semigroup $(e^{-t(A+iA)})_{t \geq 0}$ on $H + iH$, from which we obtain $e^{-tA}H \subseteq D(A)$ for $t > 0$. We prove that every $u \in D(\mathfrak{a})$ can be approximated in $(D(\mathfrak{a}), \|.\|_\mathfrak{a})$ by $e^{-tA}u$. We have

$$\|e^{-tA}u - u\|_\mathfrak{a}^2 = \Re < e^{-tA}\mathcal{A}u - \mathcal{A}u, e^{-tA}u - u > + \|e^{-tA}u - u\|^2$$
$$\leq \|e^{-tA}\mathcal{A}u - \mathcal{A}u\|_{D(\mathfrak{a})'}\|e^{-tA}u - u\|_\mathfrak{a} + \|e^{-tA}u - u\|^2.$$

Hence

$$\|e^{-tA}u - u\|_\mathfrak{a}^2 \leq \|e^{-tA}\mathcal{A}u - \mathcal{A}u\|_{D(\mathfrak{a})'}^2 + 2\|e^{-tA}u - u\|^2.$$

The strong continuity of $(e^{-tA})_{t \geq 0}$ on $D(\mathfrak{a})'$ and of $(e^{-tA})_{t \geq 0}$ on H imply

$$\|e^{-tA}u - u\|_\mathfrak{a} \to 0 \text{ as } t \to 0.$$

This proves the lemma. □

It is seen in this proof that $e^{-tA}H \subseteq D(A) \subseteq D(\mathfrak{a})$ for $t > 0$ and that the restriction of $(e^{-tA})_{t \geq 0}$ is a strongly continuous semigroup on $(D(\mathfrak{a}), \|.\|_\mathfrak{a})$.

If the Hilbert space H is complex, then Theorem 1.55 and the same arguments as in the above proof show that

$$\|e^{-zA}u - u\|_{\mathfrak{a}} \to 0 \text{ as } z \to 0, z \in \Sigma(\psi)$$

for every $u \in D(\mathfrak{a})$ and every $\psi \in [0, \frac{\pi}{2} - \arctan M)$.

If for each $t \geq 0$, $T(t)$ denotes the restriction of e^{-tA} to $D(\mathfrak{a})$, then $(T(t))_{t\geq 0}$ is a strongly continuous semigroup on $(D(\mathfrak{a}), \|.\|_{\mathfrak{a}})$. It is also holomorphic on the sector $\Sigma(\frac{\pi}{2} - \arctan M)$ when the space H is complex. If $-B$ denotes the generator of $T(t)_{t\geq 0}$, then B is the part of A in $D(\mathfrak{a})$, that is,

$$D(B) = \{u \in D(A), Au \in D(\mathfrak{a})\}, \quad Bu = Au \text{ for all } u \in D(B).$$

1.5 CORRESPONDENCE BETWEEN FORMS, OPERATORS, AND SEMI-GROUPS

Let $\mathfrak{a}$ be a sesquilinear form on H which satisfies the standard assumptions (1.2)–(1.5). One associates with $\mathfrak{a}$ an operator A and a semigroup $(e^{-tA})_{t\geq 0}$. In this section we show that there is a unique correspondence between sesquilinear forms and a class of operators and semigroups.

The first result shows that $\mathfrak{a}$ can be described completely by its associated semigroup $(e^{-tA})_{t\geq 0}$.

LEMMA 1.56 *Let $\mathfrak{a}$ be a densely defined, accretive, continuous, and closed sesquilinear form. Let $u \in H$. Then $u \in D(\mathfrak{a})$ if and only if*

$$\sup_{t>0} \frac{1}{t} \Re(u - e^{-tA}u; u) < \infty.$$

In addition, for every $u, v \in D(\mathfrak{a})$

$$\mathfrak{a}(u, v) = \lim_{t\downarrow 0} \frac{1}{t}(u - e^{-tA}u; v).$$

Proof. Let $u, v \in D(\mathfrak{a})$. By Theorem 1.55 we have

$$\frac{1}{t}(u - e^{-tA}u; v) = \frac{1}{t} < u - e^{-tA}u, v > .$$

Since $u \in D(\mathfrak{a}) = D(\mathcal{A})$, we have

$$\frac{1}{t} < u - e^{-tA}u, v > \to < \mathcal{A}u, v > = \mathfrak{a}(u, v) \text{ as } t \to 0.$$

This proves the last assertion of the lemma. In particular,

$$\frac{1}{t}(u - e^{-tA}u; u) \to \mathfrak{a}(u, u) \text{ for all } u \in D(\mathfrak{a}).$$

Assume now that $u \in H$ is such that $\sup_{t>0} \frac{1}{t}\Re(u - e^{-tA}u; u) < \infty$. For $\lambda > 0$, we write for simplicity $(\lambda I + A)^{-1} = R(\lambda)$. We have

$$
\begin{aligned}
\Re\mathfrak{a}(\lambda R(\lambda)u; \lambda R(\lambda)u) &= \Re\lambda(AR(\lambda)u; \lambda R(\lambda)u) \\
&= \Re\lambda(u - \lambda R(\lambda)u; \lambda R(\lambda)u) \\
&\leq \Re\lambda(u - \lambda R(\lambda)u; u) \\
&= \Re\int_0^\infty \lambda^2 e^{-\lambda t}(u - e^{-tA}u; u)dt \\
&\leq \sup_{t>0}\frac{1}{t}\Re(u - e^{-tA}u; u)\int_0^\infty t\lambda^2 e^{-\lambda t}dt \\
&= \sup_{t>0}\frac{1}{t}\Re(u - e^{-tA}u; u)\int_0^\infty se^{-s}ds.
\end{aligned}
$$

It follows now that $\lambda R(\lambda)u$ is bounded uniformly with respect to λ in $(D(\mathfrak{a}), \|.\|_\mathfrak{a})$ (recall that $\lambda R(\lambda)$ is a contraction operator on H by Proposition 1.22). In addition, $\lambda R(\lambda)$ converges strongly to the identity operator in H as $\lambda \to +\infty$ (this can be seen by again using Proposition 1.22 and $\|u - \lambda R(\lambda)u\| = \|R(\lambda)Au\| \leq \lambda^{-1}\|Au\|$ for $u \in D(A)$. The desired convergence then follows by the density of $D(A)$ in H). By Lemma 1.32, we deduce that $u \in D(\mathfrak{a})$. $\qquad\square$

A natural question is how to recognize in Hilbert spaces those operators or semigroups that are associated with sesquilinear forms. In the next results, we describe such operators and semigroups.

If A is the operator associated with a densely defined, accretive, continuous, and closed sesquilinear form $\mathfrak{a}$, then $I + A$ is sectorial (cf. Lemma 1.9) and A is m-accretive (cf. Proposition 1.22). The next result shows that these properties characterize operators that are associated with sesquilinear forms.

THEOREM 1.57 *Let A be an m-accretive operator on a complex Hilbert space H. Assume that $I+A$ is sectorial. Then there exists a unique sesquilinear form $\mathfrak{a}$ which is densely defined, accretive, continuous, and closed and such that A is the operator associated with $\mathfrak{a}$.*

Proof. Define the form

$$\mathfrak{b}(u, v) := (Au; v), \quad D(\mathfrak{b}) = D(A).$$

By Lemma 1.29, the form $\mathfrak{b}$ is closable. Let A_1 be the operator associated with the closure $\mathfrak{a} := \overline{\mathfrak{b}}$ of $\mathfrak{b}$.[4] By assumption, $I + A$ is invertible. By Proposition 1.22, $I + A_1$ is invertible, too. Thus, $I + A_1$ is an extension of $I + A$ and both operators are invertible. This implies that $A = A_1$ and hence A is the operator associated with the form $\mathfrak{a}$.

The uniqueness of $\mathfrak{a}$ follows from Lemma 1.25. $\square$

If we assume in the previous theorem that A is sectorial, then the associated form $\mathfrak{a}$ is sectorial. This follows from the above proof, since then $\mathfrak{b}$ is sectorial and this also holds for the closure $\mathfrak{a} = \overline{\mathfrak{b}}$.

Note also that if A is self-adjoint, then $\mathfrak{a}$ is a symmetric form.

Theorem 1.53 asserts that the semigroup generated by (minus) the operator associated with a sesquilinear form is holomorphic and $e^{-z}e^{-zA}$ is a contraction operator for every z in some sector. Here we give a converse to that result.

THEOREM 1.58 *Let $(T(t))_{t \geq 0}$ be a contraction semigroup acting on a complex Hilbert space H. Assume that this semigroup is holomorphic on the sector $\Sigma(\psi)$ (for some $\psi \in (0, \frac{\pi}{2})$) and such that for every $z \in \Sigma(\psi)$, $e^{-z}T(z)$ is a contraction operator on H. Then the generator of $(T(t))_{t \geq 0}$ is (minus) the operator associated with a densely defined, accretive, continuous, and closed sesquilinear form.*

Proof. Denote by $-A$ the generator of the contraction semigroup $(T(t))_{t \geq 0}$. The operator A is accretive by Theorem 1.49. Now fix $\theta \in (0, \psi)$. The semigroup $(T(te^{i\theta})e^{-te^{i\theta}})_{t \geq 0}$ is contractive on H and its generator is $-e^{i\theta}(I + A)$. Thus, $e^{i\theta}(I+A)$ is accretive and hence $\Re(e^{i\theta}(I+A)u; u) \geq 0$ for every $u \in D(A)$. This gives $\Im(Au; u) \leq \frac{1}{\tan\theta}\Re((I + A)u; u)$. For similar reasons, $\Re(e^{-i\theta}(I+A)u; u) \geq 0$ and thus $-\Im(Au; u) \leq \frac{1}{\tan\theta}\Re((I+A)u; u)$. It follows that $I+A$ is a sectorial operator. The proof is finished by applying Theorem 1.57. $\square$

Remark. If we assume that $e^{-\alpha z}T(z)$ is a contraction operator for every $z \in \Sigma(\psi)$, then we obtain an operator A such that $\alpha I + A$ is sectorial. In particular, if $T(z)$ is a contraction for all $z \in \Sigma(\psi)$, then A is sectorial.

Applying Lemma 1.56, we can reformulate the above theorem as follows.

THEOREM 1.59 *Let $(T(t))_{t \geq 0}$ be a contraction semigroup acting on a complex Hilbert space H. Assume that this semigroup is holomorphic on the sector $\Sigma(\psi)$ (for some $\psi \in (0, \frac{\pi}{2})$) and such that for every $z \in \Sigma(\psi)$, $e^{-z}T(z)$*

---

[4] A_1 is the Friedrich extension of A.

is a contraction operator on H. Then the form given by

$$\mathfrak{a}(u, v) := \lim_{t \downarrow 0} \frac{1}{t}(u - T(t)u; v),$$

$$D(\mathfrak{a}) := \left\{ u \in H, \sup_{t > 0} \frac{1}{t}\Re(u - e^{-tA}u; u) < \infty \right\},$$

is densely defined, accretive, continuous, and closed and $(T(t))_{t \geq 0}$ is its associated semigroup.

Notes

The material of this chapter is known. See Davies [Dav80], Kato [Kat80], Lions [Lio61], Reed-Simon [ReSi80] or Tanabe [Tan79]. Our presentation is however different and some proofs are simplified by using semigroups from the beginning. One of the aims of this chapter is to give a systematic account of the interplay between forms, operators, and semigroups.

Sections 1.1 and 1.2. Bounded sesquilinear forms on Hilbert spaces can be found in many textbooks on Functional Analysis. For the Lax-Milgram lemma see, e.g., Brezis [Bre92], Lions [Lio61], Yosida [Yos65].

An exhaustive study of sectorial forms can be found in [Kat80]. In [ReSi80] closed sectorial forms are called strictly accretive. Note also that the notion of a sectorial form in [ReSi80] is slightly different from ours. Operators associated with forms are called regularly accretive in [Tan79].

Proposition 1.13 is sometimes considered as the definition of a closable form. Example 1.2.1 of a symmetric form which is not closable is borrowed from [Kat80]. At this point, we mention the following remarkable result proved by Simon [Sim78] (see also Reed-Simon [ReSi80]).

Theorem. *Let $\mathfrak{a}$ be a symmetric non-negative form on a Hilbert space H. Then, there exists a largest closable symmetric form $\mathfrak{a}_r$ that is smaller than $\mathfrak{a}$.*

In this theorem, a symmetric form $\mathfrak{b}$ is said to be smaller than $\mathfrak{a}$ if $D(\mathfrak{a}) \subseteq D(\mathfrak{b})$ and $\mathfrak{b}(u, u) \leq \mathfrak{a}(u, u)$ for all $u \in D(\mathfrak{a})$.

Proposition 1.18 and related results can also be found in [Kat80]. Theorem 1.19 is often called the KLMN theorem. The version given here can be found in [Kat80] (where it is formulated for sectorial forms). This theorem was proved in various versions by Kato [Kat55], Lions [Lio61], Lax-Milgram [LaMi54]. See also Nelson [Nel64] and Reed-Simon [ReSi75].

Section 1.3. Semigroup theory and its various applications is a well documented subject; hence we make only brief comments and give some more references. The fundamental Hille-Yosida generation theorem was proved in 1948 and became the starting point of the subsequent theory of semigroups. Because of applications to equations of different types and because of interactions with other fields of analysis and probability, the theory has seen important developments. Several textbooks on semigroups are now available. Systematic treatments of semigroups as well as applications to many equations can be found in Arendt et al. [ABHN01], Clement et al. [CHADP87], Davies [Dav80], Engel-Nagel [EnNa99], Fattorini [Fat83], Goldstein [Gol85], Hille-Phillips [HiPh57], Kato [Kat80], Nagel et al. [Nag86], Pazy [Paz83], Robinson [Rob96], Yosida [Yos65]. For holomorphic semigroups and applications to parabolic problems, see Amann [Ama95] and Lunardi [Lun95]. Applications to Volterra equations are given in Engel-Nagel [EnNa99] and Prüss [Prü93].

We gave the definition of accretive operators only in the Hilbert space case. The reason why we considered only operators on Hilbert spaces is to make connection with operators that are associated with accretive forms. The definition makes sense for operators acting on Banach spaces. This was introduced by Phillips [Phi59] (he used the terminology of dissipative operator; which means that $-A$ is accretive). A linear operator A acting on a Banach space E is called accretive (or $-A$ is dissipative) if $\Re < Au, u^* > \geq 0$ for all $u \in D(A)$ and some u^* in the subdifferential of the norm $\|.\|_E$ of E at x. That is, u^* is in the dual space E' and such that $\|u^*\|_{E'} \leq 1$ and $< u, u^* > = \|u\|_E$, where $< .,. >$ denotes the pairing between E and E'.

Theorem 1.49 is the Hilbert space version of the well-known Lumer-Phillips theorem proved in [LuPh61]. The latter holds for accretive operators on any Banach space. Note also that similar theorems hold for nonlinear accretive operators, see Bénilan-Crandall-Pazy [BCP90].

Theorem 1.50 is due to Kato and is taken from Okazawa [Oka80], where Banach space versions are also given.

Sections 1.4 and 1.5. Related results to Theorems 1.53 and 1.54 can be found in Goldstein [Gol85] and Kato [Kat80]. Theorem 1.55 can be found in a different form in Tanabe [Tan79], but here we give a more precise angle of holomorphy in $D(\mathfrak{a})'$. Lemma 1.56 is an extension to the nonsymmetric case of a well-known result for symmetric forms (in the latter case, it is usually proved by using the spectral theorem for self-adjoint operators). The "if" part of this lemma is shown in Albeverio, Röckner, and Stannat [ARS95] and the "only if" part in Ouhabaz [Ouh92a] (see also [Ouh96]). Finally, Theorems 1.57 and 1.58 are implicit in [Kat80] and [Tan79].

Chapter Two

CONTRACTIVITY PROPERTIES

Let H be a Hilbert space over $\mathbb{K} = \mathbb{R}$ or $\mathbb{C}$. Denote by $\mathfrak{a}$ a sesquilinear form on H. We assume that $\mathfrak{a}$ is densely defined, accretive, continuous, and closed (see (1.2)–(1.5)). Denote by A its associated operator. We have seen in the previous chapter that $-A$ generates a strongly continuous semigroup $(e^{-tA})_{t \geq 0}$ on H. Assume now that $H = L^2(X, \mu, \mathbb{C})$, where (X, μ) is a σ-finite measure space. Several properties of the semigroup like positivity, L^p-contractivity, domination, and so on can be characterized in terms of the operator A. However, in most applications, one does not precisely know the operator A. Typical situations where this occurs are when A is an elliptic operator with measurable coefficients and acts on $L^2(\Omega)$, where Ω is any open subset of $\mathbb{R}^n$ (see Chapter 4). Thus, characterizations in terms of the generator cannot be applied in several situations. On the contrary, in most situations one knows the form $\mathfrak{a}$.[1] Therefore, criteria for properties of the semigroup $(e^{-tA})_{t \geq 0}$ would be more useful and powerful if they are given in terms of the form.

In the present chapter, we give criteria in terms of the form $\mathfrak{a}$ for positivity, irreducibility, and L^∞-contractivity of the semigroup $(e^{-tA})_{t \geq 0}$. We also study the domination property of semigroups by using the associated forms. The results are in the spirit of the famous Beurling-Deny criteria. The latter characterize the sub-Markovian property of semigroups associated with symmetric forms. We will consider forms that are not necessarily symmetric and recover the Beurling-Deny criteria. The method used here works also for semigroups acting on vector-valued functions. The approach is based on criteria for invariance of closed convex sets of H under the action of the semigroup. These criteria hold in a general setting. The previously mentioned properties of the semigroup are obtained as particular cases, by choosing the appropriate convex set.

It should be emphasized that all the results in the present chapter hold in both complex or real spaces. We do not distinguish the two cases unless we mention this explicitly. As in the previous chapter, we write real and imaginary parts of elements of $\mathbb{K}$ without assuming $\mathbb{K} = \mathbb{C}$. In the case

---

[1] Usually, one starts by defining the form, hence its expression and domain are known. The associated operator A is given by Definition 1.21. In several situations, one cannot describe A precisely.

of real Hilbert spaces, the real part $\Re$ should of course be omitted in the statements.

2.1 INVARIANCE OF CLOSED CONVEX SETS

Let H be a complex or real Hilbert space with norm and scalar product denoted by $\|.\|$ and $(.;.)$. Let $\mathfrak{a}$ be a densely defined, accretive, continuous, and closed sesquilinear form on H. Denote by A and $(e^{-tA})_{t\geq 0}$ the operator and semigroup associated with $\mathfrak{a}$. Let C be a non-empty closed convex subset of H and denote by $\mathcal{P}$ the projection of H onto C. Recall that for every $f \in H, \mathcal{P}f \in C$ and satisfies

$$\|f - \mathcal{P}f\| = \min_{g \in C} \|f - g\|.$$

It is an elementary fact that $\mathcal{P}f$ is characterized by

$$[\, h = \mathcal{P}f \,] \Leftrightarrow [\, h \in C \text{ and } \Re(f - h; g - h) \leq 0 \text{ for all } g \in C \,]. \quad (2.1)$$

This section is devoted to characterizations, in terms of the form $\mathfrak{a}$, of the invariance of the convex set C under the action of the semigroup $(e^{-tA})_{t\geq 0}$. That is, the property

$$e^{-tA}C \subseteq C \text{ for all } t \geq 0,$$

by which we mean $e^{-tA}u \in C$ for every $u \in C$ and every $t \geq 0$.

We first show the following

PROPOSITION 2.1 *The following assertions are equivalent:*
1) $e^{-tA}C \subseteq C$ *for all* $t \geq 0$.
2) $\lambda(\lambda I + A)^{-1}C \subseteq C$ *for all* $\lambda > 0$.

Proof. Assume that for all $t \geq 0, e^{-tA}C \subseteq C$. Let $\lambda > 0$ and $u \in C$. Since the resolvent is the Laplace transform of the semigroup, we have

$$\lambda(\lambda I + A)^{-1}u = \lambda \int_0^\infty e^{-\lambda t}e^{-tA}u\,dt.$$

Assume for a contradiction that $\lambda(\lambda I + A)^{-1}u \notin C$. By the Hahn-Banach theorem, there exist a constant $\alpha \in \mathbb{R}$ and a linear continuous functional ϕ on H such that

$$\Re\phi(\lambda(\lambda I + A)^{-1}u) > \alpha \geq \Re\phi(g) \text{ for all } g \in C.$$

Applying this with $g = e^{-tA}u$ gives

$$\Re\phi(\lambda(\lambda I + A)^{-1}u) > \alpha$$

$$= \int_0^\infty \lambda e^{-\lambda t}\alpha dt$$

$$\geq \Re\phi\left(\int_0^\infty \lambda e^{-\lambda t}e^{-tA}u dt\right)$$

$$= \Re\phi(\lambda(\lambda I + A)^{-1}u)$$

which is not possible.

Conversely, assume that for every $\lambda > 0$, $\lambda(\lambda I + A)^{-1}$ leaves $\mathcal{C}$ invariant. Hence $(\lambda(\lambda + A)^{-1})^n$ leaves $\mathcal{C}$ invariant for every $n \in \mathbb{N}$. Using the exponential formulae

$$e^{-tA}u = \lim_{n\to\infty}\left(I + \frac{t}{n}A\right)^{-n} u \text{ for all } u \in H$$

and the fact that $\mathcal{C}$ is closed, one obtains assertion 1). $\square$

The following theorem gives criteria in terms of the form for the invariance of the closed convex set $\mathcal{C}$ under the semigroup $(e^{-tA})_{t\geq 0}$.

THEOREM 2.2 *Let $\mathfrak{a}$ be a densely defined, accretive, continuous, and closed sesquilinear form on H. The following assertions are equivalent:*
1) $e^{-tA}\mathcal{C} \subseteq \mathcal{C}$ for all $t \geq 0$.
2) $\mathcal{P}(D(\mathfrak{a})) \subseteq D(\mathfrak{a})$ and $\Re\mathfrak{a}(\mathcal{P}u, u - \mathcal{P}u) \geq 0$ for all $u \in D(\mathfrak{a})$.
3) $\mathcal{P}(D(\mathfrak{a})) \subseteq D(\mathfrak{a})$ and $\Re\mathfrak{a}(u, u - \mathcal{P}u) \geq 0$ for all $u \in D(\mathfrak{a})$.
4) There exists a core D of $\mathfrak{a}$ such that $\mathcal{P}(D) \subseteq D(\mathfrak{a})$ and $\Re\mathfrak{a}(\mathcal{P}u, u - \mathcal{P}u) \geq 0$ for all $u \in D$.

Proof. We show that 1) implies 2).

Let us write $R(\lambda) := (\lambda I + A)^{-1}$ for $\lambda > 0$. Note that $R(\lambda)H \subseteq D(A)$ and $AR(\lambda) = I - \lambda R(\lambda)$. Fix $u \in D(\mathfrak{a})$ and observe that

$$\Re\mathfrak{a}(\lambda R(\lambda)\mathcal{P}u, \lambda R(\lambda)\mathcal{P}u)$$

$$= \Re\lambda(AR(\lambda)\mathcal{P}u; \lambda R(\lambda)\mathcal{P}u)$$

$$= \Re\lambda(\mathcal{P}u - \lambda R(\lambda)\mathcal{P}u; \lambda R(\lambda)\mathcal{P}u - \mathcal{P}u) + \Re\lambda(AR(\lambda)\mathcal{P}u; \mathcal{P}u)$$

$$\leq \Re\lambda(AR(\lambda)\mathcal{P}u; \mathcal{P}u)$$

$$= \Re\lambda(\mathcal{P}u - \lambda R(\lambda)\mathcal{P}u; \mathcal{P}u - u) + \Re\lambda(AR(\lambda)\mathcal{P}u; u).$$

Now by Proposition 2.1, we have $\lambda R(\lambda)\mathcal{P}u \in \mathcal{C}$. Hence

$$\Re(\mathcal{P}u - \lambda R(\lambda)\mathcal{P}u; \mathcal{P}u - u) \leq 0$$

by (2.1). It follows from this and the continuity of the form $\mathfrak{a}$ that

$$\Re\mathfrak{a}(\lambda R(\lambda)\mathcal{P}u, \lambda R(\lambda)\mathcal{P}u) \leq \Re\lambda(AR(\lambda)\mathcal{P}u; u)$$
$$= \Re\mathfrak{a}(\lambda R(\lambda)\mathcal{P}u, u)$$
$$\leq M\|\lambda R(\lambda)\mathcal{P}u\|_{\mathfrak{a}}\|u\|_{\mathfrak{a}}$$
$$\leq \frac{1}{2}\|\lambda R(\lambda)\mathcal{P}u\|_{\mathfrak{a}}^{2} + \frac{M^{2}}{2}\|u\|_{\mathfrak{a}}^{2}$$
$$= \frac{1}{2}\Re\mathfrak{a}(\lambda R(\lambda)\mathcal{P}u, \lambda R(\lambda)\mathcal{P}u) + \frac{1}{2}\|\lambda R(\lambda)\mathcal{P}u\|^{2}$$
$$+ \frac{M^{2}}{2}\|u\|_{\mathfrak{a}}^{2}.$$

We have then shown that for $\lambda > 0$,

$$\Re\mathfrak{a}(\lambda R(\lambda)\mathcal{P}u, \lambda R(\lambda)\mathcal{P}u) \leq \|\lambda R(\lambda)\mathcal{P}u\|^{2} + M^{2}\|u\|_{\mathfrak{a}}^{2}.$$

Since $\lambda R(\lambda)$ is a contraction operator on H (see Proposition 1.22), we obtain

$$\Re\mathfrak{a}(\lambda R(\lambda)\mathcal{P}u, \lambda R(\lambda)\mathcal{P}u) \leq \|\mathcal{P}u\|^{2} + M^{2}\|u\|_{\mathfrak{a}}^{2}.$$

This inequality implies in particular that $(\lambda R(\lambda)\mathcal{P}u)_{\lambda>0}$ is bounded (with respect to λ) in the Hilbert space $(D(\mathfrak{a}), \|.\|_{\mathfrak{a}})$. On the other hand $\lambda R(\lambda)\mathcal{P}u$ converges in H to $\mathcal{P}u$ as $\lambda \to \infty$. Lemma 1.32 can be used to conclude that $\mathcal{P}u \in D(\mathfrak{a})$. Thus, $\mathcal{P}D(\mathfrak{a}) \subseteq D(\mathfrak{a})$.

By Lemma 1.56, we have for every $u \in D(\mathfrak{a})$

$$\Re\mathfrak{a}(\mathcal{P}u, u - \mathcal{P}u) = \lim_{t \to 0} \Re\frac{1}{t}(\mathcal{P}u - e^{-tA}\mathcal{P}u; u - \mathcal{P}u).$$

But for all $t > 0$,

$$\Re(\mathcal{P}u - e^{-tA}\mathcal{P}u; u - \mathcal{P}u) \geq 0,$$

because of (2.1) and the fact that $e^{-tA}\mathcal{P}u \in \mathcal{C}$. This gives $\Re\mathfrak{a}(\mathcal{P}u, u - \mathcal{P}u) \geq 0$ for all $u \in D(\mathfrak{a})$ and proves assertion 2).

Assume now that 2) holds. Since the form $\mathfrak{a}$ is accretive,

$$\Re\mathfrak{a}(u - \mathcal{P}u, u - \mathcal{P}u) \geq 0 \text{ for all } u \in D(\mathfrak{a}).$$

Hence,

$$\Re\mathfrak{a}(u, u - \mathcal{P}u) = \Re\mathfrak{a}(u - \mathcal{P}u, u - \mathcal{P}u) + \Re\mathfrak{a}(\mathcal{P}u, u - \mathcal{P}u) \geq 0,$$

which is assertion 3).

We show that 3) implies 1). Given $u \in \mathcal{C}$ we apply 3) to $\lambda R(\lambda)u$, where $\lambda > 0$ is fixed. We have

$$
\begin{aligned}
0 \leq \Re\mathfrak{a}(\lambda R(\lambda)u, \lambda R(\lambda)u - \mathcal{P}\lambda R(\lambda)u) \\
= \lambda \Re(u - \lambda R(\lambda)u; \lambda R(\lambda)u - \mathcal{P}\lambda R(\lambda)u) \\
= \lambda \Re(u - \mathcal{P}\lambda R(\lambda)u; \lambda R(\lambda)u - \mathcal{P}\lambda R(\lambda)u) \\
+ \lambda \Re(\mathcal{P}\lambda R(\lambda)u - \lambda R(\lambda)u; \lambda R(\lambda)u - \mathcal{P}\lambda R(\lambda)u) \\
\leq \lambda \Re(u - \mathcal{P}\lambda R(\lambda)u; \lambda R(\lambda)u - \mathcal{P}\lambda R(\lambda)u).
\end{aligned}
$$

Since $u \in \mathcal{C}$, one has by (2.1)

$$
\Re(u - \mathcal{P}\lambda R(\lambda)u; \lambda R(\lambda)u - \mathcal{P}\lambda R(\lambda)u) \leq 0.
$$

Therefore, it follows from the previous inequalities that

$$
\|\mathcal{P}\lambda R(\lambda)u - \lambda R(\lambda)u\| = 0.
$$

Hence, $\lambda R(\lambda)u \in \mathcal{C}$ for all $\lambda > 0$ and all $u \in \mathcal{C}$. Assertion 1) is then obtained by applying Proposition 2.1.

Obviously, 2) implies 4). Now, we prove that 4) implies 2).

Given $u \in D(\mathfrak{a})$ and let $(u_n)_n \in D$ be a sequence which converges to u for the norm $\|.\|_\mathfrak{a}$. We apply 4) and the continuity assumption to obtain

$$
\begin{aligned}
\Re\mathfrak{a}(\mathcal{P}u_n, \mathcal{P}u_n) = \Re\mathfrak{a}(\mathcal{P}u_n, \mathcal{P}u_n - u_n) + \Re\mathfrak{a}(\mathcal{P}u_n, u_n) \\
\leq \Re\mathfrak{a}(\mathcal{P}u_n, u_n) \\
\leq M \|\mathcal{P}u_n\|_\mathfrak{a} \|u_n\|_\mathfrak{a} \\
\leq \frac{1}{2}\|\mathcal{P}u_n\|_\mathfrak{a}^2 + \frac{M^2}{2}\|u_n\|_\mathfrak{a}^2 \\
= \frac{1}{2}\Re\mathfrak{a}(\mathcal{P}u_n, \mathcal{P}u_n) + \frac{1}{2}\|\mathcal{P}u_n\|^2 + \frac{M^2}{2}\|u_n\|_\mathfrak{a}^2.
\end{aligned}
$$

This implies that

$$
\Re\mathfrak{a}(\mathcal{P}u_n, \mathcal{P}u_n) \leq \|\mathcal{P}u_n\|^2 + M^2\|u_n\|_\mathfrak{a}^2,
$$

from which it follows that $(\mathcal{P}u_n)_n$ is a bounded sequence in $(D(\mathfrak{a}), \|.\|_\mathfrak{a})$ (note that $(\mathcal{P}u_n)_n$ is bounded in H, since $\mathcal{P}$ is continuous). We conclude by Lemma 1.32 that $\mathcal{P}u \in D(\mathfrak{a})$. In addition, the same lemma and assertion 4) give

$$
\Re\mathfrak{a}(\mathcal{P}u, \mathcal{P}u) \leq \liminf \Re\mathfrak{a}(\mathcal{P}u_{n_k}, \mathcal{P}u_{n_k}) \leq \liminf \Re\mathfrak{a}(\mathcal{P}u_{n_k}, u_{n_k}).
$$

But

$$
\mathfrak{a}(\mathcal{P}u_{n_k}, u_{n_k}) - \mathfrak{a}(\mathcal{P}u, u) = \mathfrak{a}(\mathcal{P}u_{n_k} - \mathcal{P}u, u) + \mathfrak{a}(\mathcal{P}u_{n_k}, u_{n_k} - u).
$$

The second term on the right-hand side converges to 0 since (u_{n_k}) converges to u for the norm $\|.\|_\mathfrak{a}$ and the sequence $(\|\mathcal{P}u_{n_k}\|_\mathfrak{a})$ is bounded. The first term also converges to 0. This follows from the fact that $v \mapsto \mathfrak{a}(v, u)$ is a continuous linear functional on $D(\mathfrak{a})$ and $(\mathcal{P}u_{n_k})$ converges weakly in $D(\mathfrak{a})$ to $\mathcal{P}u$. Thus, $\liminf \Re\mathfrak{a}(\mathcal{P}u_{n_k}, u_{n_k}) = \Re\mathfrak{a}(\mathcal{P}u, u)$. This together with the previous inequalities give assertion 2). $\qquad\square$

Remark.   The assertions in the previous theorem are also equivalent to:
3′) There exists a core D of $\mathfrak{a}$ such that $\mathcal{P}(D) \subseteq D$ and $\Re\mathfrak{a}(u, u - \mathcal{P}u) \geq 0$ for all $u \in D$.
Indeed, assume that 3′) holds and fix $u \in D$ and $\varepsilon > 0$. Define $u_\varepsilon := \mathcal{P}u + \varepsilon(u - \mathcal{P}u)$. We have $u_\varepsilon \in D$ and $\mathcal{P}u_\varepsilon = \mathcal{P}u$. Applying 3′) to u_ε gives

$$\Re\mathfrak{a}(\mathcal{P}u, u - \mathcal{P}u) + \varepsilon\mathfrak{a}(u - \mathcal{P}u, u - \mathcal{P}u) \geq 0.$$

Letting $\varepsilon \to 0$ yields assertion 4) of the previous theorem.

In the case where the form $\mathfrak{a}$ is symmetric, the criteria of Theorem 2.2 can be given in terms of the quadratic form. More precisely,

THEOREM 2.3 *Assume that $\mathfrak{a}$ is a densely defined, symmetric, accretive, and closed form on H. The following assertions are equivalent:*
1) $e^{-tA}\mathcal{C} \subseteq \mathcal{C}$ for all $t \geq 0$.
2) $\mathcal{P}(D(\mathfrak{a})) \subseteq D(\mathfrak{a})$ and $\mathfrak{a}(\mathcal{P}u, \mathcal{P}u) \leq \mathfrak{a}(u, u)$ for all $u \in D(\mathfrak{a})$.
3) There exists a core D of $\mathfrak{a}$ such that $\mathcal{P}(D) \subseteq D(\mathfrak{a})$ and $\mathfrak{a}(\mathcal{P}u, \mathcal{P}u) \leq \mathfrak{a}(u, u)$ for all $u \in D$.

Proof. Assume that 3) is satisfied. Let $u \in D(\mathfrak{a})$ and $(u_n)_n \in D$ be a sequence which converges to u with respect to the norm $\|.\|_\mathfrak{a}$. By assumption, we have for every n

$$\mathfrak{a}(\mathcal{P}u_n, \mathcal{P}u_n) \leq \mathfrak{a}(u_n, u_n).$$

Hence, $(\mathcal{P}u_n)_n$ is a bounded sesquence in $(D(\mathfrak{a}), \|.\|_\mathfrak{a})$. Since $\mathcal{P}u_n$ converges to $\mathcal{P}u$ in H, it follows from Lemma 1.32 that $\mathcal{P}u \in D(\mathfrak{a})$ and

$$\begin{aligned}\mathfrak{a}(\mathcal{P}u, \mathcal{P}u) &\leq \liminf \mathfrak{a}(\mathcal{P}u_n, \mathcal{P}u_n) \\ &\leq \liminf \mathfrak{a}(u_n, u_n) \\ &= \mathfrak{a}(u, u).\end{aligned}$$

This shows that assertion 2) holds.

Assume now that 2) holds. By the Cauchy-Schwarz inequality, we have for every $u \in D(\mathfrak{a})$

$$\begin{aligned}
\Re\mathfrak{a}(u, u - \mathcal{P}u) &= \mathfrak{a}(u, u) - \Re\mathfrak{a}(u, \mathcal{P}u) \\
&\geq \mathfrak{a}(u, u) - \mathfrak{a}(u, u)^{1/2}\mathfrak{a}(\mathcal{P}u, \mathcal{P}u)^{1/2} \\
&\geq 0.
\end{aligned}$$

We apply Theorem 2.2 to obtain assertion 1).

Assume that 1) holds. By the previous theorem, we have $\mathcal{P}u \in D(\mathfrak{a})$ for every $u \in D(\mathfrak{a})$ and $\Re\mathfrak{a}(\mathcal{P}u, u - \mathcal{P}u) \geq 0$. Using this, we can write

$$\begin{aligned}
\mathfrak{a}(\mathcal{P}u, \mathcal{P}u) &= \Re\mathfrak{a}(\mathcal{P}u, \mathcal{P}u - u) + \Re\mathfrak{a}(\mathcal{P}u, u) \\
&\leq \Re\mathfrak{a}(\mathcal{P}u, u) \\
&\leq \mathfrak{a}(\mathcal{P}u, \mathcal{P}u)^{1/2}\mathfrak{a}(u, u)^{1/2},
\end{aligned}$$

which gives assertion 2). This proves the theorem since 2) implies 3). $\qquad\square$

2.2 POSITIVE AND L^p-CONTRACTIVE SEMIGROUPS

In this section, we apply the criteria obtained in the previous section to study positivity and L^p-contractivity properties of semigroups. We will state the results when the L^p-spaces in consideration are complex but all the results are valid if the spaces are real as mentioned in the previous section.

We assume in this section that $H = L^2(X, \mu, \mathbb{C})$, where (X, μ) is a σ-finite measure space. Let

$$H_{\mathbb{R}} := L^2(X, \mu, \mathbb{R})$$

be the subset of H of real-valued functions and

$$H^+ := \{f \in H, f(x) \geq 0 \ \mu \text{ a.e.} x \in X\}$$

the cone of non-negative functions.

DEFINITION 2.4 *Let $(T(t))_{t \geq 0}$ be a strongly continuous semigroup on H. We say that $(T(t))_{t \geq 0}$ is real if $T(t)H_{\mathbb{R}} \subseteq H_{\mathbb{R}}$ for all $t \geq 0$. The semigroup $(T(t))_{t \geq 0}$ is positive if $T(t)H^+ \subseteq H^+$ for all $t \geq 0$.*

For a given $u \in H$, we denote by $\overline{u} := \Re u - i\Im u$ the conjugate function of u. By $|u|$ we denote the absolute value of u (i.e., the function $x \to |u(x)| := \sqrt{u(x)\overline{u(x)}}$) and by sign u the function defined by

$$\operatorname{sign} u(x) = \begin{cases} \frac{u(x)}{|u(x)|} & \text{if } u(x) \neq 0, \\ 0 & \text{if } u(x) = 0. \end{cases} \tag{2.2}$$

For $u, v \in H_{\mathbb{R}}$, we set $u^+ := \sup\{u, 0\}, u^- := \sup\{-u, 0\}, u \wedge v := \inf\{u, v\}, u \vee v := \sup\{u, v\}$ and $1 \wedge u := \inf\{1, u\}$.

Remark. All the inequalities (and equalities) between measurable functions of H are understood in the μ a.e. sense. We often omit writing μ a.e.

Let $\mathfrak{a}$ be a densely defined, accretive, continuous, and closed sesquilinear form on $L^2(X, \mu, \mathbb{C})$. Denote again by A its associated operator and by $(e^{-tA})_{t \geq 0}$ the semigroup generated by $-A$ on $L^2(X, \mu, \mathbb{C})$. Note that the assumption that $\mathfrak{a}$ is densely defined can be removed in all the results below (except Theorems 2.9 and 2.10, which must be reformulated differently for non-densely defined forms). See Section 2.6 below.

PROPOSITION 2.5 *The following assertions are equivalent:*
1) The semigroup $(e^{-tA})_{t \geq 0}$ is real.
2) $u \in D(\mathfrak{a}) \implies \Re u \in D(\mathfrak{a})$ and $\mathfrak{a}(\Re u, \Im u) \in \mathbb{R}$.
3) $u \in D(\mathfrak{a}) \implies \bar{u} \in D(\mathfrak{a})$ and $\mathfrak{a}(u, v) \in \mathbb{R}$ for every $u, v \in D(\mathfrak{a}) \cap H_{\mathbb{R}}$.
4) There exists a core D of $\mathfrak{a}$ such that $\Re u \in D(\mathfrak{a})$ and $\mathfrak{a}(\Re u, \Im u) \in \mathbb{R}$ for all $u \in D$.

Proof. The fact that the semigroup $(e^{-tA})_{t \geq 0}$ is real is equivalent to the fact that $(e^{-tA})_{t \geq 0}$ leaves invariant the closed convex set $\mathcal{C} = H_{\mathbb{R}}$. The projection onto $H_{\mathbb{R}}$ is given by $\mathcal{P}u = \Re u$. By Theorem 2.2, assertion 1) is equivalent to

$$u \in D(\mathfrak{a}) \implies \Re u \in D(\mathfrak{a}) \text{ and } \Re \mathfrak{a}(\Re u, i\Im u) \geq 0.$$

By the same theorem, it is enough to consider u in a core D of $\mathfrak{a}$. Applying the latter inequality to $-\Re u + i\Im u$ shows that $\mathfrak{a}(\Re u, \Im u) \in \mathbb{R}$. This shows the equivalence of the assertions 1), 2), and 4). The equivalence of 2) and 3) is obvious. □

THEOREM 2.6 *The following assertions are equivalent:*
1) The semigroup $(e^{-tA})_{t \geq 0}$ is positive.
2) $u \in D(\mathfrak{a}) \implies (\Re u)^+ \in D(\mathfrak{a}), \mathfrak{a}(\Re u, \Im u) \in \mathbb{R}$ and $\mathfrak{a}((\Re u)^+, (\Re u)^-) \leq 0$.
3) There exists a core D of $\mathfrak{a}$ such that $(\Re u)^+ \in D(\mathfrak{a}), \mathfrak{a}(\Re u, \Im u) \in \mathbb{R}$ and $\mathfrak{a}((\Re u)^+, (\Re u)^-) \leq 0$ for all $u \in D$.
4) The semigroup $(e^{-tA})_{t \geq 0}$ is real and for every $u \in D(\mathfrak{a}) \cap H_{\mathbb{R}}$ we have $u^+ \in D(\mathfrak{a})$ and $\mathfrak{a}(u^+, u^-) \leq 0$.

Proof. Consider the convex set $\mathcal{C} = H^+$. The projection $\mathcal{P}$ is given by $\mathcal{P}u = (\Re u)^+$.

The equivalence of the first three assertions follows from Theorem 2.2 and Proposition 2.5, since a positive semigroup is in particular real. Clearly, each of these assertions implies 4). Conversely, if 4) holds, then $\Re u \in D(\mathfrak{a}) \cap H_\mathbb{R}$ and $\mathfrak{a}(\Re u, \Im u) \in \mathbb{R}$ for all $u \in D(\mathfrak{a})$ by the previous proposition. Applying then 4) to $\Re u$ gives assertion 2).   □

THEOREM 2.7 *Assume that the form $\mathfrak{a}$ is symmetric, densely defined, accretive, and closed. The following assertions are equivalent:*
1) The semigroup $(e^{-tA})_{t \geq 0}$ is positive.
2) $u \in D(\mathfrak{a}) \Longrightarrow (\Re u)^+ \in D(\mathfrak{a})$ and $\mathfrak{a}((\Re u)^+, (\Re u)^+) \leq \mathfrak{a}(u, u)$.
3) There exists a core D of $\mathfrak{a}$ such that $(\Re u)^+ \in D(\mathfrak{a})$ and $\mathfrak{a}((\Re u)^+, (\Re u)^+) \leq \mathfrak{a}(u, u)$ for all $u \in D$.
4) The semigroup $(e^{-tA})_{t \geq 0}$ is real and for every $u \in D(\mathfrak{a}) \cap H_\mathbb{R}$, we have $|u| \in D(\mathfrak{a})$ and $\mathfrak{a}(|u|, |u|) \leq \mathfrak{a}(u, u)$.

Proof. Applying Theorem 2.3 with $\mathcal{C} = H^+$ and $\mathcal{P}u = (\Re u)^+$ gives the equivalence of the first three assertions.

For $u \in D(\mathfrak{a}) \cap H_\mathbb{R}$, one writes $u = u^+ - u^-$ and $|u| = u^+ + u^-$. Thus, $u^+ \in D(\mathfrak{a})$ for all $u \in D(\mathfrak{a}) \cap H_\mathbb{R}$ if and only if $|u| \in D(\mathfrak{a})$ for all $u \in D(\mathfrak{a}) \cap H_\mathbb{R}$. Since the inequality $\mathfrak{a}(|u|, |u|) \leq \mathfrak{a}(u, u)$ is equivalent to $\mathfrak{a}(u^+, u^-) \leq 0$, the equivalence of assertions 1) and 4) follows from Theorem 2.6.   □

DEFINITION 2.8 *Let $(e^{-tA})_{t \geq 0}$ be the semigroup associated with the form $\mathfrak{a}$ on $H = L^2(X, \mu, \mathbb{C})$. We say that $(e^{-tA})_{t \geq 0}$ is irreducible if for every $t > 0$ and every nonzero function $f \in H^+$, we have*

$$e^{-tA} f(x) > 0 \text{ for } \mu \text{a.e.} x \in X.$$

As for positivity, we wish to characterize in terms of the form $\mathfrak{a}$ the irreducibility of its associated semigroup $(e^{-tA})_{t \geq 0}$. In order to do this, we have to reformulate the irreducibility property in terms of invariance of closed convex sets. This reformulation is given in the following result.

THEOREM 2.9 *Let $\mathfrak{a}$ be a densely defined, accretive, continuous, and closed form on $L^2(X, \mu, \mathbb{C})$ and assume that its associated semigroup $(e^{-tA})_{t \geq 0}$ is positive. The following assertions are equivalent:*
1) $(e^{-tA})_{t \geq 0}$ is irreducible.
2) If the semigroup $(e^{-tA})_{t \geq 0}$ leaves $L^2(\Omega, \mu, \mathbb{C})$ invariant for some subset Ω of X, then either $\mu(\Omega) = 0$ or $\mu(X \setminus \Omega) = 0$.

Note that in this theorem, $L^2(\Omega, \mu, \mathbb{C})$ is seen as the subspace of $L^2(X, \mu, \mathbb{C})$ of functions which are zero (μ a.e.) on $X \setminus \Omega$. Given $f \in L^2(\Omega, \mu, \mathbb{C})$, we denote again by f the function in $L^2(X, \mu, \mathbb{C})$ which takes the value 0 (μ

a.e.) on $X \setminus \Omega$ and coincides (μ a.e.) with f on Ω.

Proof. Assume that assertion 1) holds. If $(e^{-tA})_{t \geq 0}$ leaves invariant a subspace $L^2(\Omega, \mu, \mathbb{C})$, then $e^{-tA}f$ is 0 (μ a.e.) on $X \setminus \Omega$ for all $t > 0$ and $f \in L^2(\Omega, \mu, \mathbb{C})$. If $\mu(\Omega) > 0$, then we can apply this to a nontrivial $f \geq 0$. We then obtain from 1) that $\mu(X \setminus \Omega) = 0$.

Assume that 2) is satisfied. Let $0 \leq f \in L^2(X, \mu, \mathbb{C})$ be a nontrivial function and let $\tau > 0$. Set $\Omega := \{e^{-\tau A}f = 0\}$ (the set on which $e^{-\tau A}f$ is 0, μ a.e.) and assume for a contradiction that $\mu(\Omega) > 0$. We have $(e^{-\tau A}f; \phi) = 0$ for every $\phi \in L^2(\Omega, \mu, \mathbb{C})$. By strong continuity of $(e^{-tA})_{t \geq 0}$, we can find a sequence $t_n \in (0, \tau)$ such that $\|e^{-t_n A}f - f\|_2 \leq 2^{-n}$. Set

$$f_n := e^{-t_n A}f \text{ and } g_n := f - \sum_{k \geq n}(f - f_k)^+.$$

We have

$$g_n \leq f - (f - f_m)^+ = \inf\{f, f_m\} \leq f_m \text{ for all } m \geq n.$$

Using the positivity of the semigroup, we obtain for all $m \geq n$, and all $0 \leq \phi \in L^2(\Omega, \mu, \mathbb{C})$,

$$0 \leq (e^{-(\tau-t_m)A}g_n^+; \phi) \leq (e^{-(\tau-t_m)A}f_m; \phi)$$
$$= (e^{-\tau A}f; \phi)$$
$$= 0.$$

Hence we have for every $m \geq n$,

$$(e^{-(\tau-t_m)A}g_n^+; \phi) = 0.$$

The semigroup $(e^{-tA})_{t \geq 0}$ is holomorphic on H (see Chapter 1).[2] Hence we obtain from the above equality that

$$(e^{-tA}g_n^+; \phi) = 0 \text{ for all } t \geq 0.$$

Letting $n \to \infty$, we obtain

$$(e^{-tA}f; \phi) = 0 \text{ for all } t \geq 0, \phi \in L^2(\Omega, \mu, \mathbb{C}). \tag{2.3}$$

Setting $h := e^{-\tau A}f$, we have from (2.3)

$$(e^{-tA}h; \phi) = 0 \text{ for all } t \geq 0, \phi \in L^2(\Omega, \mu, \mathbb{C}). \tag{2.4}$$

---

[2]If H is real, one argues similarly by using the complexification procedure, as explained in the previous chapter

Given now $0 \leq g \in L^2(X \setminus \Omega, \mu, \mathbb{C})$ and fix $\phi \in L^2(\Omega, \mu, \mathbb{C})$. We may write $g = (g - nh)^+ + \inf\{g, nh\}$. It follows from (2.4) that

$$|(e^{-tA}(\inf\{g, nh\}); \phi)| \leq n(e^{-tA}h; |\phi|) = 0.$$

This implies that

$$(e^{-tA}g; \phi) = (e^{-tA}(g - nh)^+; \phi).$$

using the facts that $g \in L^2(X \setminus \Omega, \mu, \mathbb{C})$ and $\Omega = \{h = 0\}$, we see that $(g - nh)^+ \to 0$ in $L^2(X, \mu, \mathbb{C})$ as $n \to \infty$. Taking the limit in the previous equation yields

$$(e^{-tA}g; \phi) = 0 \text{ for all } t \geq 0.$$

Since this is true for all $\phi \in L^2(\Omega, \mu, \mathbb{C})$ and all $g \in L^2(X \setminus \Omega, \mu, \mathbb{C})$, we conclude that $(e^{-tA})_{t \geq 0}$ leaves $L^2(X \setminus \Omega, \mu, \mathbb{C})$ invariant. Assertion 2) and the hypothesis $\mu(\Omega) > 0$ imply that $\mu(X \setminus \Omega) = 0$. This and (2.3) for $t = 0$, then give $f = 0$ (μ a.e. on X), which is a contradiction. $\square$

We can apply the previous criteria for invariance of closed convex sets to characterize the irreducibility of semigroups. The projection onto the closed convex set $L^2(\Omega, \mu, \mathbb{C})$ is given by $\mathcal{P}u = \chi_\Omega u$, where χ_Ω denotes the characteristic function of Ω. Theorem 2.2 shows that $(e^{-tA})_{t \geq 0}$ leaves $L^2(\Omega, \mu, \mathbb{C})$ invariant if and only if

$$\chi_\Omega u \in D(\mathfrak{a}) \text{ and } \Re\mathfrak{a}(\chi_\Omega u, \chi_{X \setminus \Omega} u) \geq 0 \text{ for all } u \in D(\mathfrak{a}).$$

By the same theorem, it is enough to check this condition for u in some core of $\mathfrak{a}$.

Using this and Theorem 2.9, we obtain the following criterion for irreducibility.

THEOREM 2.10 *Let $\mathfrak{a}$ be a densely defined, accretive, continuous, and closed form on $L^2(X, \mu, \mathbb{C})$. Assume that its associated semigroup $(e^{-tA})_{t \geq 0}$ is positive. The following assertions are equivalent:*
1) $(e^{-tA})_{t \geq 0}$ is an irreducible semigroup.
2) If $\Omega \subseteq X$ is such that $\chi_\Omega u \in D(\mathfrak{a})$ and $\Re\mathfrak{a}(\chi_\Omega u, \chi_{X \setminus \Omega} u) \geq 0$ for all $u \in D(\mathfrak{a})$, then either $\mu(\Omega) = 0$ or $\mu(X \setminus \Omega) = 0$.
3) If $\Omega \subseteq X$ is such that $\chi_\Omega u \in D(\mathfrak{a})$ and $\Re\mathfrak{a}(\chi_\Omega u, \chi_{X \setminus \Omega} u) \geq 0$ for all u in a core D of $\mathfrak{a}$, then either $\mu(\Omega) = 0$ or $\mu(X \setminus \Omega) = 0$.

Note that if the form $\mathfrak{a}$ is local, i.e., $\mathfrak{a}(u, v) = 0$ for all $u, v \in D(\mathfrak{a})$ which have disjoint supports, then the condition $\Re\mathfrak{a}(\chi_\Omega u, \chi_{X \setminus \Omega} u) \geq 0$ is automatically satisfied. Hence for local forms, the irreducibility criterion is reduced to the question of whether or not characteristic functions operate on $D(\mathfrak{a})$. More precisely,

COROLLARY 2.11 *Let $\mathfrak{a}$ and $(e^{-tA})_{t\geq 0}$ be as in the previous theorem. Assume in addition that the form $\mathfrak{a}$ is local. The following assertions are equivalent:*
1) $(e^{-tA})_{t\geq 0}$ *is an irreducible semigroup.*
2) If $\Omega \subseteq X$ is such that $\chi_\Omega(D(\mathfrak{a})) \subseteq D(\mathfrak{a})$, then either $\mu(\Omega) = 0$ or $\mu(X \setminus \Omega) = 0$.
3) If $\Omega \subseteq X$ is such that $\chi_\Omega(D) \subseteq D(\mathfrak{a})$ for some core D of $\mathfrak{a}$, then either $\mu(\Omega) = 0$ or $\mu(X \setminus \Omega) = 0$.

We now turn to another property of the semigroup. We want to extend the contraction semigroup $(e^{-tA})_{t\geq 0}$, initially defined on $L^2(X, \mu, \mathbb{C})$ to other $L^p(X, \mu, \mathbb{C})$ spaces. For this reason, we study L^∞-contractivity, which we introduce in the following definition.

DEFINITION 2.12 *We say that* $(e^{-tA})_{t\geq 0}$ *is L^∞-contractive if for every $t \geq 0$ and every $u \in L^2(X, \mu, \mathbb{C}) \cap L^\infty(X, \mu, \mathbb{C})$,*

$$\|e^{-tA}u\|_{L^\infty(X,\mu,\mathbb{C})} \leq \|u\|_{L^\infty(X,\mu,\mathbb{C})}.$$

If the semigroup $(e^{-tA})_{t\geq 0}$ *is both positive and L^∞-contractive, we say that it is sub-Markovian.*

Recall the notation $1 \wedge u := \inf\{1, u\}$ and sign u is the function defined by (2.2).

THEOREM 2.13 *Let $\mathfrak{a}$ be a densely defined, accretive, continuous, and closed form on $L^2(X, \mu, \mathbb{C})$. The following assertions are equivalent:*
1) The semigroup $(e^{-tA})_{t\geq 0}$ is L^∞-contractive.
2) $u \in D(\mathfrak{a}) \Longrightarrow (1 \wedge |u|)\text{sign } u \in D(\mathfrak{a})$ *and* $\mathfrak{Ra}(u, (|u| - 1)^+\text{sign } u) \geq 0.$
3) $u \in D(\mathfrak{a}) \Longrightarrow (1 \wedge |u|)\text{sign } u \in D(\mathfrak{a})$ *and*
$\mathfrak{Ra}((1 \wedge |u|)\text{sign } u, (|u| - 1)^+\text{sign } u) \geq 0.$
3′) There exists a core D of $\mathfrak{a}$ such that $(1 \wedge |u|)\text{sign } u \in D(\mathfrak{a})$ and
$\mathfrak{Ra}((1 \wedge |u|)\text{sign } u, (|u| - 1)^+\text{sign } u) \geq 0$ *for all $u \in D$.*

Proof. Note that L^∞-contractivity is equivalent to the fact that the semigroup leaves invariant the closed convex set

$$\mathcal{C} = \{u \in L^2(X, \mu, \mathbb{C}), |u| \leq 1 \ (\mu \text{ a.e.})\}.$$

The projection onto this convex set is given by $\mathcal{P}u = (1 \wedge |u|)\text{sign } u$. Note also that

$$u - (1 \wedge |u|)\text{sign } u = (|u| - 1)^+\text{sign } u.$$

Applying Theorems 2.2 we obtain the above result. $\qquad\qquad\square$

Using the same proof and applying Theorem 2.3, we obtain in the particular case of symmetric forms

THEOREM 2.14 *Assume that the form* $\mathfrak{a}$ *is symmetric. The following assertions are equivalent:*
1) The semigroup $(e^{-tA})_{t\geq0}$ *is* L^∞*-contractive.*
2) $u \in D(\mathfrak{a}) \Longrightarrow (1 \wedge |u|)\text{sign } u \in D(\mathfrak{a})$ *and*
   $\mathfrak{a}((1 \wedge |u|)\text{sign } u, (1 \wedge |u|)\text{sign } u) \leq \mathfrak{a}(u, u).$
3) There exists a core D *of* $\mathfrak{a}$ *such that* $(1 \wedge |u|)\text{sign } u \in D(\mathfrak{a})$ *and*
   $\mathfrak{a}((1 \wedge |u|)\text{sign } u, (1 \wedge |u|)\text{sign } u) \leq \mathfrak{a}(u, u)$ *for all* $u \in D.$

We have assumed in Theorem 2.13 that $\mathfrak{a}$ is accretive. In the next result, we show that actually this assumption is not needed in the equivalence of assertions 1) and 3). This refinement is of some interest and will be applied to uniformly elliptic operators in Chapter 4.

THEOREM 2.15 *Assume that the form* $\mathfrak{a}$ *is densely defined and there exists a constant* $w \in \mathbb{R}$ *such that the form*

$$(\mathfrak{a} + w)(u, v) := \mathfrak{a}(u, v) + w(u; v), \quad u, v \in D(\mathfrak{a})$$

is accretive, continuous, and closed. The following assertions are equivalent:
1) The semigroup $(e^{-tA})_{t\geq0}$ *is* L^∞*-contractive.*
2) $u \in D(\mathfrak{a}) \Longrightarrow (1 \wedge |u|)\text{sign } u \in D(\mathfrak{a})$ *and*
   $\Re\mathfrak{a}((1 \wedge |u|)\text{sign } u, (|u| - 1)^+\text{sign } u) \geq 0.$

Proof. We can assume that $w \geq 0$; otherwise the result is already proved. Assume that 1) holds. This implies that the semigroup $(e^{-t(A+w)})_{t\geq0}$ associated with the form $\mathfrak{a} + w$ is L^∞-contractive, too. Thus by the previous theorem we obtain

$$(1 \wedge |u|)\text{sign } u \in D(\mathfrak{a}) \quad \text{for all } u \in D(\mathfrak{a}).$$

Now by Lemma 1.56, we have

$\Re\mathfrak{a}((1 \wedge |u|)\text{sign } u, (|u| - 1)^+\text{sign } u)$

$= \lim\limits_{t \to 0} \dfrac{1}{t}\Re \displaystyle\int_X [(1 \wedge |u|)\text{sign } u - e^{-tA}((1 \wedge |u|)\text{sign } u)](|u| - 1)^+\text{sign } \overline{u}d\mu$

$= \lim\limits_{t \to 0} \dfrac{1}{t}\Re \displaystyle\int_X (|u| - 1)^+[1 - \text{sign } (\overline{u})e^{-tA}((1 \wedge |u|)\text{sign } u)]d\mu$

$\geq 0.$

Here we use the fact that $|e^{-tA}((1 \wedge |u|\text{sign } u))| \leq 1$ to obtain the last inequality.
Assume now that 2) holds. Let $u \in L^2(X, \mu, \mathbb{C})$ be such that $|u| \leq 1$. Set

$$\phi(t) = \frac{1}{2} \int_X [(|e^{-tA}u| - 1)^+]^2 d\mu.$$

A simple calculation shows that $\frac{d}{dt}|e^{-tA}u| = \Re(-\text{sign}\,(\overline{e^{-tA}u})Ae^{-tA}u)$ at each $t > 0$ (see Proposition 4.4 below). Thus, we have for every $t > 0$,

$$\phi'(t) = \Re \int_X -Ae^{-tA}u(|e^{-tA}u| - 1)^+ \text{sign}\,(\overline{e^{-tA}u})d\mu$$

$$= -\Re\mathfrak{a}(e^{-tA}u, (|e^{-tA}u| - 1)^+ \text{sign}\,(e^{-tA}u)).$$

The second equality makes sense because of 2) and the fact that $e^{-tA}u \in D(\mathfrak{a})$ for $t > 0$ (see the previous chapter). Using

$$e^{-tA}u = (1 \wedge |e^{-tA}u|)\text{sign}\,(e^{-tA}u) + (|e^{-tA}u| - 1)^+ \text{sign}\,(e^{-tA}u)$$

and assertion 2) we obtain

$$\phi'(t) \leq -\Re\mathfrak{a}((|e^{-tA}u| - 1)^+ \text{sign}\,(e^{-tA}u), (|e^{-tA}u| - 1)^+ \text{sign}\,(e^{-tA}u)).$$

This and the accretivity assumption of the form $\mathfrak{a} + w$ imply that $\phi'(t) \leq 2w\phi(t)$ for all $t > 0$. Since $\lim_{t \to 0} \phi(t) = 0$, we obtain $\phi(t) = 0$ for all $t > 0$. This gives $|e^{-tA}u| \leq 1$, which is the L^∞-contractivity property. $\square$

If our starting contraction semigroup $(e^{-tA})_{t \geq 0}$ on $L^2(X, \mu, \mathbb{C})$ satisfies the L^∞-contractivity property, then using the Riesz-Thorin interpolation theorem, we can extend each operator e^{-tA} from $L^2(X, \mu, \mathbb{C}) \cap L^p(X, \mu, \mathbb{C})$ to a contraction operator on $L^p(X, \mu, \mathbb{C})$ for $2 \leq p \leq \infty$. We denote again by e^{-tA} this extension to $L^p(X, \mu, \mathbb{C})$. A density argument shows that the family $(e^{-tA})_{t \geq 0}$ defines a strongly continuous semigroup of contractions on $L^p(X, \mu, \mathbb{C})$ for each p with $2 \leq p < \infty$. By duality, the adjoint semigroup $(e^{-tA^*})_{t \geq 0}$ is strongly continuous and contractive on $L^p(X, \mu, \mathbb{C})$ for $1 < p \leq 2$. In addition, it also defines a strongly continuous contractive semigroup on $L^1(X, \mu, \mathbb{C})$. In order to show the strong continuity, we prove that

$$\lim_{t \to 0} \|e^{-tA^*}\chi_B - \chi_B\|_1 = 0$$

for every measurable subset B of finite measure, where χ_B denotes the indicator function of B.

By Hölder's inequality, we have

$$\left| \int_B |e^{-tA^*}\chi_B|d\mu - \int_B \chi_B d\mu \right| \leq \mu(B)^{1/2}\|e^{-tA^*}\chi_B - \chi_B\|_2.$$

Thus, the strong continuity on $L^2(X, \mu, \mathbb{C})$ implies

$$\lim_{t \to 0} \left[\int_B |e^{-tA^*}\chi_B|d\mu - \int_B \chi_B d\mu \right] = 0.$$

Since

$$\|\chi_B\|_1 \geq \|e^{-tA^*}\chi_B\|_1 = \int_{X\setminus B} |e^{-tA^*}\chi_B| d\mu + \int_B |e^{-tA^*}\chi_B| d\mu,$$

it follows that

$$\lim_{t \to 0} \int_{X\setminus B} |e^{-tA^*}\chi_B| d\mu = 0.$$

Writing

$$\|e^{-tA^*}\chi_B - \chi_B\|_1 = \int_B |e^{-tA^*}\chi_B - \chi_B| d\mu + \int_{X\setminus B} |e^{-tA^*}\chi_B| d\mu$$

$$\leq \mu(B)^{1/2}\|e^{-tA^*}\chi_B - \chi_B\|_2 + \int_{X\setminus B} |e^{-tA^*}\chi_B| d\mu,$$

we see that $\lim_{t \to 0} \|e^{-tA^*}\chi_B - \chi_B\|_1 = 0$. The strong continuity of the semigroup $(e^{-tA^*})_{t\geq0}$ on $L^1(X, \mu, \mathbb{C})$ follows then by a density argument.

Recall that $(e^{-tA^*})_{t\geq0}$ is the semigroup associated with the adjoint form $\mathfrak{a}^*$ (cf. Proposition 1.24). Using this and the previous results, we can characterize the fact that the semigroup $(e^{-tA^*})_{t\geq0}$ is L^∞-contractive (or equivalently, that $(e^{-tA})_{t\geq0}$ is L^1-contractive). We have

COROLLARY 2.16 *Under the assumptions of Theorem 2.13, the following assertions are equivalent:*
1) The semigroup $(e^{-tA^*})_{t\geq0}$ is L^∞-contractive.
2) For each $t > 0$, e^{-tA} is a contraction operator on $L^p(X, \mu, \mathbb{C})$ for all p, $1 \leq p \leq 2$.
3) $u \in D(\mathfrak{a}) \implies (1 \wedge |u|)\text{sign } u \in D(\mathfrak{a})$ and $\Re\mathfrak{a}((|u| - 1)^+\text{sign } u, u) \geq 0$.
4) $u \in D(\mathfrak{a}) \implies (1 \wedge |u|)\text{sign } u \in D(\mathfrak{a})$ and
   $\Re\mathfrak{a}((|u| - 1)^+\text{sign } u, (1 \wedge |u|)\text{sign } u) \geq 0$.
4′) There exists a core D of $\mathfrak{a}$ such that $(1 \wedge |u|)\text{sign } u \in D(\mathfrak{a})$ and
   $\Re\mathfrak{a}((|u| - 1)^+\text{sign } u, (1 \wedge |u|)\text{sign } u) \geq 0$ for all $u \in D$.

In order to extend the semigroup $(e^{-tA})_{t\geq0}$ to a contraction semigroup on $L^p(X, \mu, \mathbb{C})$ for all $p, 1 \leq p < \infty$, one has to check the L^∞-contractivity for both semigroups $(e^{-tA})_{t\geq0}$ and $(e^{-tA^*})_{t\geq0}$. By the previous results, this is equivalent to the validity of the following three conditions:
i) $u \in D(\mathfrak{a}) \implies (1 \wedge |u|)\text{sign } u \in D(\mathfrak{a})$.
ii) $\Re\mathfrak{a}((|u| - 1)^+\text{sign } u, (1 \wedge |u|)\text{sign } u) \geq 0$ for all $u \in D(\mathfrak{a})$.
iii) $\Re\mathfrak{a}((1 \wedge |u|)\text{sign } u, (|u| - 1)^+\text{sign } u) \geq 0$ for all $u \in D(\mathfrak{a})$.
Note also that it enough to check these properties on any core of $\mathfrak{a}$.

COROLLARY 2.17 *Under the assumptions of Theorem 2.15, the following assertions are equivalent:*
1) The semigroup $(e^{-tA})_{t \geq 0}$ is sub-Markovian.
2) $u \in D(\mathfrak{a}) \Longrightarrow (\Re u)^+, (1 \wedge |u|)\text{sign } u \in D(\mathfrak{a}), \mathfrak{a}(\Re u, \Im u) \in \mathbb{R},$
$\mathfrak{a}((\Re u)^+, (\Re u)^-) \leq 0,$ and $\Re\mathfrak{a}((1 \wedge |u|)\text{sign } u, (|u| - 1)^+\text{sign } u) \geq 0.$
3) $(e^{-tA})_{t \geq 0}$ is positive, $1 \wedge u \in D(\mathfrak{a}),$ and $\mathfrak{a}(1 \wedge u, (u-1)^+) \geq 0$ for all $u \in D(\mathfrak{a}) \cap H^+.$

Proof. The equivalence of 1) and 2) follows from Theorems 2.6 and 2.15. Theorem 2.6 shows that 2) implies 3). It remains to prove that 3) implies 1). For $u \in D(\mathfrak{a})$, we have $(\Re u)^+ \in D(\mathfrak{a})$ because of the positivity of $(e^{-tA})_{t \geq 0}$ (cf. Theorem 2.6). Applying then 3) to $(\Re u)^+$ yields

$$1 \wedge (\Re u)^+ \in D(\mathfrak{a}) \text{ and } \mathfrak{a}(1 \wedge (\Re u)^+, ((\Re u)^+ - 1)^+) \geq 0 \text{ for all } u \in D(\mathfrak{a}).$$

By Theorem 2.2, this implies that $(e^{-tA})_{t \geq 0}$ leaves invariant the convex set $C = \{u \in L^2(X, \mu, \mathbb{C}), 0 \leq u \leq 1\}$ (the projection onto this convex set is given by $\mathcal{P}u = 1 \wedge (\Re u)^+$). $\qquad\square$

We have seen that for symmetric forms all the above criteria can be given in terms of the quadratic form. Thus, as a corollary, we obtain the well-known Beurling-Deny criteria.

COROLLARY 2.18 *Assume that the form $\mathfrak{a}$ is symmetric on the real Hilbert space $H = L^2(X, \mu, \mathbb{R})$. The following properties are equivalent:*
1) $(e^{-tA})_{t \geq 0}$ is positive.
2) $u \in D(\mathfrak{a}) \Longrightarrow |u| \in D(\mathfrak{a})$ and $\mathfrak{a}(|u|, |u|) \leq \mathfrak{a}(u, u).$
Assume now that $(e^{-tA})_{t \geq 0}$ is positive. Then the following assertions are equivalent:
3) $(e^{-tA})_{t \geq 0}$ is sub-Markovian.
4) $0 \leq u \in D(\mathfrak{a}) \Longrightarrow 1 \wedge u \in D(\mathfrak{a})$ and $\mathfrak{a}(1 \wedge u, 1 \wedge u) \leq \mathfrak{a}(u, u).$

2.3 DOMINATION OF SEMIGROUPS

We turn now to another property. Let $\mathfrak{a}$ and $\mathfrak{b}$ be two sesquilinear forms on $H = L^2(X, \mu, \mathbb{C})$ and satisfying the standard assumptions (1.2)–(1.5) as in the previous sections. Denote by A and B their associated operators, respectively. Denote by $(e^{-tA})_{t \geq 0}$ and $(e^{-tB})_{t \geq 0}$ their associated semigroups on H. We want to characterize in terms of the forms the domination property

$$|e^{-tA} f| \leq e^{-tB} |f| \text{ for all } f \in H \text{ and } t \geq 0.$$

Here and in the rest of this section, all the inequalities are understood in the μ a.e. sense.

When this inequality holds, we will say that $(e^{-tA})_{t \geq 0}$ is dominated by $(e^{-tB})_{t \geq 0}$. This property plays an important role in several situations, e.g., in estimates of heat kernels or in spectral theory (properties like compactness or Hilbert-Schmidt carry over from e^{-tB} to e^{-tA}).

We first introduce the following definition.

DEFINITION 2.19 *Let U and V be two subspaces of H. We shall say that U is an ideal of V if the following two conditions are satisfied:*
1) $u \in U \implies |u| \in V$.
2) If $u \in U$ and $v \in V$ are such that $|v| \leq |u|$, then the product $v$sign $u \in U$.

In the first result, we show that positivity of the semigroup implies that $D(\mathfrak{a})$ is an ideal of itself. We have

PROPOSITION 2.20 *Let $\mathfrak{a}$ be densely defined, acrretive, continuous, and closed form on $H = L^2(X, \mu, \mathbb{C})$. Consider the following assertions:*
1) The semigroup $(e^{-tA})_{t \geq 0}$ is positive.
2) $D(\mathfrak{a})$ is an ideal of itself and

$$\Re\mathfrak{a}(u, |v|\text{sign } u) \geq \mathfrak{a}(|u|, |v|) \text{ for all } u, v \in D(\mathfrak{a}) \text{ such that } |v| \leq |u|.$$

Then 1) implies 2). The converse is true if $(e^{-tA})_{t \geq 0}$ is real.

Proof. Assume that assertion 1) holds. Then for all $t \geq 0$ and all $f \in H$, we have

$$|e^{-tA}f| \leq e^{-tA}|f|. \tag{2.5}$$

Indeed, if f is real-valued then (2.5) follows from the positivity of e^{-tA} and the obvious inequalities $f \leq |f|$ and $-f \leq |f|$. For general f, it follows from

$$|f| = \sup\{\Re(e^{i\theta}f), \, 0 \leq \theta \leq 2\pi\}.$$

Define now on $H \times H$, the sesquilinear form

$$\mathfrak{c}(U_0, U_1) := \mathfrak{a}(u_0, u_1) + \mathfrak{a}(v_0, v_1),$$
$$D(\mathfrak{c}) := D(\mathfrak{a}) \times D(\mathfrak{a}), \, U_0 = (u_0, v_0), U_1 = (u_1, v_1).$$

It is easy to see that the sesquilinear form $\mathfrak{c}$ is densely defined, accretive, continuous, and closed. The semigroup $(T(t))_{t \geq 0}$ generated by (minus) the operator associated with $\mathfrak{c}$ on $H \times H$ is given by

$$T(t) = \begin{pmatrix} e^{-tA} & 0 \\ 0 & e^{-tA} \end{pmatrix} \text{ for all } t \geq 0.$$

Now it is clear from (2.5) that $(e^{-tA})_{t\geq 0}$ is positive if and only if $(T(t))_{t\geq 0}$ leaves invariant the following closed convex set of $H \times H$,

$$C = \{(u, v) \in H \times H, |u| \leq v\}. \tag{2.6}$$

Using (2.1), one shows that the projection $\mathcal{P}$ of $H \times H$ onto C is given by

$$\mathcal{P}(u, v) = \frac{1}{2}\left(\left[|u| + |u| \wedge \Re v\right]^+ \text{sign } u, \left[|u| \vee \Re v + \Re v\right]^+\right). \tag{2.7}$$

Theorem 2.2 implies that $\mathcal{P}(u, v) \in D(\mathfrak{a}) \times D(\mathfrak{a})$ for all $(u, v) \in D(\mathfrak{a}) \times D(\mathfrak{a})$. This implies that $D(\mathfrak{a})$ is an ideal of itself.

Now by Lemma 1.56, we have

$$\Re\mathfrak{a}(u, |v|\text{sign } u) = \lim_{t\to 0} \frac{1}{t}\Re(|u| - \text{sign}(\overline{u})e^{-tA}u; |v|)$$

$$\geq \limsup_{t\to 0} \frac{1}{t}(|u| - e^{-tA}|u|; |v|)$$

$$= \mathfrak{a}(|u|, |v|).$$

Conversely, assume that $(e^{-tA})_{t\geq 0}$ is real and that $D(\mathfrak{a})$ is an ideal of itself. Thus, $|u| \in D(\mathfrak{a})$ for every $u \in D(\mathfrak{a})$. This implies that $u^+ \in D(\mathfrak{a})$ for every $u \in D(\mathfrak{a}) \cap H_{\mathbb{R}}$. Apply now assertion 2) with $v = u^+$ to obtain

$$\mathfrak{a}(u, u^+) \geq \mathfrak{a}(|u|, u^+),$$

that is,

$$\mathfrak{a}(u^-, u^+) \leq 0.$$

By Theorem 2.6, this implies that the semigroup $(e^{-tA})_{t\geq 0}$ is positive. $\square$

Remark. The positivity of $(e^{-tA})_{t\geq 0}$ implies that $|v|\text{sign } u \in D(\mathfrak{a})$ for all $u, v \in D(\mathfrak{a})$ such that $|v| \leq |u|$. This follows directly from the fact that $D(\mathfrak{a})$ is an ideal of itself.

We come now to the domination property. We consider two densely defined, accretive, continuous, and closed sesquilinear forms $\mathfrak{a}$ and $\mathfrak{b}$ on $H = L^2(X, \mu, \mathbb{C})$. We denote by $(e^{-tA})_{t\geq 0}$ and $(e^{-tB})_{t\geq 0}$ their associated semigroups, respectively.

THEOREM 2.21 *Assume that the semigroup $(e^{-tB})_{t\geq 0}$ is positive. The following assertions are equivalent:*
1) $|e^{-tA}f| \leq e^{-tB}|f|$ for all $t \geq 0$ and all $f \in H$.
2) $D(\mathfrak{a})$ is an ideal of $D(\mathfrak{b})$ and $\Re\mathfrak{a}(u, |v|\text{sign } u) \geq \mathfrak{b}(|u|, |v|)$ for all $(u, v) \in D(\mathfrak{a}) \times D(\mathfrak{b})$ such that $|v| \leq |u|$.
3) $D(\mathfrak{a})$ is an ideal of $D(\mathfrak{b})$ and $\Re\mathfrak{a}(u, v) \geq \mathfrak{b}(|u|, |v|)$ for all $u, v \in D(\mathfrak{a})$ such that $u\overline{v} \geq 0$.

Proof. We define on $H \times H$ the sesquilinear form

$$\mathfrak{c}(U_0, U_1) = \mathfrak{a}(u_0, u_1) + \mathfrak{b}(v_0, v_1),$$
$$D(\mathfrak{c}) = D(\mathfrak{a}) \times D(\mathfrak{b}), U_0 = (u_0, v_0), U_1 = (u_1, v_1).$$

The semigroup $(T(t))_{t \geq 0}$, associated with $\mathfrak{c}$ on $H \times H$ is given by

$$T(t) = \begin{pmatrix} e^{-tA} & 0 \\ 0 & e^{-tB} \end{pmatrix}.$$

Clearly, 1) is equivalent to the fact that $T(t)\mathcal{C} \subseteq \mathcal{C}$ for all $t \geq 0$, where $\mathcal{C}$ is the convex set given by (2.6).

Assume that 1) holds. By Theorem 2.2, we have $\mathcal{P}(D(\mathfrak{a}) \times D(\mathfrak{b})) \subseteq D(\mathfrak{a}) \times D(\mathfrak{b})$. Using (2.7), we obtain that $D(\mathfrak{a})$ is an ideal of $D(\mathfrak{b})$.

Let $u, v \in D(\mathfrak{a})$ be such that $u\bar{v} \geq 0$. In particular, $u\bar{v} = |u||v|$. By Lemma 1.56, we have

$$\Re\mathfrak{a}(u, v) = \lim_{t \to 0} \Re \frac{1}{t}(u - e^{-tA}u; v)$$

$$= \lim_{t \to 0} \Re \frac{1}{t}(|u| - \text{sign}(\bar{v})e^{-tA}u; |v|)$$

$$\geq \limsup_{t \to 0} \Re \frac{1}{t}(|u| - e^{-tB}|u|; |v|)$$

$$= \mathfrak{b}(|u|, |v|).$$

This proves 3).

To prove that 3) implies 2), we pick $u \in D(\mathfrak{a}), v \in D(\mathfrak{b})$ such that $|v| \leq |u|$. Hence $|v|\text{sign } u \in D(\mathfrak{a})$ and $u|v|\text{sign}(\bar{u}) \geq 0$. So by assertion 3),

$$\Re\mathfrak{a}(u, |v|\text{sign } u) \geq \mathfrak{b}(|u|, |v\text{sign } u|) = \mathfrak{b}(|u|, |v|).$$

Finally, we prove that 2) implies 1). Let $(u, v) \in D(\mathfrak{a}) \times D(\mathfrak{b})$. Since $|u| \in D(\mathfrak{b})$ and the semigroup $(e^{-tB})_{t \geq 0}$ is positive, we obtain $(|u| + |u| \wedge \Re v)^+ \in D(\mathfrak{b})$ (see Theorem 2.6 and Proposition 2.20). For the same reason, $(|u| \vee \Re v + \Re v)^+ \in D(\mathfrak{b})$. Since $D(\mathfrak{a})$ is an ideal of $D(\mathfrak{b})$, we obtain $(|u| + |u| \wedge \Re v)^+ \text{sign } u \in D(\mathfrak{a})$. Thus, $\mathcal{P}(D(\mathfrak{a}) \times D(\mathfrak{b})) \subseteq D(\mathfrak{a}) \times D(\mathfrak{b}) = D(\mathfrak{c})$, where $\mathcal{P}$ is give by (2.7).

Since $u \in D(\mathfrak{a})$ and $\frac{1}{2}(|u| + |u| \wedge \Re v)^+ \in D(\mathfrak{b})$, we apply 2) to obtain

$$\Re\mathfrak{a}\left(u, u - \frac{1}{2}(|u| + |u| \wedge \Re v)^+\text{sign } u\right)$$

$$= \Re\mathfrak{a}\left(u, [|u| - \frac{1}{2}(|u| + |u| \wedge \Re v)^+]\text{sign } u\right)$$

$$\geq \mathfrak{b}\left(|u|, |u| - \frac{1}{2}(|u| + |u| \wedge \Re v)^+\right).$$

Hence, for every $(u, v) \in D(\mathfrak{a}) \times D(\mathfrak{b})$,

$$\Re\mathfrak{c}((u, v), (u, v) - \mathcal{P}(u, v)) \geq \mathfrak{b}\left(|u|, |u| - \frac{1}{2}(|u| + |u| \wedge \Re v)^+\right)$$

$$+ \mathfrak{b}\left(v, v - \frac{1}{2}(|u| \vee \Re v + \Re v)^+\right)$$

$$= \Re\mathfrak{d}((|u|, v), (|u|, v) - \mathcal{P}(|u|, v)),$$

where $\mathfrak{d}$ is the form given by

$$\mathfrak{d}(U_0, U_1) = \mathfrak{b}(u_0, u_1) + \mathfrak{b}(v_0, v_1),$$
$$D(\mathfrak{d}) = D(\mathfrak{b}) \times D(\mathfrak{b}), U_0 = (u_0, v_0), U_1 = (u_1, v_1).$$

The semigroup associated with $\mathfrak{d}$ is given by $S(t) = \begin{pmatrix} e^{-tB} & 0 \\ 0 & e^{-tB} \end{pmatrix}$ $(t \geq 0)$. The positivity of $(e^{-tB})_{t \geq 0}$ implies that $S(t)\mathcal{C} \subseteq \mathcal{C}$ for all $t \geq 0$ (again $\mathcal{C}$ is given by (2.6)). Thus, by Theorem 2.2

$$\Re\mathfrak{d}((|u|, v), (|u|, v) - \mathcal{P}(|u|, v)) \geq 0.$$

Thus, we have proved

$$\Re\mathfrak{c}((u, v), (u, v) - \mathcal{P}(u, v)) \geq 0.$$

We again apply Theorem 2.2 to conclude that $T(t)\mathcal{C} \subseteq \mathcal{C}$ for all $t \geq 0$. $\quad\square$

COROLLARY 2.22 *Assume that $\mathfrak{a}$ and $\mathfrak{b}$ are restrictions of a form $\mathfrak{s}$ whose associated semigroup is positive. Assume in addition that $(e^{-tB})_{t \geq 0}$ is positive. Then the following assertions are equivalent:*
1) $(e^{-tA})_{t \geq 0}$ is dominated by $(e^{-tB})_{t \geq 0}$.
2) $D(\mathfrak{a})$ is an ideal of $D(\mathfrak{b})$.

Proof. Let $(u, v) \in D(\mathfrak{a}) \times D(\mathfrak{b})$ be such that $|v| \leq |u|$. By assumption, the semigroup associated with the form $\mathfrak{s}$ is positive, hence by Proposition 2.20

$$\Re\mathfrak{s}(u, |v|\text{sign } u) \geq \mathfrak{s}(|u|, |v|).$$

Using this, we have

$$\Re\mathfrak{a}(u, |v|\text{sign } u) = \Re\mathfrak{s}(u, |v|\text{sign } u)$$
$$\geq \mathfrak{s}(|u|, |v|)$$
$$= \mathfrak{b}(|u|, |v|).$$

The conclusion follows from Theorem 2.21. $\quad\square$

We consider now the situation where the two semigroups $(e^{-tA})_{t\geq 0}$ and $(e^{-tB})_{t\geq 0}$ are positive. Clearly, in this case, it is enough to consider only non-negative f in the domination property. This leads to a simplification in the characterization of the domination. In particular, the ideal property coincides now with the known notion of lattice ideal in Banach lattice spaces.[3] More precisely, we have

PROPOSITION 2.23 *Assume that the semigroups $(e^{-tA})_{t\geq 0}$ and $(e^{-tB})_{t\geq 0}$ are both positive. If $(e^{-tA})_{t\geq 0}$ is dominated by $(e^{-tB})_{t\geq 0}$ then $D(\mathfrak{a}) \subseteq D(\mathfrak{b})$. Moreover, the following are equivalent:*
1) $D(\mathfrak{a})$ is an ideal of $D(\mathfrak{b})$.
2) If $0 \leq v \leq u, u \in D(\mathfrak{a})$ and $v \in D(\mathfrak{b})$ then $v \in D(\mathfrak{a})$.
3) If $|v| \leq |u|, u \in D(\mathfrak{a})$ and $v \in D(\mathfrak{b})$ then $v \in D(\mathfrak{a})$.

Proof. Let $u \in D(\mathfrak{a})$. Since $(e^{-tA})_{t\geq 0}$ is positive, we have $\Re u, (\Re u)^+ \in D(\mathfrak{a})$ (cf. Theorem 2.6). Since $D(\mathfrak{a})$ is an ideal of $D(\mathfrak{b})$, we have $(\Re u)^+ = |(\Re u)^+| \in D(\mathfrak{b})$. The same argument applied to $-u$, yields then $\Re u \in D(\mathfrak{b})$. Applying this with iu if necessary, we conclude that $u \in D(\mathfrak{b})$. Thus, $D(\mathfrak{a}) \subseteq D(\mathfrak{b})$.

Assume now that assertion 2) is satisfied. Let $(u, v) \in D(\mathfrak{a}) \times D(\mathfrak{b})$ such that $|v| \leq |u|$. We have to show that $v\,\mathrm{sign}\, u \in D(\mathfrak{a})$. In fact, $\Re(v)\mathrm{sign}\, u \in D(\mathfrak{b})$ since $D(\mathfrak{b})$ is an ideal of itself by Proposition 2.20. By the same proposition, one has $|u| \in D(\mathfrak{a})$. Since $0 \leq (\Re(v)\mathrm{sign}\, u)^+ \leq |u|$, and that $(\Re(v)\mathrm{sign}\, u))^+ \in D(\mathfrak{b})$ by Theorem 2.6, we obtain from assertion 2) that $(Re(v)\mathrm{sign}\, u))^+ \in D(\mathfrak{a})$. Using also this for $-v$ and iv if necessary, we obtain $v\,\mathrm{sign}\, u \in D(\mathfrak{a})$. This shows that 2) implies 1).

Finally, we show that 1) implies 3). For, let $|v| \leq |u|$ with $u \in D(\mathfrak{a})$ and $v \in D(\mathfrak{b})$. Then $v\,\mathrm{sign}\, u \in D(\mathfrak{a}) \subseteq D(\mathfrak{b})$. But $|v\,\mathrm{sign}\, u| \leq |\overline{u}|$, and this implies again that $v\,\mathrm{sign}\, u\,\mathrm{sign}\,\overline{u} \in D(\mathfrak{a})$, that is, $v \in D(\mathfrak{a})$. $\square$

THEOREM 2.24 *Assume that the two semigroups $(e^{-tA})_{t\geq 0}$ and $(e^{-tB})_{t\geq 0}$ are positive. The following assertions are equivalent:*
1) $(e^{-tA})_{t\geq 0}$ is dominated by $(e^{-tB})_{t\geq 0}$.
2) $D(\mathfrak{a})$ is an ideal of $D(\mathfrak{b})$ and $\mathfrak{a}(u, v) \geq \mathfrak{b}(u, v)$ for all $u, v \in D(\mathfrak{a}) \cap H^+$.

Proof. Theorem 2.21 shows that 1) implies 2).
Conversely, let $u, v \in D(\mathfrak{a})$ be such that $u\overline{v} \geq 0$. Because of (2.5) we can apply Theorem 2.21 with $\mathfrak{a}$ in place of $\mathfrak{b}$ and then obtain

$$\Re\mathfrak{a}(u, v) \geq \mathfrak{a}(|u|, |v|).$$

This and assertion 2) imply

$$\Re\mathfrak{a}(u, v) \geq \mathfrak{a}(|u|, |v|) \geq \mathfrak{b}(|u|, |v|).$$

---

[3]This is why we used the terminology of "ideal" in Definition 2.19.

The conclusion follows again from Theorem 2.21.   □

Remark.   1) The results in this section are true if instead of assuming that both forms are accretive we merely assume that $\Re \mathfrak{a}(u, u) \geq -c_1 \|u\|_H^2$ and $\Re \mathfrak{b}(u, u) \geq -c_2 \|u\|_H^2$ with c_1 and c_2 two positive constants. In fact, let $c = \max \{c_1, c_2\}$; then the forms $(\mathfrak{a} + c)$ and $(\mathfrak{b} + c)$ are accretive. It is clear that $(e^{-t A})_{t \geq 0}$ is dominated by $(e^{-t B})_{t \geq 0}$ if and only if $(e^{-ct} e^{-t A})_{t \geq 0}$ is dominated by $(e^{-ct} e^{-t B})_{t \geq 0}$.

2) The assumption that the forms are densely defined can be removed in all the above results. See Section 2.6 below.

3) The method presented in this section can be used in other circumstances. For example, to characterize the property

$$\phi(e^{-t A} f) \leq e^{-t B} \phi(f)$$

for some convex function $\phi : \mathbb{R} \to \mathbb{R}$, one can apply Theorem 2.2 to the semigroup $\begin{pmatrix} e^{-t A} & 0 \\ 0 & e^{-t B} \end{pmatrix}$ and the convex set

$$\mathcal{C} = \{(u, v) \in H \times H, \phi(u) \leq v\}.$$

2.4 OPERATIONS ON THE FORM-DOMAIN

A normal contraction on $\mathbb{C}$ is a function $p : \mathbb{C} \to \mathbb{C}$ which satisfies

$$p(0) = 0 \text{ and } |p(x) - p(y)| \leq |x - y| \text{ for all } x, y \in \mathbb{C}.$$

Given $u \in H = L^2(X, \mu, \mathbb{C})$, we define $p(u)$ to be the composition function $p(u)(x) = p(u(x))$. It is well known that normal contractions operate on the Sobolev space $H^1(\mathbb{R}^n)$, that is, $p(u) \in H^1(\mathbb{R}^n)$ for every normal contraction p and every $u \in H^1(\mathbb{R}^n)$. More generally, it is well known that normal contractions operate on the domain of any symmetric Dirichlet form (the latter means a symmetric form whose semigroup is sub-Markovian). We prove this in the next result. We also show a similar result for forms whose associated semigroup is merely L^∞-contractive.

THEOREM 2.25 *Assume that $\mathfrak{a}$ is a densely defined, accretive, and closed symmetric form on $H = L^2(X, \mu, \mathbb{C})$. The following assertions are equivalent:*

1) The semigroup $(e^{-t A})_{t \geq 0}$ is sub-Markovian.
2) $p(D(\mathfrak{a})) \subseteq D(\mathfrak{a})$ and $\mathfrak{a}(p(u), p(u)) \leq \mathfrak{a}(u, u)$ for every $u \in D(\mathfrak{a})$ and every normal contraction p.

Proof. Choosing the functions $p(z) = (\Re z)^+$ and $p(z) = (1 \wedge |z|)\text{sign } z$ and applying Theorems 2.7 and 2.14, we see that 2) implies 1).

Assume now that $(e^{-tA})_{t \geq 0}$ is sub-Markovian. Using Lemma 1.56, we see that it is enough to prove that for every $t > 0$ and every $u \in H$,

$$(p(u) - e^{-tA}p(u); p(u)) \leq (u - e^{-tA}u; u).$$

By a density argument, it is enough to prove this inequality for functions $u = \sum_{i=1}^{n} \alpha_i \chi_{A_i}$, where A_i are disjoint measurable subsets of X such that $\mu(A_i) < \infty$. Here χ_{A_i} denotes the characteristic function of A_i. Thus, we need to prove that

$$\sum_{i,j=1}^{n} (\chi_{A_i} - e^{-tA}\chi_{A_i}; \chi_{A_j})p(\alpha_i)\overline{p(\alpha_j)} \leq \sum_{i,j=1}^{n} (\chi_{A_i} - e^{-tA}\chi_{A_i}; \chi_{A_j})\alpha_i\overline{\alpha_j}.$$

Set

$$b_{ij} := (\chi_{A_i} - e^{-tA}\chi_{A_i}; \chi_{A_j}), \quad \lambda_i := (\chi_{A_i}; \chi_{A_i}), \quad a_{ij} := (e^{-tA}\chi_{A_i}; \chi_{A_j}).$$

Since e^{-tA} is self-adjoint, we have

$$\sum_{i,j} b_{ij}p(\alpha_i)\overline{p(\alpha_j)} = \sum_{i<j} a_{ij}|p(\alpha_i) - p(\alpha_j)|^2 + \sum_{j}[\lambda_j - \sum_i a_{ij}]|p(\alpha_j)|^2.$$

The sub-Markovian property of the semigroup implies that $a_{ij} \geq 0$ and $\sum_{i=1}^{n} a_{ij} \leq \lambda_j$. Using this and the fact that p is a normal contraction we obtain the desired inequality. $\square$

In the next proposition we show that if $(e^{-tA})_{t \geq 0}$ is L^∞-contractive, then it is dominated by a sub-Markovian semigroup. More precisely,

PROPOSITION 2.26 *Assume that $(T(t))_{t \geq 0}$ is strongly continuous semigroup on $L^2(X, \mu, \mathbb{C})$ which is contractive on $L^p(X, \mu, \mathbb{C})$ for every $p \in [1, \infty]$. Then there exists a sub-Markovian semigroup $(T^{\odot}(t))_{t \geq 0}$ which satisfies*

$$|T(t)u| \leq T^{\odot}(t)|u| \text{ for all } t \geq 0, u \in L^p(X, \mu, \mathbb{C}).$$

Moreover, any other semigroup which dominates $(T(t))_{t \geq 0}$ also dominates the semigroup $(T^{\odot}(t))_{t \geq 0}$.[4]

Proof. For each $t \geq 0$, the operator $T(t)$ is a contraction on $L^1(X, \mu, \mathbb{C})$. Thus by a result of Chacon and Krengel [ChKr64], it has a modulus operator $|T(t)| \in \mathcal{L}(L^1(X, \mu, \mathbb{C}))$, defined for $0 \leq f \in L^1(X, \mu, \mathbb{C})$ by

$$|T(t)|f := \sup_{|g| \leq f} |T(t)g| \text{ for } 0 \leq f \in L^1(X, \mu, \mathbb{C}).$$

---

[4] $(T^{\odot}(t))_{t \geq 0}$ is called the modulus semigroup of $(T(t))_{t \geq 0}$.

Kubokawa [Kub75][5] has shown that the semigroup $(T^\odot(t))_{t\geq 0}$ can be defined in $L^1(X, \mu, \mathbb{C})$ as follows:

$$T^\odot(r)f := \lim_n \left| T\left(\frac{1}{2^n}\right) \right|^{[2^n r]} f \text{ for } r \in \mathbb{Q}^+ \text{ and } 0 \leq f \in L^1(X, \mu, \mathbb{C}),$$

where $[.]$ denotes the integer part. For real $t \geq 0$, one puts

$$T^\odot(t)f = \lim_{r \to t, r \in \mathbb{Q}^+} T^\odot(r)f.$$

Now it is clear that $T^\odot(t)$ satisfies $0 \leq T^\odot(t)f \leq 1$ for $0 \leq f \leq 1$. Hence $(T^\odot(t))_{t\geq 0}$ is a sub-Markovian semigroup.

Assume that $(S(t))_{t\geq 0}$ is a positive semigroup acting on $L^p(X, \mu, \mathbb{C})$ for some $p \in [1, \infty)$, and which dominates $(T(t))_{t\geq 0}$. Then for every $f \in L^1 \cap L^p$,

$$\left| T\left(\frac{1}{2^n}\right) \right| |f| = \sup_{|g| \leq |f|} \left| T\left(\frac{1}{2^n}\right) g \right|$$

$$\leq S\left(\frac{1}{2^n}\right) |f|.$$

Hence, it follows from the definition of $T^\odot(r)$ that

$$T^\odot(r)|f| \leq S(r)|f|$$

for each $r \in \mathbb{Q}^+$. This implies the last claim of the proposition.   □

The next result shows how normal contractions operate on domains of forms whose semigroup is merely L^∞-contractive.

THEOREM 2.27 *Assume that the form $\mathfrak{a}$ is densely defined, accretive, continuous, and closed. If the semigroups $(e^{-tA})_{t\geq 0}$ and its adjoint $(e^{-tA^*})_{t\geq 0}$ are both L^∞-contractive, then:*
$p(|u|)\text{sign } u \in D(\mathfrak{a})$ for every $u \in D(\mathfrak{a})$ and every normal contraction p.

Proof. If $(e^{-tA})_{t\geq 0}$ and its adjoint $(e^{-tA^*})_{t\geq 0}$ are L^∞-contractive, then the semigroup $(e^{-tB})_{t\geq 0}$ associated to the symmetric part $\mathfrak{b} = \frac{1}{2}(\mathfrak{a} + \mathfrak{a}^*)$ is L^∞-contractive. This can be seen by applying Theorem 2.13 to both forms $\mathfrak{a}$ and $\mathfrak{a}^*$. It follows now that $(e^{-tB})_{t\geq 0}$ acts as a contraction semigroup on $L^q(X, \mu, \mathbb{C})$ for every $q \in [1, \infty)$. By the previous proposition, there exists a modulus semigroup $(T^\odot(t))_{t\geq 0}$ of $(e^{-tB})_{t\geq 0}$. We have also seen in the proof of the latter proposition that $(T^\odot(t))_{t\geq 0}$ is a contraction semigroup on $L^q(X, \mu, \mathbb{C})$ for $1 \leq q < \infty$.

---

[5]See also Kipnis [Kip74].

The semigroup $(T^{\odot}(t))_{t\geq 0}$ dominates $(e^{-tB})_{t\geq 0}$ on $L^2(X,\mu,\mathbb{C})$; then by taking the adjoints, we obtain that $(T^{\odot}(t)^*)_{t\geq 0}$ dominates $(e^{-tB})_{t\geq 0}$. Hence, $(T^{\odot}(t)^*)_{t\geq 0}$ dominates $(T^{\odot}(t))_{t\geq 0}$. We deduce from this that $T^{\odot}(t)$ is self-adjoint for each $t \geq 0$. This implies that the generator of $(T^{\odot}(t)^*)_{t\geq 0}$, say $-C$, is a self-adjoint operator. Let $\mathfrak{c}$ be the symmetric non-negative (i.e., accretive) form whose associated operator is C ($\mathfrak{c}$ exists by Theorem 1.57). Theorem 2.21 asserts that $D(\mathfrak{a})(= D(\mathfrak{b}))$ is an ideal of $D(\mathfrak{c})$. Now let $u \in D(\mathfrak{a})$ and p be a normal contraction. Since $|u| \in D(\mathfrak{c})$, we have $p(|u|) \in D(\mathfrak{c})$ by Theorem 2.25. But $|p(|u|)| \leq |u|$ and using again the fact that $D(\mathfrak{a})$ is an ideal of $D(\mathfrak{c})$, we obtain $p(|u|)\operatorname{sign} u \in D(\mathfrak{a})$. $\qquad\square$

THEOREM 2.28 *Let $\mathfrak{a}$ and $(e^{-tA})_{t\geq 0}$ be as in the previous theorem.*
1) If $(e^{-tA})_{t\geq 0}$ and $(e^{-tA^*})_{t\geq 0}$ are sub-Markovian, then $uv \in D(\mathfrak{a})$ for all $u,v \in D(\mathfrak{a}) \cap L^\infty(X,\mu,\mathbb{C})$.
2) If $(e^{-tA})_{t\geq 0}$ and $(e^{-tA^*})_{t\geq 0}$ are L^∞-contractive, then $u|v| \in D(\mathfrak{a})$ for all $u,v \in D(\mathfrak{a}) \cap L^\infty(X,\mu,\mathbb{C})$.

Proof. Let $\mathfrak{b}, \mathfrak{c}, B$, and C be as in the proof of Theorem 2.27. Note that if $(e^{-tA})_{t\geq 0}$ and $(e^{-tA^*})_{t\geq 0}$ are sub-Markovian, then $(e^{-tB})_{t\geq 0}$ is sub-Markovian, too; and if $(e^{-tA})_{t\geq 0}$ and $(e^{-tA^*})_{t\geq 0}$ are L^∞-contractive, then so is $(e^{-tB})_{t\geq 0}$ (apply Theorems 2.6 and 2.13).

1) Using $uv = \frac{1}{2}(u+v)^2 - \frac{1}{2}u^2 - \frac{1}{2}v^2$, it is enough to prove that $u^2 \in D(\mathfrak{a}) = D(\mathfrak{b})$ for all $u \in D(\mathfrak{a}) \cap L^\infty(X,\mu,\mathbb{C})$. In order to show this, we use the same idea as in the proof of Theorem 2.25. Let $u = \sum_{i=1}^n \alpha_i \chi_{A_i}$, where A_i are disjoint measurable subsets with finite measure. Set $b_{ij} := (\chi_{A_i} - e^{-tB}\chi_{A_i}; \chi_{A_j})$, $\lambda_i := (\chi_{A_i}; \chi_{A_i})$ and $a_{ij} := (e^{-tB}\chi_{A_i}; \chi_{A_j})$. Since e^{-tB} is self-adjoint, we have

$$(u^2 - e^{-tB}u^2, u^2) = \sum_{i,j} b_{ij}\alpha_i^2\overline{\alpha_j}^2$$

$$= \sum_{i<j} a_{ij}|\alpha_i^2 - \alpha_j^2|^2 + \sum_j \left[\lambda_j - \sum_i a_{ij}\right]|\alpha_j^2|^2$$

$$\leq 4\sup_i |\alpha_i|^2 \sum_{i<j} a_{ij}|\alpha_i - \alpha_j|^2$$

$$+ \sup_i |\alpha_i|^2 \sum_j \left[\lambda_j - \sum_i a_{ij}\right]|\alpha_j|^2.$$

To obtain the last inequality, we have used the facts that $a_{ij} \geq 0$ and $\lambda_j - \sum_i a_{ij} \geq 0$ which follow from the sub-Markovian property of $(e^{-tB})_{t\geq 0}$. Thus, we have proved

$$(u^2 - e^{-tB}u^2; u^2) \leq 4\|u\|_\infty^2 (u - e^{-tB}u; u).$$

This inequality extends to all $u \in L^2(X, \mu, \mathbb{C}) \cap L^\infty(X, \mu, \mathbb{C})$. An application of Lemma 1.56 shows that $u^2 \in D(\mathfrak{b}) = D(\mathfrak{a})$ for all $u \in D(\mathfrak{a}) \cap L^\infty(X, \mu, \mathbb{C})$.

2) As in the proof of Theorem 2.27, we construct a symmetric form $\mathfrak{c}$ whose associated semigroup is sub-Markovian and such that $D(\mathfrak{a})(= D(\mathfrak{b}))$ is an ideal of $D(\mathfrak{c})$. Now if $u, v \in D(\mathfrak{a}) \cap L^\infty(X, \mu, \mathbb{C})$, then $|u|, |v| \in D(\mathfrak{c}) \cap L^\infty(X, \mu, \mathbb{C})$. In addition, assertion 1) gives $|u||v| \in D(\mathfrak{c})$. Using now

$$|u||v| \leq \|v\|_\infty |u|$$

and the fact that $D(\mathfrak{a})$ is an ideal of $D(\mathfrak{c})$, we obtain assertion 2). $\qquad \square$

2.5 SEMIGROUPS ACTING ON VECTOR-VALUED FUNCTIONS

In this section, we analyze semigroups acting on vector-valued functions. Let us first fix notation. Let K be a Hilbert space with scalar product $< ., . >_K$ and associated norm $|.|_K$. Let $H := L^2(X, \mu, K)$ be the space of measurable functions f such that

$$\|f\| := \left[\int_X |f(x)|_K^2 d\mu(x) \right]^{1/2} < \infty.$$

Here (X, μ) is a σ-finite measure space. As above we denote by $\|.\|$ and $(.;.)$ the norm and scalar product of H, that is,

$$(u; v) := \int_X < u(x), v(x) >_K d\mu(x) \text{ for all } u, v \in H$$

and $\|u\| = \sqrt{(u; u)}$.

Let $\tilde{\mathfrak{a}}$ be a densely defined, accretive, continuous, and closed sesquilinear form on H. We denote by $\tilde{A}$ the operator associated with $\tilde{\mathfrak{a}}$. The operator $-\tilde{A}$ generates a contraction semigroup $(e^{-t\tilde{A}})_{t \geq 0}$ on $H = L^2(X, \mu, K)$.

In this section we study the question of extending the semigroup $e^{-t\tilde{A}}$ to a contraction semigroup on $L^p(X, \mu, K)$ for $p \neq 2$. In order to do this, we will study two properties. The first property is L^∞-contractivity:

$$\|e^{-t\tilde{A}} u\|_{L^\infty(X,\mu,K)} \leq \|u\|_{L^\infty(X,\mu,K)},$$

for all $u \in L^2(X, \mu, K) \cap L^\infty(X, \mu, K)$ and all $t \geq 0$.

As in the scalar case, if $(e^{-t\tilde{A}})_{t \geq 0}$ is L^∞-contractive then it extends to a strongly continuous semigroup on $L^p(X, \mu, K)$ for $2 \leq p < \infty$. The same conclusion holds for $1 \leq p \leq 2$ if the adjoint semigroup $(e^{-t\tilde{A}^*})_{t \geq 0}$

is L^∞-contractive. If both $(e^{-t\tilde{A}})_{t\geq 0}$ and $(e^{-t\tilde{A}^*})_{t\geq 0}$ are L^∞-contractive, then the semigroup $(e^{-t\tilde{A}})_{t\geq 0}$ extends to a strongly continuous semigroup on $L^p(X, \mu, K), 1 \leq p < \infty$. The strong continuity on $L^1(X, \mu, K)$ can be shown by similar arguments as in the scalar case (see Section 2.2).

There is a second way which allows us to extend $(e^{-t\tilde{A}})_{t\geq 0}$ to a contraction semigroup on $L^p(X, \mu, K)$ spaces. It consists of finding a sub-Markovian semigroup $(e^{-tA})_{t\geq 0}$ acting on $L^2(X, \mu, \mathbb{R})$ which satisfies the domination property

$$|e^{-t\tilde{A}}u(.)|_K \leq e^{-tA}|u|_K(.) \text{ for all } t \geq 0, u \in H.$$

As above, if this property holds, we say that $(e^{-t\tilde{A}})_{t\geq 0}$ is dominated by $(e^{-tA})_{t\geq 0}$.

We have studied both properties in previous sections in the scalar-valued case $K = \mathbb{C}$ or $\mathbb{R}$. We will see here that the same results hold in the vector-valued setting.

We define sign u as in the scalar case by replacing the absolute value by the norm of K. That is

$$\text{sign } u(x) = \begin{cases} \frac{u(x)}{|u(x)|_K} & \text{if } u(x) \neq 0, \\ 0 & \text{if } u(x) = 0, \end{cases}$$

for $u \in H = L^2(X, \mu, K)$. We also use the notation $|u|_K$ for the function $x \mapsto |u(x)|_K$.

THEOREM 2.29 *The following assertions are equivalent:*
1) The semigroup $(e^{-t\tilde{A}})_{t\geq 0}$ is L^∞-contractive.
2) $u \in D(\tilde{a}) \implies (1 \wedge |u|_K)\text{sign } u \in D(\tilde{a})$ and
$\Re\tilde{a}(u, u - (1 \wedge |u|_K)\text{sign} u) \geq 0.$
3) $u \in D(\tilde{a}) \implies (1 \wedge |u|_K)\text{sign } u \in D(\tilde{a})$ and
    $\Re\tilde{a}((1 \wedge |u|_K)\text{sign } u, u - (1 \wedge |u|_K)\text{sign } u) \geq 0.$
3´) There exists a core D of $\tilde{a}$ such that $(1 \wedge |u|_K)\text{sign } u \in D(\tilde{a})$ and
    $\Re\tilde{a}((1 \wedge |u|_K)\text{sign } u, u - (1 \wedge |u|_K)\text{sign } u) \geq 0$ *for all $u \in D$.*
    If the form $\tilde{a}$ is symmetric, then the above assertions are equivalent to each of the following:
4) $u \in D(\tilde{a}) \implies (1 \wedge |u|_K)\text{sign } u \in D(\tilde{a})$ and
    $\tilde{a}((1 \wedge |u|_K)\text{sign } u, (1 \wedge |u|_K)\text{sign } u) \leq \tilde{a}(u, u).$
4´) There exists a core D of $\tilde{a}$ such that $(1 \wedge |u|_K)\text{sign } u \in D(\tilde{a})$ and
    $\tilde{a}((1 \wedge |u|_K)\text{sign } u, (1 \wedge |u|_K)\text{sign } u) \leq \tilde{a}(u, u)$ *for all $u \in D$.*

Proof. It is easy to see that L^∞-contractivity is equivalent to the fact that for every $u \in L^2(X, \mu, K) \cap L^\infty(X, \mu, K)$,

$$|u|_K \leq 1 \implies |e^{-t\tilde{A}}u|_K \leq 1 \text{ for all } t \geq 0.$$

The latter property is equivalent to the fact that the semigroup $(e^{-t\tilde{A}})_{t\geq 0}$ leaves invariant the closed convex set (of H) given by

$$\mathcal{C} = \{u \in L^2(X, \mu, K), |u|_K \leq 1\}.$$

The projection of $L^2(X, \mu, K)$ onto $\mathcal{C}$ is given by $\mathcal{P}u = (1 \wedge |u|_K)\text{sign } u$. Applying Theorems 2.2 and 2.3 we obtain the above result.   □

Now let $\mathfrak{a}$ be a densely defined, accretive, continuous, and closed form on $L^2(X, \mu, \mathbb{R})$. Denote by $(e^{-tA})_{t\geq 0}$ its associated semigroup. We have

THEOREM 2.30 *Assume that $(e^{-tA})_{t\geq 0}$ is positive. The following assertions are equivalent:*
1) $(e^{-t\tilde{A}})_{t\geq 0}$ is dominated by $(e^{-tA})_{t\geq 0}$.
2) $D(\tilde{\mathfrak{a}})$ is an ideal of $D(\mathfrak{a})$ and $\Re\tilde{\mathfrak{a}}(u, |f|\text{sign } u) \geq \mathfrak{a}(|u|_K, |f|)$ for all $(u, f) \in D(\tilde{\mathfrak{a}}) \times D(\mathfrak{a})$ such that $|f| \leq |u|_K$.
3) $D(\tilde{\mathfrak{a}})$ is an ideal of $D(\mathfrak{a})$ and $\Re\tilde{\mathfrak{a}}(u, v) \geq \mathfrak{a}(|u|_K, |v|_K)$ for all $u, v \in D(\tilde{\mathfrak{a}})$ such that $< u, v >_K = |u|_K |v|_K$.

Here, we say that $D(\tilde{\mathfrak{a}})$ is an ideal of $D(\mathfrak{a})$ if the following two properties are satisfied:
i) $u \in D(\tilde{\mathfrak{a}}) \implies |u|_K \in D(\mathfrak{a})$.
ii) If $u \in D(\tilde{\mathfrak{a}})$, $f \in D(\mathfrak{a})$ are such that $|f| \leq |u|_K$, then $|f|\text{sign } u \in D(\tilde{\mathfrak{a}})$.

Proof. Again the proof is very similar to the case of scalar-valued functions. Suppose that assertion 1) holds. Define the form $\mathfrak{c}$ on $\mathfrak{H} := L^2(X, \mu, K) \times L^2(X, \mu, \mathbb{R})$ by

$$\mathfrak{c}(U, V) = \tilde{\mathfrak{a}}(u, v) + \mathfrak{a}(f, g), \quad U = (u, f), V = (v, g) \in D(\mathfrak{c}) = D(\tilde{\mathfrak{a}}) \times D(\mathfrak{a}).$$

The semigroup generated by $\mathfrak{c}$ on $\mathfrak{H}$ is given by

$$e^{-t\mathfrak{c}} := \begin{pmatrix} e^{-t\tilde{A}} & 0 \\ 0 & e^{-tA} \end{pmatrix} \quad \text{for all } t \geq 0.$$

The domination property is equivalent to the fact that the semigroup $e^{-t\mathfrak{c}}$ leaves invariant the closed convex set

$$\mathcal{C} := \{(u, f) \in L^2(X, \mu, K) \times L^2(X, \mu, \mathbb{R}), |u|_K \leq f\}.$$

The projection $\mathcal{P}$ onto $\mathcal{C}$ is given by

$$\mathcal{P}(u, f) = \frac{1}{2}\left(\left[|u|_K + (\Re f) \wedge |u|_K\right]^+ \text{sign } u, \left[|u|_K \vee \Re f + \Re f\right]^+\right).$$

Theorem 2.2, together with the fact that $\Re f, (\Re f)^+ \in D(\mathfrak{a})$ for all $f \in D(\mathfrak{a})$ due to the positivity of $(e^{-tA})_{t\geq 0}$ (see Theorem 2.6) imply that $D(\tilde{\mathfrak{a}})$

is an ideal of $D(\mathfrak{a})$. For $u, v \in D(\tilde{\mathfrak{a}})$ such that $< u, v >_K = |u|_K |v|_K$ we have by Lemma 1.56

$$
\begin{aligned}
\Re\tilde{\mathfrak{a}}(u, v) &= \lim_{t \to 0} \frac{1}{t} \Re(u - e^{-t\tilde{A}} u; v) \\
&= \lim_{t \to 0} \frac{1}{t} \Re \int_X < u(x) - e^{-t\tilde{A}} u(x), v(x) >_K d\mu(x) \\
&\geq \limsup_{t \to 0} \frac{1}{t} \int_X [|u(x)|_K - |e^{-t\tilde{A}} u(x)|_K] |v(x)|_K d\mu(x) \\
&\geq \limsup_{t \to 0} \frac{1}{t} \int_X [|u(x)|_K - e^{-tA} |u|_K(x)] |v(x)|_K d\mu(x) \\
&= \mathfrak{a}(|u|_K, |u|_K).
\end{aligned}
$$

This shows assertion 3).

If 3) holds, then for $(u, f) \in D(\tilde{\mathfrak{a}}) \times D(\mathfrak{a})$ such that $|f| \leq |u|_K$ we have $|f| \operatorname{sign} u \in D(\tilde{\mathfrak{a}})$ and $< u, |f| \operatorname{sign} u >_K = |u|_K ||f|$. Applying 3) to u and $|f| \operatorname{sign} u$ yields 2).

The proof that 2) implies 1) is exactly the same as in the scalar case. The idea is to show that

$$
\Re\mathfrak{c}((u, f), (u, f) - \mathcal{P}(u, v)) \geq 0
$$

for all $(u, f) \in D(\mathfrak{c}) = D(\tilde{\mathfrak{a}}) \times D(\mathfrak{a})$ and apply Theorem 2.2. $\square$

Remark.   1) If we assume in addition that $(e^{-tA})_{t \geq 0}$ is sub-Markovian, then the domination property allows us to extend $(e^{-t\tilde{A}})_{t \geq 0}$ to a strongly continuous semigroup on $L^p(X, \mu, K)$ for all $2 \leq p < \infty$.

2) We have written the results of this section in the case where K is a Hilbert space, but we could include the case where K is a vector bundle over X equipped with an inner product $< ., . >_x$ on each fiber K_x. The inner product $< ., . >_x$ depends measurably on x and each K_x is a Hilbert space with respect to the norm $|.|_x := < ., . >_x^{\frac{1}{2}}$. The scalar product of $H = L^2(X, \mu, K)$ is given by $(u; v) = \int_X < u(x), v(x) >_x dm(x)$. In this situation, L^∞-contractivity is equivalent to the fact that the semigroup $(e^{-t\tilde{A}})_{t \geq 0}$ leaves invariant the closed convex set

$$
\mathcal{C} := \{u \in L^2(X, m, K), |u(x)|_x \leq 1 \ (\mu \text{ a.e.})\}.
$$

We then obtain Theorem 2.29 with $|.|_K$ replaced by $|.|_x$ in the statements. Similarly, Theorem 2.30 holds in this setting, too.

Consider as above, a sesquilinear form $\tilde{\mathfrak{a}}$ on $L^2(X, \mu, K)$ and denote by $(e^{-t\tilde{A}})_{t \geq 0}$ its associated contraction semigroup. Let $\mathcal{R} : X \to \mathcal{L}(K)$ be a field of bounded linear operators on K and assume that the map $x \mapsto <$ $\mathcal{R}(x)u(x), u(x) >_K$ is measurable for all $u \in L^2(X, \mu, K)$. We define the perturbed form $\tilde{\mathfrak{b}} := \tilde{\mathfrak{a}} + \mathcal{R}$ by

$$\tilde{\mathfrak{b}}(u, v) = \tilde{\mathfrak{a}}(u, v) + \int_X < \mathcal{R}(x)u(x), v(x) >_K d\mu(x),$$

$$D(\tilde{\mathfrak{b}}) = \left\{ u \in D(\tilde{\mathfrak{a}}), \int_X | < \mathcal{R}(x)u(x), u(x) >_K | d\mu(x) < \infty \right\}$$

(in the sequel, we shall write for simplicity $< \mathcal{R}(x)u, u >_K$ instead of $< \mathcal{R}(x)u(x), u(x) >_K$). We assume that the form $\tilde{\mathfrak{b}}$ is densely defined, accretive, continuous, and closed. Denote by $(e^{-t\tilde{B}})_{t \geq 0}$ its associated semigroup on $L^2(X, \mu, K)$. Note that if the family $\mathcal{R}(x)$ consists of compact self-adjoint operators on K then we can write (analogous to Schrödinger operators) $\mathcal{R} = \mathcal{R}_+ - \mathcal{R}_-$, where $< \mathcal{R}_\pm(x)u, u >_K \geq 0$. This can be seen by putting $\mathcal{R}_+(x)\xi = \mathcal{R}(x)\xi$ if ξ belongs to the eigenspace corresponding to some non-negative eigenvalue of $\mathcal{R}(x)$ and $= 0$ otherwise, and $\mathcal{R}_- = (-\mathcal{R})_+$. Now the questions of continuity and closability of $\tilde{\mathfrak{b}}$ can be studied more easily because this form can be seen as the perturbation of $\tilde{\mathfrak{a}} + \mathcal{R}_+$ (which is the sum of two accretive forms) by $-\mathcal{R}_-$, which is assumed to be ($\tilde{\mathfrak{a}} + \mathcal{R}_+$)-bounded with bound less than 1.

PROPOSITION 2.31 *Suppose that $(e^{-t\tilde{A}})_{t \geq 0}$ is L^∞-contractive. Then the semigroup $(e^{-t\tilde{B}})_{t \geq 0}$ is L^∞-contractive if and only if*

$$\Re \int_X < \mathcal{R}(x)u, u >_K \frac{1}{|u|_K^2}(|u|_K - 1)^+ d\mu$$
$$\geq -\Re\tilde{\mathfrak{a}}((1 \wedge |u|_K)\operatorname{sign} u, (|u|_K - 1)^+\operatorname{sign} u),$$

for all $u \in D(\tilde{\mathfrak{b}})$. In particular, if $\Re < \mathcal{R}(x)u, u >_K \geq 0$ (for a.e. $x \in X$ and all $u \in L^2(X, \mu, K)$), then $(e^{-t\tilde{B}})_{t \geq 0}$ is L^∞-contractive.

The proof is a simple application of Theorem 2.29 and the obvious remark that

$$< \mathcal{R}(x)(1 \wedge |u|_K)\operatorname{sign} u, u - (1 \wedge |u|_K)\operatorname{sign} u >_K$$
$$= < \mathcal{R}(x)u, u >_K \frac{1}{|u|_K^2}(|u|_K - 1)^+.$$

Note that if $\tilde{\mathfrak{a}}$ is symmetric and the $\mathcal{R}(x)$ are self-adjoint operators on K, then the inequality in this proposition can be replaced by

$$\int_X < \mathcal{R}(x)u, u >_K (1 - \frac{1}{|u|_K^2}(1 \wedge |u|_K)^2)d\mu$$
$$\geq \tilde{\mathfrak{a}}((1 \wedge |u|_K)\text{sign } u, (1 \wedge |u|_K)\text{sign } u) - \tilde{\mathfrak{a}}(u, u).$$

Now let $(e^{-tA})_{t\geq 0}$ be the semigroup on $L^2(X, \mu, \mathbb{R})$ associated with a form $\mathfrak{a}$ and let $V : X \to \mathbb{R}$ be a measurable function. Define as above the form $\mathfrak{b} := \mathfrak{a}+V$, which we assume to satisfy the usual conditions. Denote by $(e^{-tB})_{t\geq 0}$ its associated semigroup on $L^2(X, \mu, \mathbb{R})$. Since V is real-valued, if $(e^{-tA})_{t\geq 0}$ is positivity preserving then the same holds for $(e^{-tB})_{t\geq 0}$. This can be seen by applying Theorem 2.6, since $\mathfrak{a}(u^+, u^-) + (Vu^+, u^-) = \mathfrak{a}(u^+, u^-) \leq 0$.

PROPOSITION 2.32 *Suppose that $(e^{-t\tilde{A}})_{t\geq 0}$ is dominated by $(e^{-tA})_{t\geq 0}$ and suppose that*

$$\Re < \mathcal{R}(x)u(x), u(x) >_K \geq V(x)|u(x)|_K^2 \ \ for \ all \ u \in D(\tilde{\mathfrak{b}}).$$

Then $(e^{-t\tilde{B}})_{t\geq 0}$ is dominated by $(e^{-tB})_{t\geq 0}$.

Proof. By Theorem 2.30, if $(e^{-t\tilde{A}})_{t\geq 0}$ is dominated by $(e^{-tA})_{t\geq 0}$, then $D(\tilde{\mathfrak{a}})$ is an ideal of $D(\mathfrak{a})$ and

$$\Re\tilde{\mathfrak{a}}(u, |f|\text{sign } u) \geq \mathfrak{a}(|u|_K, |f|)$$

for all $(u, f) \in D(\tilde{\mathfrak{a}}) \times D(\mathfrak{a})$ such that $|f| \leq |u|_K$.

Now it is easy to see that this implies that $D(\tilde{\mathfrak{b}})$ is an ideal of $D(\mathfrak{b})$ and this inequality holds if we replace $\tilde{\mathfrak{a}}$ and $\mathfrak{a}$ by $\tilde{\mathfrak{b}}$ and $\mathfrak{b}$, respectively. We apply Theorem 2.30 again to conclude. $\square$

If, for example, the $\mathcal{R}(x)$ are self-adjoint compact operators on K, then the above proposition can be applied by taking $V(x)$ to be the smallest eigenvalue of $\mathcal{R}(x)$.

These results can be applied to study systems of evolution equations in L^p spaces. For a system of n equations, $K = \mathbb{C}^n$ and $\tilde{A}$ (the operator associated with the form $\tilde{\mathfrak{a}}$) and $\mathcal{R}(x)$ are $n \times n$ matrix operators. The operator $\tilde{A} + \mathcal{R}$ can be studied on $L^2(X, \mu, K)$ by the sesquilinear form theory, and if $\tilde{A}$ and $\mathcal{R}$ satisfy the the assumptions of Proposition 2.31, one can then extend the original semigroup defined on $L^2(X, \mu, \mathbb{C}^n)$ to $L^p(X, \mu, \mathbb{C}^n)$. This allows one to prove existence and uniqueness of the solution to the system with initial data in $L^p(X, \mu, \mathbb{C}^n)$ for $2 \leq p < \infty$. The system on $L^p(X, \mu, \mathbb{C}^n)$ for $p \leq 2$ can be treated by duality.

2.6 SESQUILINEAR FORMS WITH NONDENSE DOMAINS

In the previous sections, we have studied several properties of semigroups in terms of their sesquilinear forms and we have assumed that the forms are densely defined. In this section, we show that we can remove this assumption. All the results of the present chapter (except Theorems 2.9, 2.10, and Corollary 2.11) hold for forms which are not densely defined.

To understand why this situation is of some interest, let us mention, for example, that in order to compare (in the domination sense) two semigroups where one is acting on $L^2(X, \mu, \mathbb{R})$ and the other on $L^2(\Omega, \mu, \mathbb{R})$, where Ω is an open set of X, then the sesquilinear form associated with the semigroup acting on $L^2(\Omega, \mu, \mathbb{R})$ can be seen as a form on $L^2(X, \mu, \mathbb{R})$ which is not densely defined.

Our approach to obtain criteria for positivity, L^∞-contractivity, and domination of semigroups is based on Theorems 2.2 and 2.3. Therefore, these criteria hold once we extend the latter two theorems to the case of forms which are not densely defined.

Let H be a general Hilbert space as in Section 2.2 and let $\mathfrak{a}$ be an accretive, continuous, and closed form. There exists an associated operator A acting on the closure (with respect with the norm of H) $\overline{D(\mathfrak{a})}$ of $D(\mathfrak{a})$. The operator $-A$ generates on $\overline{D(\mathfrak{a})}$ a semigroup which we denote by $(e^{-tA})_{t \geq 0}$. We extend this semigroup to H by putting

$$
T(t)u = \begin{cases} e^{-tA}u & \text{if} \quad u \in \overline{D(\mathfrak{a})}, \\ 0 & \text{if} \quad u \in D(\mathfrak{a})^\perp, \end{cases}
$$

where $D(\mathfrak{a})^\perp$ denotes the complement of $\overline{D(\mathfrak{a})}$ in H.

Now let $\mathcal{C}$ be a closed convex set of H and let $\mathcal{P}$ be the projection of H onto $\mathcal{C}$. We have

THEOREM 2.33 *The following assertions are equivalent:*
1) $T(t)\mathcal{C} \subseteq \mathcal{C}$ *for all* $t \geq 0$.
2) $\mathcal{P}(D(\mathfrak{a})) \subseteq D(\mathfrak{a})$ *and* $\Re\mathfrak{a}(u, u - \mathcal{P}u) \geq 0$ *for all* $u \in D(\mathfrak{a})$.
3) $\mathcal{P}(D(\mathfrak{a})) \subseteq D(\mathfrak{a})$ *and* $\Re\mathfrak{a}(\mathcal{P}u, u - \mathcal{P}u) \geq 0$ *for all* $u \in D(\mathfrak{a})$.
3´) *There exists a core* D *of* $\mathfrak{a}$ *such that* $\mathcal{P}(D) \subseteq D(\mathfrak{a})$ *and*
     $\Re\mathfrak{a}(\mathcal{P}u, u - \mathcal{P}u) \geq 0$ *for all* $u \in D$.
If the form $\mathfrak{a}$ *is symmetric, then 1) is equivalent to each of the following assertions:*
4) $\mathcal{P}(D(\mathfrak{a})) \subseteq D(\mathfrak{a})$ *and* $\mathfrak{a}(\mathcal{P}u, \mathcal{P}u) \leq \mathfrak{a}(u, u)$ *for all* $u \in D(\mathfrak{a})$.
4´) *There exists a core* D *of* $\mathfrak{a}$ *such that* $\mathcal{P}(D) \subseteq D(\mathfrak{a})$ *and* $\mathfrak{a}(\mathcal{P}u, \mathcal{P}u) \leq \mathfrak{a}(u, u)$ *for all* $u \in D$.

Proof. Denote by $\mathcal{Q}$ the orthogonal projection of H onto $\overline{D(\mathfrak{a})}$. If 1) holds, then in particular $T(0)\mathcal{C} \subseteq \mathcal{C}$, that is,

$$\mathcal{Q}\mathcal{C} \subseteq \mathcal{C}. \tag{2.8}$$

This implies that

$$\mathcal{P}(\overline{D(\mathfrak{a})}) \subseteq \overline{D(\mathfrak{a})}. \tag{2.9}$$

Indeed, for $u \in \overline{D(\mathfrak{a})}$ and every $v \in \mathcal{C}$, we have

$$\|u - \mathcal{Q}\mathcal{P}u\| = \|\mathcal{Q}u - \mathcal{Q}\mathcal{P}u\| \le \|u - \mathcal{P}u\| \le \|u - v\|.$$

But $\mathcal{Q}\mathcal{P}u \in \mathcal{C}$ because of (2.8). It follows then from the definition of the projection that $\mathcal{P}u = \mathcal{Q}\mathcal{P}u$, which gives (2.9).

Clearly, $\overline{D(\mathfrak{a})} \cap \mathcal{C}$ is a closed convex subset of $\overline{D(\mathfrak{a})}$. In addition, since $\mathcal{P}(\overline{D(\mathfrak{a})}) \subseteq \overline{D(\mathfrak{a})} \cap \mathcal{C}$ (remember (2.9)) it follows that $\overline{D(\mathfrak{a})} \cap \mathcal{C}$ is not empty. It is also clear that the projection of $\overline{D(\mathfrak{a})}$ onto $\overline{D(\mathfrak{a})} \cap \mathcal{C}$ is $\mathcal{P}$. Thus, Theorem 2.2 applied to $\mathfrak{a}$ considered as a form on $\overline{D(\mathfrak{a})}$ and to the convex set $\overline{D(\mathfrak{a})} \cap \mathcal{C}$ shows that assertion 1) implies each of the other assertions of the theorem.

Conversely, assume that one of the assertions 2), 3), 3´) (or 4), 4´) in the symmetric case) holds. Using the continuity of $\mathcal{P}$ and the fact that $\mathcal{P}(D) \subseteq D(\mathfrak{a})$ we obtain (2.9). In particular, the closed convex set $\overline{D(\mathfrak{a})} \cap \mathcal{C}$ is not empty. Theorem 2.2 applied again to $\mathfrak{a}$ and the convex set $\overline{D(\mathfrak{a})} \cap \mathcal{C}$ gives

$$e^{-tA}(\overline{D(\mathfrak{a})} \cap \mathcal{C}) \subseteq \overline{D(\mathfrak{a})} \cap \mathcal{C} \text{ for all } t \ge 0. \tag{2.10}$$

Now we show that (2.9) implies (2.8). Let $u \in \mathcal{C}$ and $v \in \overline{D(\mathfrak{a})}$. As above,

$$\|u - \mathcal{P}\mathcal{Q}u\| = \|\mathcal{P}u - \mathcal{P}\mathcal{Q}u\| \le \|u - \mathcal{Q}u\| \le \|u - v\|.$$

It follows from (2.9) that $\mathcal{P}\mathcal{Q}u \in \overline{D(\mathfrak{a})}$ and since the above inequality holds for all $v \in \overline{D(\mathfrak{a})}$, we conclude that $\mathcal{P}\mathcal{Q}u = \mathcal{Q}u$ and hence (2.8) holds.

Since $T(t)u = e^{-tA}\mathcal{Q}u$ for all $u \in H$, we conclude from (2.8) and (2.10) that assertion 1) holds. This finishes the proof. $\square$

Based on this result, our approach in the previous sections to obtaining criteria for positivity, L^∞-contractivity, and the domination is still valid for forms with not necessarily dense domains. Thus all those criteria hold for these forms with $(T(t))_{t\ge0}$ (defined above) in place of $(e^{-tA})_{t\ge0}$ in the statements. As we already mentioned, Theorems 2.9, 2.10, and Corollary 2.11 have to be reformulated in this case. Theorem 2.10 and Corollary 2.11 must be written as criteria for invariance under $(T(t))_{t\ge0}$ of $L^2(\Omega, \mu, \mathbb{C})$

and not as irreducibility of $(T(t))_{t \geq 0}$.

We have shown in the previous proof that condition (2.9) is a necessary condition for the invariance of the convex set $\mathcal{C}$ under the action of $(T(t))_{t \geq 0}$. This implies in particular the following properties:

1) Assume that $H = L^2(X, \mu, \mathbb{C})$. If $(T(t))_{t \geq 0}$ is positive then $\Re u, (\Re u)^+ \in \overline{D(\mathfrak{a})}$ for every $u \in \overline{D(\mathfrak{a})}$.

2) Let $H = L^2(X, \mu, K)$, where K is any Hilbert space. If $(T(t))_{t \geq 0}$ is L^∞-contractive, then $(1 \wedge |u|_K)\text{sign } u \in \overline{D(\mathfrak{a})}$ for every $u \in \overline{D(\mathfrak{a})}$.

3) Let $\tilde{\mathfrak{a}}$ and $\mathfrak{a}$ be as in Theorem 2.30. Define as above

$$\tilde{T}(t)u = \begin{cases} e^{-t\tilde{A}}u & \text{if} \quad u \in \overline{D(\tilde{\mathfrak{a}})}, \\ 0 & \text{if} \quad u \in D(\tilde{\mathfrak{a}})^\perp. \end{cases}$$

If $(\tilde{T}(t))_{t \geq 0}$ is dominated by $(T(t))_{t \geq 0}$, then $\overline{D(\tilde{\mathfrak{a}})}$ is an ideal of $\overline{D(\mathfrak{a})}$. This follows by applying (2.9) with

$$\mathcal{C} = \{(u, f) \in L^2(X, \mu, K) \times L^2(X, \mu, \mathbb{R}), \ |u|_K \leq f\}$$

and noting that for $(u, f) \in L^2(X, \mu, K) \times L^2(X, \mu, \mathbb{R})$,

$$\mathcal{P}(u, f) = \frac{1}{2}\left(\left[|u|_K + (\Re f) \wedge |u|_K \right]^+ \text{sign } u, \ \left[|u|_K \vee \Re f + \Re f \right]^+ \right).$$

Notes

Section 2.1. This part is mainly taken from Ouhabaz [Ouh96]. Theorem 2.2 (except assertion 4)) is proved in Ouhabaz [Ouh96]. The core version in assertion 4) of Theorem 2.2 and the remark following its proof were observed by Manavi, Vogt, and Voigt [MVV01]. Theorem 2.3 (without assertion 3)) is proved in [Ouh96]. Theorem 2.2 is inspired from the work of Brezis and Pazy [BrPa70] (where an operator version can be found). The same theorem was extended by Barthélemy [Bar96] to forms that are associated with nonlinear coercive operators.

Section 2.2. The idea of deducing positivity, L^∞-contractivity, and domination criteria as corollaries of invariance of closed convex sets comes from [Ouh96]. Proposition 2.5, Theorems 2.6, 2.13 (without assertion 3′)), and Theorem 2.14 (without assertion 3)) are proved in [Ouh92b] and [Ouh96]. Theorem 2.15 is taken from Auscher et al. [ABBO00]. A more general definition of irreducibility of semigroups can be found in Nagel [Nag86]. It coincides, however, with Definition 2.8 for positive holomorphic semigroups (see again [Nag86]. Remember that the semigroup associated to a form is holomorphic; see Chapter 1). Theorem 2.9 can be deduced from a result in Schaefer [Sch74] saying that every closed ideal of

$L^2(X, \mu, \mathbb{C})$ is an $L^2(\Omega, \mu, \mathbb{C})$ for some $\Omega \subseteq X$ together with the fact that irreducibility means that the semigroup has no nontrivial closed invariant ideal (see [Nag86]).

Beurling-Deny criteria can be found in several books, e.g., Fukushima [Fuk80], Fukushima, Oshima, and Takeda [FOT94], Bouleau and Hirsch [BoHi91], Davies [Dav89], Ma and Röckner [MaRö92]. The theory of Dirichlet forms has seen important developments since the publication of the pioneering papers of Beurling and Deny [BeDe58], [BeDe59]. This is due to the rich interplay between Dirichlet forms and Markov processes. We refer the reader to the books [Fuk80], [FOT94], [BoHi91], and [MaRö92] for more information on Dirichlet forms. In [MaRö92] nonsymmetric forms are also considered and criteria are given there for the sub-Markovian property. As we want to study elliptic operators with complex coefficients as well, we give separate criteria for L^∞-contractivity without assuming positivity.

The arguments used to prove strong continuity on L^1 of $(e^{-tA^*})_{t \geq 0}$ (when $(e^{-tA})_{t \geq 0}$ is L^∞-contractive) are taken from Davies [Dav89].

Section 2.3. The results on the domination of semigroups presented here are proved in Ouhabaz [Ouh96]. The notion of ideal in Definition 2.19 was introduced in that paper. The implication 2) $\Rightarrow$ 1) in Theorem 2.24 was first proved by Stollmann and Voigt [StVo96] in the case of symmetric forms. Closed ideals in Dirichlet spaces are studied in Stollmann [Sto93]. Weak versions of Theorem 2.21 were proved by Simon [Sim79], Hess, Schrader, and Uhlenbrock [HSU77], and Bérard [Bér86]. Note that characterizations of the domination of semigroups in terms of their generators are known and hold in a more general setting of Banach spaces (see Nagel [Nag86]).

Domination of semigroups plays a crucial role in spectral theory, in particular in the proof of Cwikel-Lieb-Rozenblum eigenvalue estimates; see Rozenblum and Solomyak [RoSo97] and Rozenblum [Roz00]. Theorem 2.21 is used by Melgaard, Ouhabaz, and Rozenblum [MOR03] in order to derive eigenvalue estimates for Aharonov-Bohm hamiltonians.

Section 2.4. Theorem 2.25 and assertion 1) of Theorem 2.28 are well-known properties of Dirichlet spaces. Such results and more on Dirichlet spaces can be found in the books mentioned above [BoHi91], [Fuk80], [FOT94], [MaRö92], and [Dav89]. The proof given here for Theorem 2.25 can be found in [MaRö92] and is based on similar ideas used in the proof of Theorem XIII.51 in Reed-Simon [ReSi79]. Theorem 2.27 and assertion 2) of Theorem 2.28 seem to be new.

Section 2.5. The section on semigroups acting on vector-valued functions is taken from Ouhabaz [Ouh99] following a work by Shigikawa [Shi97]. We describe briefly another application of the results in that section. For details and definitions, see Bérard [Bér86], Chavel [Cha84], Cycon et al. [CFKS87] (Chap. 12), and Strichartz [Str86]. On a complete Riemannian manifold M, one considers the space $\Lambda^k(M)$ of C^∞ differential forms of degree $k \leq \dim M$. Let d be the exterior derivative and δ its formal adjoint. The Hodge-de Rham Laplacian $\tilde{\Delta}$ acting on differential forms of degree k is the operator $-(d\delta + \delta d)$. It has a decomposition (the Weitzenböck formula) $\tilde{\Delta} = \overline{\Delta} + \mathcal{R}^k$, where $\overline{\Delta}$ is the Bochner Laplacian which satisfies

$$|e^{-t\overline{\Delta}}w| \leq e^{-t\Delta}|w|.$$

Here Δ is the Laplacian acting on functions (note that we are in the setting described in Remark 2) following the proof of Theorem 2.30; in particular, the norm $|.|$ depends on the point). For a proof of the above domination property, see Hess, Schrader, and Uhlenbrock [HSU80], Bérard [Bér86], p. 172, or Cycon et al. [CFKS87], p. 264. If $\lambda(x)$ denotes the smallest eigenvalue of $\mathcal{R}^k(x)$, then it follows from Proposition 2.32 that

$$|e^{-t\tilde{\Delta}}w| \leq e^{-t(\Delta + \lambda(x))}|w|. \tag{2.11}$$

In particular, if λ is non-negative then $(e^{-t\tilde{\Delta}})_{t\geq 0}$ is L^∞-contractive.
The domination property (2.11) is also established in Shigikawa [Shi97]. The L^∞-contractivity (when λ is non-negative) is proved by Strichartz [Str86], who also proved negative results to the L^p-contractivity for every $p \neq 2$.

Section 2.6. This section is taken partly from Thomaschewski [Tho98] and partly from Barthelémy [Bar96].

Chapter Three

INEQUALITIES FOR SUB-MARKOVIAN

SEMIGROUPS

In this chapter we prove certain inequalities for sub-Markovian semigroups and their generators. A relationship between the sub-Markovian property and Kato type inequalities for the generator is established. This completes some of the results of Section 2.2. Note that the operators in consideration here are not necessarily associated with sesquilinear forms.

Starting with a symmetric sub-Markovian semigroup on some L^2-space, we have seen in the previous chapter that it induces strongly continuous semigroups on the scale of L^p-spaces. We prove some inequalities for the corresponding generator on L^p. As a consequence, we obtain the holomorphy of such semigroups on $L^p, 1 < p < \infty$, with a sector of holomorphy wider than the usual one obtained by the Stein interpolation theorem. In the same time, a partial description of the domain of the corresponding generator on L^p is given.

Throughout this chapter, (X, μ) will denote a σ-finite measure space. For $u \in L^p(X, \mu, \mathbb{C})$ and $v \in L^q(X, \mu, \mathbb{C})$ with $\frac{1}{p} + \frac{1}{q} = 1$, we use the notation

$$(u; v) := \int_X u(x)\overline{v(x)}d\mu(x).$$

If $u \in L^p(X, \mu, \mathbb{C})$ with $p \in [1, \infty)$, we set

$$F(u) := |u|^{p-1}\text{sign } u, \tag{3.1}$$

where sign denotes the signum function defined in (2.2). If $p = 1$, then $F(u) = \text{sign } u$. As in the previous chapter, all inequalities between measurable functions are understood in the μ a.e. sense. We use again the notation $(e^{-tA})_{t \geq 0}$ for the strongly continuous semigroup generated by the operator $-A$.

3.1 SUB-MARKOVIAN SEMIGROUPS AND KATO TYPE INEQUALITIES

We consider a strongly continuous semigroup $(e^{-tA})_{t \geq 0}$ on $L^p(X, \mu, \mathbb{C})$ for some p, the following result characterizes in terms of its generator the

L^∞-contractivity of the semigroup.

THEOREM 3.1 *Let* $(e^{-tA})_{t\geq 0}$ *be a contraction semigroup on* $L^p(X, \mu, \mathbb{C})$ *for some* $p \in [1, \infty)$. *The following assertions are equivalent:*
1) The semigroup $(e^{-tA})_{t\geq 0}$ *is* L^∞-*contractive.*
2) $\Re\, (Au; F((|u| - 1)^+ \mathrm{sign}\, u)) \geq 0$ *for all* $u \in D(A)$.
3) There exists a core D *of* A *such that* $\Re\, (Au; F((|u| - 1)^+ \mathrm{sign}\, u)) \geq 0$ *for all* $u \in D$.

Proof. We use the obvious fact that for $p > 1$

$$F((|u| - 1)^+ \mathrm{sign}\, u) = [(|u| - 1)^+]^{p-1} \mathrm{sign}\, u,$$

and for $p = 1$, the term in the right-hand side is replaced by $\chi_{\{|u|>1\}} \mathrm{sign}\, u$, where $\chi_{\{|u|>1\}}$ denotes the characteristic function of the set where $|u| > 1$.

Of course, 2) implies 3). Assume now that assertion 3) holds. By a density argument, it is clear that the inequality there extends to all $u \in D(A)$. Fix now $\lambda > 0, f \in L^p(X, \mu, \mathbb{C})$ with $|f| \leq 1$ and apply this inequality with $u = \lambda(\lambda I + A)^{-1}f$. We obtain

$$\Re \int_X A(\lambda I + A)^{-1}f((|\lambda(\lambda I + A)^{-1}f| - 1)^+)^{p-1} \mathrm{sign}\, \overline{\lambda(\lambda I + A)^{-1}f} d\mu$$

$$\geq 0.$$

In other words, the term

$$\Re \int_X [f \mathrm{sign}\, \overline{\lambda(\lambda I + A)^{-1}f} - |\lambda(\lambda I + A)^{-1}f|][(|\lambda(\lambda I + A)^{-1}f| - 1)^+]^{p-1} d\mu$$

is non-negative. Since $|f| \leq 1$, it follows that

$$\Re[f \mathrm{sign}\, \overline{\lambda(\lambda I + A)^{-1}f} - |\lambda(\lambda I + A)^{-1}f|][(|\lambda(\lambda I + A)^{-1}f| - 1)^+]^{p-1} \leq 0.$$

Thus,

$$\Re[f \mathrm{sign}\, \overline{\lambda(\lambda I + A)^{-1}f} - |\lambda(\lambda I + A)^{-1}f|][(|\lambda(\lambda I + A)^{-1}f| - 1)^+]^{p-1} = 0.$$

This implies that $|\lambda(\lambda I + A)^{-1}f| \leq 1$. Since this is true for all $\lambda > 0$, we obtain assertion 1) by the exponential formula as in the proof of Proposition 2.1.

Assume that $(e^{-tA})_{t\geq 0}$ is L^∞-contractive and let $u \in D(A)$. We have

$$\Re\, (Au; F((|u| - 1)^+ \mathrm{sign}\, u)) = \lim_{t \to 0} \frac{1}{t} \Re(u - e^{-tA}u; F((|u| - 1)^+ \mathrm{sign}\, u)).$$

Let us write $u = 1 \wedge |u| \mathrm{sign}\, u + (|u| - 1)^+ \mathrm{sign}\, u$, where $1 \wedge |u| := \inf(1, |u|)$. Thus,

$$\Re(u - e^{-tA}u)F((|u| - 1)^+ \mathrm{sign}\, \overline{u}) = I_1 + I_2$$

where

$$I_1 = \Re[1 \wedge |u| - e^{-tA}(1 \wedge |u|\text{sign } u)\text{sign } \bar{u}][(|u| - 1)^+]^{p-1}$$

and

$$I_2 = \Re[(|u| - 1)^+ - e^{-tA}((|u| - 1)^+\text{sign } u)\text{sign } \bar{u}][(|u| - 1)^+]^{p-1}.$$

Since $|1 \wedge |u|\text{sign } u| \leq 1$ and the semigroup is L^∞-contractive, we have $I_1 \geq 0$. On the other hand, the semigroup is contractive on $L^p(X, \mu, \mathbb{C})$, so it follows from Hölder's inequality that $\int_X I_2 d\mu \geq 0$. This implies assertion 2) of the theorem. □

It is clear in this proof that for positive semigroups, one may consider only non-negative $u \in D(A)$ in assertion 2). This leads to the following result.

PROPOSITION 3.2 *Assume that $(e^{-tA})_{t \geq 0}$ is a positive contraction semigroup on $L^p(X, \mu, \mathbb{C})$ for some $p \in [1, \infty)$. Then the semigroup $(e^{-tA})_{t \geq 0}$ is sub-Markovian if and only if $(Au; F((u - 1)^+)) \geq 0$ for all $0 \leq u \in D(A)$.*

The result in this proposition is in the spirit of the following well-known result which characterizes generators of positive contraction semigroups.

PROPOSITION 3.3 *Assume that $(e^{-tA})_{t \geq 0}$ is a contraction semigroup on $L^p(X, \mu, \mathbb{C})$ for some $p \in [1, \infty)$. Then $(e^{-tA})_{t \geq 0}$ is positive if and only if it is real and $(Au; F(u^+)) \geq 0$ for all real $u \in D(A)$.*

Proof. Assume that $(e^{-tA})_{t \geq 0}$ is a positive semigroup. Let $u \in D(A)$ be real. We have

$$(u - e^{-tA}u)F(u^+) = (u^+)^p - (e^{-tA}u^+)(u^+)^{p-1} + (e^{-tA}u^-)(u^+)^{p-1}$$
$$\geq (u^+)^p - (e^{-tA}u^+)(u^+)^{p-1}.$$

Again, $(u^+)^{p-1}$ has to be replaced by $\chi_{\{u>0\}}$ if $p = 1$.

The fact that the semigroup is contractive together with Hölder's inequality imply that for all $t > 0$

$$(u - e^{-tA}u; F(u^+)) \geq 0.$$

Thus,

$$(Au; F(u^+)) = \lim_{t \to 0} \frac{1}{t}(u - e^{-tA}u; F(u^+)) \geq 0.$$

Conversely, let $f \in L^p(X, \mu, \mathbb{C})$ be such that $f \leq 0$. We apply the inequality $(Au; F(u^+)) \geq 0$, with $u = \lambda(\lambda I + A)^{-1}f$ where $\lambda > 0$ is fixed. This gives

$$\int_X [f - \lambda(\lambda I + A)^{-1}f][(\lambda(\lambda I + A)^{-1}f)^+]^{p-1} d\mu \geq 0.$$

Since $f \leq 0$, we have

$$[f - \lambda(\lambda I + A)^{-1}f][(\lambda(\lambda I + A)^{-1}f)^+]^{p-1} \leq 0,$$

and hence

$$[f - \lambda(\lambda I + A)^{-1}f][(\lambda(\lambda I + A)^{-1}f)^+]^{p-1} = 0.$$

This gives $(\lambda(\lambda I + A)^{-1}f)^+ = 0$, which means that $\lambda(\lambda I + A)^{-1}$ is positive. This implies the positivity of the semigroup. $\square$

In the rest of this section, we will consider only positive semigroups. Thus, we can assume without restriction that our L^p-spaces are real. Indeed, if the semigroup $(e^{-tA})_{t\geq 0}$ is positive, then clearly $\Re u \in D(A)$ for every $u \in D(A)$ and the restriction of $(e^{-tA})_{t\geq 0}$ to real parts of elements of L^p is again a strongly continuous semigroup whose generator is $-A$ with domain the set of $\Re u$ with $u \in D(A)$.

Recall also that if $(e^{-tA})_{t\geq 0}$ is sub-Markovian, then it induces a strongly continuous semigroup on L^q, for each $q \in [p, \infty)$. We will keep the same notation $(e^{-tA})_{t\geq 0}$ to denote the semigroup obtained on L^q for every $q \in [p, \infty)$.

Assume that $(e^{-tA})_{t\geq 0}$ is a strongly continuous semigroup on $L^p(X, \mu, \mathbb{R})$ for some $p \in [1, \infty)$. Let J_0 be the set of convex functions $j : \mathbb{R} \to \mathbb{R}$ such that $j(0) = 0$. We have the following Jensen's inequality.

PROPOSITION 3.4 *The semigroup $(e^{-tA})_{t\geq 0}$ is sub-Markovian if and only if it satisfies*

$$j(e^{-tA}u) \leq e^{-tA}j(u),$$

for all $u \in L^p(X, \mu, \mathbb{R}) \cap L^\infty(X, \mu, \mathbb{R}), t \geq 0$ and all $j \in J_0$. Here $j(u)$ is the function $j(u)(x) = j(u(x))$.

Proof. Note first that $j \in J_0$ is locally a Lipschitz function and hence for $u \in L^\infty(X, \mu, \mathbb{R})$, $j(u) \in L^\infty(X, \mu, \mathbb{R})$. This shows that the quantity $e^{-tA}j(u)$ makes sense.

Recall that every convex function on $\mathbb{R}$ is the supremum of affine functions on $\mathbb{R}$. Fix $j \in J_0$ and let $s \mapsto (a_n s + b_n)$ be a sequence of functions on $\mathbb{R}$ such that

$$j(s) = \sup_n (a_n s + b_n), \quad s \in \mathbb{R}.$$

Since $j(0) = 0$, it follows that $b_n \leq 0$ for all $n \geq 0$. Thus, if $(e^{-tA})_{t\geq 0}$ is sub-Markovian then $b_n \leq e^{-tA}b_n$. Hence, for every $u \in L^p(X,\mu,\mathbb{R}) \cap L^\infty(X,\mu,\mathbb{R})$, we have the pointwise inequality

$$a_n e^{-tA}u + b_n \leq e^{-tA}(a_n u + b_n), \; t \geq 0.$$

Taking the supremum yields

$$j(e^{-tA}u) \leq \sup_n e^{-tA}(a_n u + b_n) \leq e^{-tA}j(u),$$

where we use the positivity of the semigroup to obtain the second inequality.

Conversely, choosing $j(s) = |s|$, we obtain for every $t \geq 0$ and $u \in L^p(X,\mu,\mathbb{R})$, $|e^{-tA}u| \leq e^{-tA}|u|$. This gives the positivity of the semigroup. Choosing $j(s) = -(1 \wedge |s|)$, we obtain $e^{-tA}(1 \wedge |u|) \leq 1 \wedge e^{-tA}|u|$, which implies the L^∞-contractivity. $\square$

Note that every function $j \in J_0$ has at each point $r \in \mathbb{R}$, left and right derivatives $j'_L(r)$ and $j'_R(r)$. Recall that

$$j'_L(r) \leq j'_R(r) \leq \frac{j(s) - j(r)}{s - r} \leq j'_L(s) \leq j'_R(s) \text{ for all } s > r. \qquad (3.2)$$

It is clear from (3.2) that the interval $[j'_L(r), j'_R(r)]$ coincides with the set of real elements w such that

$$j(s) \geq j(r) + w(s - r) \text{ for all } s \in \mathbb{R}$$

The set of such w is called the subdifferential of j at the point r and is denoted by $\partial j(r)$. For a function $u \in L^p(X,\mu,\mathbb{R})$, we will write $w \in \partial j(u)$ to mean that $w(x) \in \partial j(u(x))$ for μ a.e. $x \in X$.

Let us now denote by J_0^∞ the subset of J_0, of functions whose right and left derivatives are bounded on $\mathbb{R}$. Note that for every $j \in J_0^\infty$, there exists a constant M such that $|j(r)| \leq M|r|$ for all $r \in \mathbb{R}$. Thus,

$$j(u) \in L^p(X,\mu,\mathbb{R}) \text{ for all } j \in J_0^\infty \text{ and } u \in L^p(X,\mu,\mathbb{R}).$$

It follows in particular that for $j \in J_0^\infty$ the inequality in the previous proposition holds for all $u \in L^p(X,\mu,\mathbb{R})$.

The sub-Markovian property can be characterized in terms of Kato type inequalities. More precisely,

THEOREM 3.5 Let $(e^{-tA})_{t\geq 0}$ be a strongly continuous and positive semigroup on $L^p(X,\mu,\mathbb{R})$ for some $p \in [1,\infty)$. The following assertions are equivalent:

1) *The semigroup* $(e^{-tA})_{t\geq 0}$ *is sub-Markovian.*
2) *For every* $j \in J_0^\infty$, *we have*

$$(j(u); A^*\phi) \leq (Au; w\phi)$$

for all $0 \leq \phi \in D(A^*)$,[1] $u \in D(A)$ *and* $w \in \partial j(u)$.
3) *For every* $u \in D(A)$ *and every* $0 \leq \phi \in D(A^*)$,

$$(\chi_{\{u<1\}}Au; \phi) \leq (1 \wedge u; A^*\phi),$$

where $\chi_{\{u<1\}}$ *denotes the indicator function of the set* $\{x \in X, u(x) < 1\}$.

Proof. We show that 1) implies 2). Let $u \in D(A)$, $w \in \partial j(u)$, and $0 \leq \phi \in D(A^*)$. Since $j \in J_0^\infty$, it follows that $w \in L^\infty(X, \mu, \mathbb{R})$. Moreover, we have from the definition of $w \in \partial j(u)$ and from the previous proposition,

$$e^{-tA}j(u) \geq j(e^{-tA}u) \geq j(u) + w(e^{-tA}u - u) \text{ for all } t > 0.$$

Here the expression $e^{-tA}j(u)$ makes sense because $j(u) \in L^p(X, \mu, \mathbb{R})$ as explained above. Thus, since $0 \leq \phi$, we have for $t > 0$

$$\frac{1}{t}(u - e^{-tA}u; w\phi) = \frac{1}{t}(w(u - e^{-tA}u); \phi)$$
$$\geq \frac{1}{t}(j(u) - e^{-tA}j(u); \phi)$$
$$= \frac{1}{t}(j(u); \phi - e^{-tA^*}\phi)$$
$$= \left(j(u); \frac{1}{t}\int_0^t e^{-tA^*}A^*\phi\right).$$

The desired inequality holds by letting $t \to 0$.

Choosing $j(s) = -(1 \wedge s)$, $s \in \mathbb{R}$, we obtain that 2) implies 3).

Assume now that 3) holds. We want to show that $(e^{-tA})_{t\geq 0}$ is sub-Markovian. Using the formula

$$e^{-tA}u = \lim_{n\to\infty}\left(I + \frac{t}{n}A\right)^{-n}u \text{ for all } u \in L^p(X, \mu, \mathbb{R}),$$

it is enough to show that $\lambda(\lambda I + A)^{-1}$ is sub-Markovian for all $\lambda > 0$, large enough. Let $v \in L^p(X, \mu, \mathbb{R})$ and $0 \leq \psi \in L^q(X, \mu, \mathbb{R})$ with $\frac{1}{p} + \frac{1}{q} = 1$. We apply 3) with $u = \lambda(\lambda I + A)^{-1}v$ and $\phi = ((\lambda I + A)^{-1})^*\psi$ to obtain

$$(\chi_{\{\lambda(\lambda I+A)^{-1}v<1\}}A(\lambda I + A)^{-1}v; \lambda((\lambda I + A)^{-1})^*\psi)$$
$$\leq (1 \wedge \lambda(\lambda I + A)^{-1}v; A^*((\lambda I + A)^{-1})^*\psi).$$

---

[1]Here A^* denotes the adjoint operator of A.

This inequality can be rewritten as

$$(\chi_{\{\lambda(\lambda I+A)^{-1}v<1\}}(v - \lambda(\lambda I + A)^{-1}v); \lambda((\lambda I + A)^{-1})^*\psi)$$
$$\leq (1 \wedge \lambda(\lambda I + A)^{-1}v; \psi - \lambda((\lambda I + A)^{-1})^*\psi).$$

Note that we always have the pointwise inequality

$$1 \wedge v - 1 \wedge \lambda(\lambda I + A)^{-1}v$$
$$\leq \chi_{\{\lambda(\lambda I+A)^{-1}v<1\}}(v - \lambda(\lambda I + A)^{-1}v).$$

Using this and the fact that $\lambda(\lambda I + A)^{-1}\psi \geq 0$ (remember that the semigroup is assumed to be positive) we obtain from the preceding inequalities

$$(1 \wedge v - 1 \wedge \lambda(\lambda + A)^{-1}v; \lambda((\lambda I + A)^{-1})^*\psi)$$
$$\leq (1 \wedge \lambda(\lambda I + A)^{-1}v; \psi - \lambda((\lambda I + A)^{-1})^*\psi),$$

and hence

$$(1 \wedge v; \lambda((\lambda I + A)^{-1})^*\psi) \leq (1 \wedge \lambda(\lambda I + A)^{-1}v; \psi).$$

Since this holds for all $0 \leq \psi \in L^q(X, \mu, \mathbb{R})$, it follows that for all $v \in L^p(X, \mu, \mathbb{R})$,

$$\lambda(\lambda I + A)^{-1}(1 \wedge v) \leq 1 \wedge \lambda(\lambda I + A)^{-1}v.$$

This implies that $\lambda(\lambda I + A)^{-1}$ is sub-Markovian. $\qquad\square$

Note that the assumption of positivity of the semigroup plays an important role in the previous result. The following example shows that without positivity, the Kato type inequalities above do not imply the sub-Markovian property.

Example 3.1.1 *Consider on $L^2((0,1), dx, \mathbb{R})$ (dx denotes the Lebesgue measure) the operator*

$$Au = -u', \ D(A) = \{u \in H^1(0,1), u(0) = -u(1)\},$$

where $H^1(0,1)$ denotes the classical Sobolev space on $(0,1)$. The operator $-A$ generates the semigroup $(e^{-tA})_{t\geq 0}$ given by

$$e^{-tA}u(x) = (-1)^n u(x + t - n) \text{ if } x + t \in [n, n+1] \ (n \in \mathbb{N}).$$

Clearly, the semigroup $(e^{-tA})_{t\geq 0}$ is not sub-Markovian. We show that the operator A satisfies assertion 2) of the previous theorem. It is not hard to see that

$$D(A^*) = \{u \in H^1(0,1), u(0) = -u(1)\}, \ A^*u = u'.$$

In particular, if $\phi \in D(A^*)$ is non-negative, then $\phi(0) = \phi(1) = 0$.

Fix $0 \le \phi \in D(A^*)$ and let $u \in D(A), j \in J_0^\infty$ and $w \in \partial j(u)$. We have for a.e. $x \in (0,1)$ and δ small

$$j(u)(x + \delta) - j(u)(x) \ge w(x)(u(x + \delta) - u(x)).$$

Therefore, for $\delta > 0$

$$\int_0^1 \frac{j(u)(x + \delta) - j(u)(x)}{\delta} \phi(x) \ge \int_0^1 w(x)\phi(x) \frac{u(x + \delta) - u(x)}{\delta} dx.$$

Hence,

$$\liminf_{\delta \downarrow 0} \int_0^1 \frac{j(u)(x + \delta) - j(u)(x)}{\delta} \phi(x) \ge -(Au; w\phi).$$

On the other hand, using the obvious equality

$$\int_0^1 j(u)(x + \delta)\phi(x)dx = \int_\delta^{\delta+1} j(u)(x)\phi(x - \delta)dx$$

and the fact that $\phi(0) = \phi(1) = 0$ one obtains

$$\lim_{\delta \downarrow 0} \int_0^1 \frac{j(u)(x + \delta) - j(u)(x)}{\delta} \phi(x)dx = -\int_0^1 j(u)(x)\phi'(x)dx$$
$$= -(j(u); A^*\phi).$$

This shows that the Kato's inequality of Theorem 3.5 (assertion 2)) holds for A.

Under an additional assumption, it is possible to characterize the sub-Markovian property of the semigroup in terms of Kato's inequality (without assuming the positivity of the semigroup). We first recall the following result which characterizes the positivity in terms of Kato's inequality (see Theorem 3.8 and Corollary 3.9 in Nagel [Nag86]).

THEOREM 3.6 *Let $(e^{-tA})_{t \ge 0}$ be a strongly continuous semigroup acting on $L^p(X, \mu, \mathbb{R})$ for some $p \in [1, \infty)$. The semigroup $(e^{-tA})_{t \ge 0}$ is positive if and only if the following two assertions hold:*
i) For every $u \in D(A)$ and every $0 \le \phi \in D(A^*)$,

$$(\text{sign } u \, Au; \phi) \ge (|u|; A^*\phi).$$

ii) There exist $\phi > 0$, $\phi \in D(A^*)$ and $\lambda \in \mathbb{R}$ such that $A^*\phi \ge \lambda\phi$.

Here $\phi > 0$ means that $\phi(x) > 0$ for μ a.e. $x \in X$.

Assume now that the generator $-A$ of $(e^{-tA})_{t \geq 0}$ satisfies the inequality of assertion 3) in Theorem 3.5. That is,

$$(\chi_{\{u<1\}} Au; \phi) \leq (1 \wedge u; A^*\phi) \text{ for all } u \in D(A), 0 \leq \phi \in D(A^*).$$

Applying this inequality to nu for all $n > 0$ and noting that $1 \wedge nu = n(\frac{1}{n} \wedge u)$, we obtain

$$(\chi_{\{u<1/n\}} Au; \phi) \leq \left(\frac{1}{n} \wedge u; A^*\phi \right)$$

for all $u \in D(A), 0 \leq \phi \in D(A^*)$. Applying this inequality to $-u$ yields

$$-(\chi_{\{u>-1/n\}} Au; \phi) \leq \left(\frac{1}{n} \wedge (-u); A^*\phi \right)$$

for all $u \in D(A), 0 \leq \phi \in D(A^*)$.

The sum of the two inequalities gives

$$([\chi_{\{u<1/n\}} - \chi_{\{u>-1/n\}}] Au; \phi) \leq \left(\frac{1}{n} \wedge u + \frac{1}{n} \wedge (-u); A^*\phi \right)$$

for all $u \in D(A), 0 \leq \phi \in D(A^*)$. Now if $n \to +\infty$, we obtain Kato's inequality

$$(\text{sign } u \, Au; \phi) \geq (|u|; A^*\phi) \text{ for all } u \in D(A), 0 \leq \phi \in D(A^*).$$

As a consequence of this and Theorems 3.6 and 3.5, we have

THEOREM 3.7 *Let $(e^{-tA})_{t \geq 0}$ be a strongly continuous semigroup acting on $L^p(X, \mu, \mathbb{R})$ for some $p \in [1, \infty)$. The semigroup $(e^{-tA})_{t \geq 0}$ is sub-Markovian if and only if the following two properties hold:*
i) For every $u \in D(A)$ and every $0 \leq \phi \in D(A^*)$,

$$(\chi_{\{u<1\}} Au; \phi) \leq (1 \wedge u; A^*\phi).$$

ii) There exist $\phi > 0$, $\phi \in D(A^*)$ and $\lambda \in \mathbb{R}$ such that $A^*\phi \geq \lambda\phi$.

It should be mentioned that, in contrast to Theorem 3.1 and Proposition 3.2, the characterization of the sub-Markovian property in terms of Kato type inequalities holds without assuming that the original semigroup (defined on $L^p(X, \mu, \mathbb{R})$) is contractive.

Theorems 3.6 and 3.7 can be used to prove the sub-Markovian property for perturbed semigroups. We consider below the case of multiplicative perturbations.

Assume that $(e^{-tA})_{t\geq 0}$ is a strongly continuous semigroup acting on $L^p(X, \mu, \mathbb{R})$ for some $p \in [1, \infty)$. Let $b : X \to \mathbb{R}^+$ be a measurable function such that $b, \frac{1}{b} \in L^\infty(X, \mu, \mathbb{R})$. Assume in addition that the operator $-bA$, with domain $D(A)$, generates a strongly continuous semigroup $(e^{-tbA})_{t\geq 0}$ on $L^p(X, \mu, \mathbb{R})$. We have

PROPOSITION 3.8 *1) The semigroup* $(e^{-tA})_{t\geq 0}$ *is positive if and only if* $(e^{-tbA})_{t\geq 0}$ *is positive.*
2) The semigroup $(e^{-tA})_{t\geq 0}$ *is sub-Markovian if and only if* $(e^{-tbA})_{t\geq 0}$ *is sub-Markovian.*

Proof. Assume that the semigroup $(e^{-tA})_{t\geq 0}$ is positive. We denote by $L^q(X, \mu, \mathbb{R})$ the dual space of $L^p(X, \mu, \mathbb{R})$. Clearly,

$$D((bA)^*) = \{\psi \in L^q(X, \mu, \mathbb{R}), b\psi \in D(A^*)\}, (bA)^*\psi = A^*(b\psi).$$

Now let $0 \leq \phi \in D((bA)^*)$ and $u \in D(bA) = D(A)$. By Theorem 3.6 we have

$$(\text{sign } u \, Au; b\phi) \geq (|u|; A^*(b\phi)).$$

This means that bA satisfies the Kato inequality of Theorem 3.6. To conclude that $(e^{-tbA})_{t\geq 0}$ is positive we have to check condition ii) of the same theorem. This follows easily since if ϕ satisfies ii) for A, then $\frac{1}{b}\phi$ satisfies ii) for the operator bA.

The converse follows as above, since $A = \frac{1}{b}(bA)$.

Assume that the semigroup $(e^{-tA})_{t\geq 0}$ is sub-Markovian. Assertion 1) implies that the semigroup $(e^{-tbA})_{t\geq 0}$ is positive. In order to conclude that this semigroup is sub-Markovian, we apply Theorem 3.5. Let $0 \leq \phi \in D((bA)^*))$ and $u \in D(bA)$. We have

$$(\chi_{\{u<1\}}bAu; \phi) = (\chi_{\{u<1\}}Au; b\phi) \leq (1 \wedge u; A^*b\phi) = (1 \wedge u; (bA)^*\phi).$$

This shows that $(e^{-tbA})_{t\geq 0}$ is sub-Markovian. $\square$

3.2 FURTHER INEQUALITIES AND THE CORRESPONDING DOMAIN IN L^p

Let (X, μ) be a σ-finite measure space and assume that $\mathfrak{a}$ is a densely defined, accretive, and closed symmetric form on $L^2(X, \mu, \mathbb{C})$. Denote by A and $(e^{-tA})_{t\geq 0}$ the self-adjoint operator and the semigroup associated with $\mathfrak{a}$. As explained in the previous chapter, if $(e^{-tA})_{t\geq 0}$ is L^∞-contractive, then $(e^{-tA})_{t\geq 0}$ extends to a strongly continuous semigroup on $L^p(X, \mu, \mathbb{C})$ for $1 \leq p < \infty$ and continuous on $L^\infty(X, \mu, \mathbb{C})$ for the w^*-topology. Denote by $-A_p$ the generator of the semigroup obtained on $L^p(X,\mu,\mathbb{C}), 1 \le p <$ ∞ ($A_2 := A$). The semigroup $(e^{-tA_p})_{t\ge0}$ generated by $-A_p$ coincides with $(e^{-tA})_{t\ge0}$ on $L^2(X,\mu,\mathbb{C}) \cap L^p(X,\mu,\mathbb{C})$. This suggests that the operator A_p should have the same formal expression as A. However, it is not easy in general to describe precisely the domain of the operator A_p. The following result gives a partial description of this domain when the semigroup is sub-Markovian. It also shows some interesting inequalities satisfied by A_p.

THEOREM 3.9 *Assume that* $(e^{-tA})_{t\ge0}$ *is a symmetric sub-Markovian semigroup on* $L^2(X,\mu,\mathbb{C})$. *For every fixed* $p \in (1,\infty)$, *we have the following properties:*
1) If $f \in D(A_p)$, *then* $f|f|^{\frac{p}{2}-1} \in D(\mathfrak{a})$ *and*

$$4\frac{p-1}{p^2}\mathfrak{a}(f|f|^{\frac{p}{2}-1}, f|f|^{\frac{p}{2}-1}) \le \Re(A_p f; |f|^{p-1}\mathrm{sign}\, f),$$

$$\Re(A_p f; |f|^{p-1}\mathrm{sign}\, f) \le 2\mathfrak{a}(f|f|^{\frac{p}{2}-1}, f|f|^{\frac{p}{2}-1}).$$

2) For every $f \in D(A_p)$, *we have*

$$|\Im(A_p f; |f|^{p-1}\mathrm{sign}\, f)| \le \frac{|p-2|}{2\sqrt{p-1}}\Re(A_p f; |f|^{p-1}\mathrm{sign}\, f).$$

Proof. We show that for every fixed $t > 0$ and every $f \in L^1(X,\mu,\mathbb{C}) \cap L^\infty(X,\mu,\mathbb{C})$ with $\mu(\{x, f(x) \ne 0\}) < \infty$,

$$4\frac{p-1}{p^2}(f|f|^{\frac{p}{2}-1} - e^{-tA}f|f|^{\frac{p}{2}-1}; f|f|^{\frac{p}{2}-1}) \le \Re(f - e^{-tA}f; |f|^{p-1}\mathrm{sign}\, f),$$
$$(3.3)$$

$$\Re(f - e^{-tA}f; |f|^{p-1}\mathrm{sign}\, f) \le 2(f|f|^{\frac{p}{2}-1} - e^{-tA}f|f|^{\frac{p}{2}-1}; f|f|^{\frac{p}{2}-1}), \quad (3.4)$$

and

$$|\Im(f - e^{-tA}f; |f|^{p-1}\mathrm{sign}\, f)| \le \frac{|p-2|}{2\sqrt{p-1}}\Re(f - e^{-tA}f; |f|^{p-1}\mathrm{sign}\, f).$$
$$(3.5)$$

By density, these inequalities will hold for all $f \in D(A_p)$. The first assertion of the theorem follows then by applying Lemma 1.56, (3.3), and (3.4). The second assertion follows from (3.5).

The proof of these inequalities will be given in several steps.

We fix $t > 0$ and $f \in L^1(X,\mu,\mathbb{C}) \cap L^\infty(X,\mu,\mathbb{C})$ with $\mu(\{x, f(x) \ne 0\}) < \infty$.

(i) First, we may assume without loss of generality that $\mu(X) < \infty$. Indeed, the operator $\chi_{\{x,f(x)\ne0\}}e^{-tA}$ is sub-Markovian and clearly we may replace X by $\{x, f(x) \ne 0\}$.

(ii) We can assume that X is a compact topological space, e^{-tA} is a symmetric sub-Markovian operator which leaves invariant $C(X)$ (the space of continuous functions on X), and $f \in C(X)$. Indeed, let $\Gamma : L^\infty(X, \mu, \mathbb{C}) \to C(K)$ be the Gelfand isomorphism, where K is a compact space (the spectrum of the algebra $L^\infty(X, \mu, \mathbb{C})$). The measure μ is transported to a measure $\hat{\mu}$ on K such that

$$\int_K \Gamma(g) d\hat{\mu} = \int_X g \, d\mu.$$

We may now replace the operator e^{-tA} by $\Gamma e^{-tA} \Gamma^{-1}$. Note that $\Gamma(g)\Gamma(h) = \Gamma(gh)$ for every $g, h \in L^\infty(X, \mu, \mathbb{C})$. This implies that $\Gamma e^{-tA} \Gamma^{-1}$ is a symmetric sub-Markovian operator on $L^2(K, \hat{\mu}, \mathbb{C})$. In addition, this operator leaves invariant $C(K)$. Note also that since $\Gamma(g) = \Gamma(\text{sign } g)\Gamma(|g|)$ and $\Gamma(|g|) = |\Gamma(g)|$, the signum function is defined on $C(K)$ by sign $\Gamma(g) = \Gamma(\text{sign } g)$.

Replacing X by K, μ by $\hat{\mu}$ and e^{-tA} by $\Gamma e^{-tA} \Gamma^{-1}$, we see that we have to show (3.3)-(3.5) with $\Gamma(f)$ in place of f.

(iii) Using (ii), e^{-tA} now becomes a symmetric sub-Markovian operator on $L^2(X, \mu, \mathbb{C})$ and leaves $C(X)$ invariant. It follows that there exists a symmetric Radon measure m_t on $X \times X$ such that[2]

$$\int_X (e^{-tA} g)\overline{h} d\mu = \int_{X \times X} g(x)\overline{h(y)} dm_t(x, y) \text{ for all } g, h \in C(X). \quad (3.6)$$

In order to prove this, we consider the subspace E of $C(X \times X)$ of functions of the type $h(x, y) = \sum_{k=1}^n f_k(x)\overline{g_k(y)}$ (where $n \in \mathbb{N}$) and the following functional on E

$$\phi(h) := \sum_{k=1}^n \int_X e^{-tA} f_k(x)\overline{g(x)} d\mu(x).$$

One checks easily that ϕ is well defined (i.e., it does not depend on the way in which we express $h \in E$). We prove that ϕ is continuous. By the sub-Markovian property, $(e^{-tA})_{t \geq 0}$ is a contraction semigroup on $C(X)$. Thus,

---

[2] see also Fukushima [Fuk80], Lemma 1.4.1.

for every h as above

$$|\phi(h)| \leq \mu(X) \sup_{x \in X} \left| \sum_k (e^{-tA} f_k)(x)\overline{g_k(x)} \right|$$

$$\leq \mu(X) \sup_y \sup_x \left| \sum_k (e^{-tA} f_k)(x)\overline{g_k(y)} \right|$$

$$= \mu(X) \sup_y \sup_x \left| e^{-tA}\left(\sum_k f_k \overline{g(y)}\right)(x) \right|$$

$$\leq \mu(X) \sup_y \sup_x \left| \sum_k f_k(x)\overline{g_k(y)} \right|$$

$$= \mu(X)\|h\|_{C(X \times X)}.$$

By density, ϕ extends to a continuous linear functional on $C(X \times X)$. We prove now that this extension is a positive functional. Let first $h(x,y) := \sum_k f_k(x)\overline{g_k(y)} \geq 0$ for all $x, y \in X$. This means that for every $y \in X$, the function $\sum_k f_k \overline{g_k(y)}$ is non-negative. By positivity of the semigroup, we have

$$\sum_k e^{-tA}(f_k)\overline{g_k(y)}(x) \geq 0 \text{ for all } t \geq 0, x, y \in X.$$

In particular,

$$\sum_k e^{-tA} f_k(x)\overline{g_k(x)} \geq 0 \text{ for all } t \geq 0, x \in X$$

and this gives $\phi(h) \geq 0$. Thus ϕ is positive on E. In order to show that its extension to $C(X \times X)$ is also positive, it is enough to prove that $\phi(|h|) \geq 0$ for all real-valued $h \in E$. Take a sequence of polynomials P_n such that $P_n(t) \geq 0$ for all $t \in [-1, 1]$ and P_n converges uniformly in $[-1, 1]$ to the function $t \to |t|$. Hence, the sequence $P_n(h/\|h\|_{C(X \times X)})$ converges in $C(X \times X)$ to $|h|/\|h\|_{C(X \times X)}$. Since, $0 \leq P_n(h/\|h\|_{C(X \times X)}) \in E$ and ϕ is positive on E, we conclude that $\phi(|h|) \geq 0$.

This proves existence of the positive measure m_t. The symmetry of m_t follows from the symmetry of the semigroup.

(iv) Let now

$$P := (f|f|^{\frac{p}{2}-1} - e^{-tA} f|f|^{\frac{p}{2}-1}; f|f|^{\frac{p}{2}-1}).$$

Using (3.6), we have

$$P = (1 - e^{-tA}1; |f|^p) + (e^{-tA}1; |f|^p) - (e^{-tA}f|f|^{\frac{p}{2}-1}; f|f|^{\frac{p}{2}-1})$$

$$= (1 - e^{-tA}1; |f|^p) + \int_{X \times X} |f(x)|^p dm_t(x, y)$$

$$- \int_{X \times X} f(x)|f(x)|^{\frac{p}{2}-1}\overline{f(y)}|f(y)|^{\frac{p}{2}-1} dm_t(x, y).$$

Using the fact that the measure $m_t(., .)$ is symmetric, we can rewrite the last equality as

$$P = (1 - e^{-tA}1; |f|^p)$$

$$+ \frac{1}{2}\int_{X \times X}\left[|f(x)|^p + |f(y)|^p - f(x)|f(x)|^{\frac{p}{2}-1}\overline{f(y)}|f(y)|^{\frac{p}{2}-1}\right.$$

$$\left. - f(y)|f(y)|^{\frac{p}{2}-1}\overline{f(x)}|f(x)|^{\frac{p}{2}-1}\right] dm_t(x, y).$$

We use the following notation. Set

$$f(x) = |f(x)|e^{i\theta(x)} = se^{i\theta(x)}, \ f(y) = le^{i\theta(y)} \text{ and } \theta = \theta(x) - \theta(y).$$

We have

$$P = (1 - e^{-tA}1; |f|^p) + \frac{1}{2}\int_{X \times X}[s^p + l^p - 2s^{\frac{p}{2}}l^{\frac{p}{2}}\cos\theta]dm_t(x, y). \quad (3.7)$$

Let $Q := (f - e^{-tA}f; |f|^{p-1}\text{sign } f)$. We have in a similar way

$$Q = (1 - e^{-tA}1; |f|^p) + (e^{-tA}1; |f|^p) - (e^{-tA}f; |f|^{p-1}\text{sign } f)$$

$$= (1 - e^{-tA}1; |f|^p) + \frac{1}{2}\int_{X \times X}\left[|f(x)|^p + |f(y)|^p\right.$$

$$\left. - f(x)|f(y)|^{p-2}\overline{f(y)} - f(y)|f(x)|^{p-2}\overline{f(x)}\right] dm_t(x, y).$$

Hence, using the same notation as above, we obtain

$$\Re Q = (1 - e^{-tA}1, |f|^p) + \frac{1}{2}\int_{X \times X}[s^p + l^p - (s^{p-1}l + sl^{p-1})\cos\theta]dm_t(x, y).$$

$$(3.8)$$

From (3.7), (3.8) and the Lemma below, we obtain (3.3) and (3.4).

From the previous expression of Q we also have

$$|\Im Q| \le \frac{1}{2}\int_{X \times X}|s^{p-1}l - sl^{p-1}||\sin\theta|dm_t(x, y).$$

Using this and (3.8) we deduce (3.5) by applying again the next lemma. □

LEMMA 3.10 *Let* $0 \leq s, l < \infty$ *and* $\theta \in \mathbb{R}$. *Fix* $p \in (1, \infty)$. *The following three inequalities hold:*

$$4\frac{p-1}{p^2}[s^p + l^p - 2s^{\frac{p}{2}}l^{\frac{p}{2}}\cos\theta] \leq s^p + l^p - (s^{p-1}l + sl^{p-1})\cos\theta,$$

$$s^p + l^p - (s^{p-1}l + sl^{p-1})\cos\theta \leq 2[s^p + l^p - 2s^{\frac{p}{2}}l^{\frac{p}{2}}\cos\theta],$$

$$|\sin\theta||s^{p-1}l - sl^{p-1}| \leq \frac{|p-2|}{2\sqrt{p-1}}[s^p + l^p - (s^{p-1}l + sl^{p-1})\cos\theta].$$

Proof. Note first that

$$2s^{\frac{p}{2}}l^{\frac{p}{2}} \leq s^{p-1}l + sl^{p-1}. \tag{3.9}$$

This follows from the elementary inequality $2a.b \leq a^2\varepsilon + \frac{1}{\varepsilon}b^2$ by choosing $\varepsilon = \frac{l}{s}$.

The inequality (3.9) implies easily that the function

$$F(u) := 4\frac{p-1}{p^2}[s^p + l^p - 2s^{\frac{p}{2}}l^{\frac{p}{2}}u] - (s^p + l^p) + (s^{p-1}l + sl^{p-1})u$$

is nondecreasing. Hence, $F(u) \leq F(1)$ for all $u \in [-1, 1]$. Thus, we merely need to prove that $F(1) \leq 0$ to obtain the first inequality. Assume, for example, that $l < s$ and use the Cauchy-Schwarz inequality to obtain

$$\frac{4}{p^2}(s^{\frac{p}{2}} - l^{\frac{p}{2}})^2 = \left(\int_l^s v^{\frac{p}{2}-1}dv\right)^2 \leq \left|\int_l^s dv\right|\left|\int_l^s v^{p-2}dv\right|$$

$$= \frac{1}{p-1}(s-l)(s^{p-1} - l^{p-1}).$$

This gives the desired inequality $F(1) \leq 0$.

Now we prove the second inequality. We want to show that

$$F_{s,l} := s^p + l^p - 2s^{\frac{p}{2}}l^{\frac{p}{2}}\cos\theta - 2s^{\frac{p}{2}}l^{\frac{p}{2}}\cos\theta + (s^{p-1}l + sl^{p-1})\cos\theta \geq 0.$$

If $\cos\theta \geq 0$, then this inequality follows from (3.9). If $\cos\theta < 0$, then

$$F_{s,l} \geq -\cos\theta[s^p + l^p + 4s^{\frac{p}{2}}l^{\frac{p}{2}} - (s^{p-1}l + sl^{p-1})]$$

$$= -\cos\theta[(s-l)(s^{p-1} - l^{p-1}) + 4s^{\frac{p}{2}}l^{\frac{p}{2}}]$$

$$\geq 0.$$

Finally, we show the third inequality. Consider the following function with variable $u \in [-1, 1]$,

$$F(u) := \frac{(s^{p-1}l - sl^{p-1})^2(1 - u^2)}{[s^p + l^p - (s^{p-1}l + sl^{p-1})u]^2}.$$

The third inequality of the lemma means that $F(u) \leq \frac{(p-2)^2}{4(p-1)}$ for all $u \in [-1, 1]$.

We first optimize $F(u)$ with respect to u. A simple calculation shows that the maximum of $F(u)$ is obtained when $u = u_1 = \frac{s^{p-1}l + sl^{p-1}}{s^p + l^p}$. Thus

$$F(u) \leq F(u_1) = \frac{(s^{p-2} - l^{p-2})^2 s^2 l^2}{(s^p + l^p)^2 - (s^{p-1}l + sl^{p-1})^2}.$$

Now we have

$$(s^{p-2} - l^{p-2})^2 = (p-2)^2 \left(\int_l^s v^{p-3} dv \right)^2$$

$$\leq (p-2)^2 \left| \int_l^s v^{-3} dv \right| \left| \int_l^s v^{2p-3} dv \right|$$

$$= \frac{(p-2)^2}{4(p-1)} \frac{(s^2 - l^2)(s^{2p-2} - l^{2p-2})}{s^2 l^2}$$

$$= \frac{(p-2)^2}{4(p-1)} \frac{(s^p + l^p)^2 - (s^{p-1}l + sl^{p-1})^2}{s^2 l^2}.$$

This shows that $F(u_1) \leq \frac{(p-2)^2}{4(p-1)}$ and finishes the proof of the lemma. □

We gave in the foregoing theorem a partial description of the domain of the generator on $L^p(X, \mu, \mathbb{C})$ when the semigroup is symmetric. For nonsymmetric semigroups we have a less precise result. Namely,

THEOREM 3.11 *Assume that $\mathfrak{a}$ is a densely defined, non-negative, continuous, and closed form on $L^2(X, \mu, \mathbb{C})$. Assume that the associated semigroup $(e^{-tA})_{t\geq 0}$ is sub-Markovian. For each $p \in [2, \infty)$, the following assertion holds:*

$$f \in D(A_p) \Rightarrow |f|^{\frac{p}{2}} \in D(\mathfrak{a}) \text{ and } \mathfrak{a}(|f|^{\frac{p}{2}}, |f|^{\frac{p}{2}}) \leq \frac{p}{2}\Re(A_p f; |f|^{p-1}\text{sign } f).$$

Proof. Fix $p \in [2, \infty)$ and let $f \in D(A_p)$. Since the semigroup $(e^{-tA})_{t\geq 0}$ is positive, we have

$$|e^{-tA} f| \leq e^{-tA}|f|.$$

By the sub-Markovian property and Proposition 3.4, we deduce that

$$|e^{-tA} f|^{\frac{p}{2}} \leq e^{-tA}|f|^{\frac{p}{2}}.$$

Thus,

$$\frac{1}{t}(|f|^{\frac{p}{2}} - e^{-tA}|f|^{\frac{p}{2}}; |f|^{\frac{p}{2}}) \leq \frac{1}{t}(|f|^{\frac{p}{2}} - |e^{-tA} f|^{\frac{p}{2}}; |f|^{\frac{p}{2}}).$$

The following elementary inequality holds:

$$|a|^{\frac{p}{2}} - |b|^{\frac{p}{2}} \le \frac{p}{2}(|a| - |b|)|a|^{\frac{p}{2}-1} \text{ for all } a, b \in \mathbb{C}, p \in [2, \infty).$$

Hence,

$$(|f|^{\frac{p}{2}} - |e^{-tA}f|^{\frac{p}{2}}; |f|^{\frac{p}{2}}) \le \frac{p}{2}(|f| - |e^{-tA}f|; |f|^{p-1}).$$

We have proved

$$\frac{1}{t}(|f|^{\frac{p}{2}} - e^{-tA}|f|^{\frac{p}{2}}; |f|^{\frac{p}{2}}) \le \frac{p}{2}\frac{1}{t}(|f| - |e^{-tA}f|; |f|^{p-1}).$$

Since $f \in D(A_p)$, then $\frac{1}{t}(f - e^{-tA}f)$ converges in $L^p(X, \mu, \mathbb{C})$ to $A_p f$ (as $t \downarrow 0$). Hence, by composition $\frac{1}{t}(|f| - |e^{-tA}f|)$ converges to $\Re(A_p f \operatorname{sign} \overline{f})$. Consequently,

$$\sup_{t>0} \frac{1}{t}(|f|^{\frac{p}{2}} - e^{-tA}|f|^{\frac{p}{2}}; |f|^{\frac{p}{2}}) < \infty.$$

We conclude by Lemma 1.56 that $|f|^{\frac{p}{2}} \in D(\mathfrak{a})$ and that

$$\mathfrak{a}(|f|^{\frac{p}{2}}, |f|^{\frac{p}{2}}) = \lim_{t \to 0} \frac{1}{t}(|f|^{\frac{p}{2}} - e^{-tA}|f|^{\frac{p}{2}}; |f|^{\frac{p}{2}})$$

$$\le \frac{p}{2}\lim_{t \to 0} \frac{1}{t}(|f| - |e^{-tA}f|; |f|^{p-1})$$

$$= \frac{p}{2}\Re(A_p f; |f|^{p-1}\operatorname{sign} f).$$

This proves the theorem. □

3.3 L^p-HOLOMORPHY OF SUB-MARKOVIAN SEMIGROUPS

We start with a semigroup $(e^{-tA})_{t \ge 0}$ which is bounded on $L^2(X, \mu, \mathbb{C})$ and on $L^{p_0}(X, \mu, \mathbb{C})$ for some $p_0 \in (2, \infty]$. As above, we may write $-A_p$ to denote the generator of the corresponding semigroup on $L^p(X, \mu, \mathbb{C})$ for $p \in [2, p_0)$. Assume that on $L^2(X, \mu, \mathbb{C})$, the semigroup $(e^{-tA})_{t \ge 0}$ is bounded holomorphic on the sector $\Sigma(\psi) := \{z \in \mathbb{C}, z \ne 0 \text{ and } |\arg z| < \psi\}$ for some $\psi \in (0, \frac{\pi}{2}]$. The holomorphy of the semigroup carries over from $L^2(X, \mu, \mathbb{C})$ to $L^p(X, \mu, \mathbb{C})$ $(2 \le p < p_0)$. However the sector of holomorphy depends on p, in general. More precisely, the following classical result holds.

PROPOSITION 3.12 *For each $p \in [2, p_0)$, the semigroup generated by $-A_p$ is bounded holomorphic on the sector $\Sigma(\psi_p)$ where $\psi_p = \psi(\frac{1}{p} - \frac{1}{p_0})(\frac{1}{2} - \frac{1}{p_0})^{-1}$.*

Proof. Fix $\theta \in (-\psi, \psi)$ and $r > 0$. Define on $L^2(X, \mu, \mathbb{C})$, the operator $T_z := e^{-Are^{i\theta}z}$. By the holomorphy assumption, $z \mapsto T_z$ is bounded and holomorphic on $0 < Rez < 1$. In addition, for z such that $\Re z = 1$, $\|T_z f\|_2 \leq M_\theta \|f\|_2$, and if $\Re z = 0$, $\|T_z f\|_{p_0} \leq M \|f\|_{p_0}$, where M and M_θ are positive constants. By the Stein interpolation theorem, we conclude that $T_t \in \mathcal{L}(L^p(X, \mu, \mathbb{C}))$ for p and $t \in [0, 1]$ such that $\frac{1}{p} = \frac{t}{2} + \frac{1-t}{p_0}$. In other words, e^{-zA} extends to a bounded operator on $L^p(X, \mu, \mathbb{C})$ for z such that $|\arg z| \leq \psi(\frac{1}{p} - \frac{1}{p_0})(\frac{1}{2} - \frac{1}{p_0})^{-1}$. For $z = t > 0$ this extension coincides with e^{-tA_p}. By a density argument one shows that $e^{-(z+z')A} = e^{-zA}e^{-z'A}$ as operators on $L^p(X, \mu, \mathbb{C})$ (for $z, z' \in \Sigma(\psi_p)$). The strong continuity is shown as follows. Fix $\nu \in (0, \psi_p)$ and $f \in L^{p_0}(X, \mu, \mathbb{C}) \cap L^2(X, \mu, \mathbb{C})$. We write the classical interpolation inequality

$$\|e^{-zA}f - f\|_p \leq \|e^{-zA}f - f\|_{p'}^s \|e^{-zA}f - f\|_2^{1-s} \quad \left(\frac{1}{p} = \frac{s}{2} + \frac{1-s}{p'} \right),$$

where $p' \in (p, p_0)$ is chosen such that e^{-zA} is uniformly bounded on $L^{p'}(X, \mu, \mathbb{C})$ for $z \in \Sigma(\nu)$. By the strong continuity on $L^2(X, \mu, \mathbb{C})$, we obtain $\|e^{-zA}f - f\|_p \to 0$ as $z \to 0$, $z \in \Sigma(\nu)$. By a density argument this holds for every $f \in L^p(X, \mu, \mathbb{C})$.

Finally, the holomorphy of $z \mapsto e^{-zA}$ on $L^p(X, \mu, \mathbb{C})$ (for $z \in \Sigma(\nu)$ with ν as above) can be shown by applying Vitali's theorem as in the proof of Theorem 6.16 (or Corollary 7.5 below). $\qquad \square$

Note that the above proof gives a similar result if we start with a semigroup which is holomorphic on $L^q(X, \mu, \mathbb{C})$ in place of $L^2(X, \mu, \mathbb{C})$.

Assume now that the operator A is associated with a non-negative, densely defined, continuous, and closed form. Then, the semigroup $(e^{-t(A+1)})_{t\geq 0}$ is bounded holomorphic on $L^2(X, \mu, \mathbb{C})$ on some sector $\Sigma(\psi)$ (cf. Theorem 1.52). If in addition, $(e^{-t(A+w)})_{t\geq 0}$ is L^∞-contractive for some constant w, then the induced semigroup $(e^{-tA_p})_{t\geq 0}$ on $L^p(X, \mu, \mathbb{C})$ is holomorphic on the sector $\Sigma(\frac{2}{p}\psi)$ for all p with $2 \leq p < \infty$. If the adjoint semigroup $(e^{-t(A^*+w)})_{t\geq 0}$ is L^∞-contractive, then one obtains by duality the holomorphy of the semigroup on $L^p(X, \mu, \mathbb{C})$ for $1 < p \leq 2$ on the sector $\Sigma(\psi(2 - \frac{2}{p}))$. In the particular case where A is self-adjoint and the semigroup $(e^{-tA})_{t\geq 0}$ is L^∞-contractive, this holds with $\psi = \frac{\pi}{2}$.

Using Theorem 3.9, we obtain a better result on the holomorphy for symmetric sub-Markovian semigroups. More precisely,

THEOREM 3.13 *Assume that A is the self-adjoint operator associated with a non-negative closed form on $L^2(X, \mu, \mathbb{C})$. Assume that the semigroup $(e^{-tA})_{t\geq 0}$ is sub-Markovian. Then for each $p \in (1, \infty)$, the semigroup*

$(e^{-tA_p})_{t\geq 0}$ *is holomorphic on the sector* $\Sigma(\frac{\pi}{2} - \arctan\frac{|p-2|}{2\sqrt{p-1}})$. *In addition,*
$\|e^{-zA_p}\|_{\mathcal{L}(L^p(X,\mu,\mathbb{C}))} \leq 1$ *for all* $z \in \Sigma(\frac{\pi}{2} - \arctan\frac{|p-2|}{2\sqrt{p-1}})$.

The proof is a simple adaptation of the arguments given in the proof of Theorem 1.54. The sectoriality condition (1.27) is now replaced by

$$|\Im(A_p f; |f|^{p-1}\text{sign } f)| \leq \frac{|p-2|}{2\sqrt{p-1}}\Re(A_p f; |f|^{p-1}\text{sign } f),$$

which we proved in Theorem 3.9.

Remark.   1) It is an elementary fact that the sector of holomorphy obtained in this result is larger than the one obtained in Proposition 3.12.

2) The previous results on the holomorphy does not give any information for the extreme case $p = 1$. We will see later, as a consequence of Gaussian upper bounds for heat kernels, that semigroups associated with several classes of uniformly elliptic operators are holomorphic on $\mathbb{C}^+$ on L^p for all p, $1 \leq p < \infty$ (see Theorem 6.16).

In the general setting, the above theorem is optimal. The most significant example showing this is given by the Ornstein-Uhlenbeck operator, that is, the operator A associated with the form

$$\mathfrak{a}(u,v) = \sum_{k=1}^{d} \int_{\mathbb{R}^d} D_k u D_k v e^{-x^2/2} dx.$$

It follows from Theorem 1 in Weissler [Wei79] or Theorem 1.1 (and Note 1.2) in Epperson [Epp89] that the maximal sector on which the semigroup $(e^{-tA})_{t\geq 0}$ is holomorphic on $L^p(\mathbb{R}^d, e^{-x^2/2}dx)$ is $\Sigma(\frac{\pi}{2} - \arctan\frac{|p-2|}{2\sqrt{p-1}})$. This semigroup is not holomorphic on $L^1(\mathbb{R}^d, e^{-x^2/2}dx)$.

As a consequence of the previous theorem and Theorem 1.45, we have the following information on the spectrum $\sigma(A_p) := \mathbb{C} \setminus \rho(A_p)$ of the operator A_p in $L^p(X, \mu, \mathbb{C})$.

COROLLARY 3.14 *Under the assumptions of the previous theorem, we have for each $p \in (1, \infty)$, the spectrum of A_p satisfies*

$$\sigma(A_p) \subseteq \left\{ z \in \mathbb{C}, \; |\arg z| \leq \arctan\frac{|p-2|}{2\sqrt{p-1}} \right\} \cup \{0\}.$$

Notes
Section 3.1. Theorem 3.1 is proved in Ouhabaz [Ouh93]. Proposition 3.2 is taken

from [Ouh93] and can also be found in Ma and Röckner [MaRö92] and Jacob [Jac01]. Results of the same type are proved for the so-called completely monotone (nonlinear) operators by Bénilan and Crandall [BéCr91]. Proposition 3.3 is well-known (see, e.g., Nagel [Nag86]). An operator B acting on L^p and such that $(Bu, F(u^+)) \leq 0$ for all $u \in D(B)$ is called dispersive.

Jensen's inequality for sub-Markovian semigroups as stated in Proposition 3.4 is taken from Arendt and Bénilan [ArBé92]. The fact that 1) $\Leftrightarrow$ 2) in Theorem 3.5 is also proved in Arendt and Bénilan [ArBé92]. 1) $\Leftrightarrow$ 3) of the same theorem as well as Theorem 3.7 are taken from Ouhabaz [Ouh93].

The original Kato's inequality is the distributional inequality

$$(\text{sign } \overline{f})\Delta f \leq \Delta |f|$$

for $f \in L^1_{\text{loc}}$ such that $\Delta f \in L^1_{\text{loc}}$. It was proved by Kato [Kat73] and was a powerful tool in his proof of essential self-adjointness of Schrödinger operators on $C^\infty_c(\mathbb{R}^d)$. Theorem 3.6 is due to Arendt [Are84]. An exhaustive study of Kato's inequality as stated in Theorem 3.6 as well as its relation to positivity of semigroups can be found in Nagel [Nag86].

Section 3.2. This section follows mainly the works of Liskevich and Semenov [LiSe93] and [LiSe96], from which Theorem 3.9 is taken. Assertion 1) of Theorem 3.9 was first proved by Varopoulos [Var85] for non-negative $f \in D(A_p)$. The proof given here is taken from Nagel and Voigt [NaVo96] and Liskevich and Semenov [LiSe96]. The representation (3.6) in Part iii) of the proof is well-known and can be found in Fukushima [Fuk80] and Davies [Dav80].

Section 3.3. Proposition 3.12 is well-known. The proof used here is taken from Davies [Dav89] (Theorem 1.4.2) and Reed and Simon [ReSi75] (Theorem X.55). Theorem 3.13 is due to Liskevich and Perelmuter [LiPe95] (see also [LiSe96]). This theorem was proved previously by Bakry [Bak89] and Okazawa [Oka91] in some special cases.

Chapter Four

UNIFORMLY ELLIPTIC OPERATORS ON DOMAINS

This chapter is devoted to second-order uniformly elliptic operators considered on domains of $\mathbb{R}^d$ and subject to various boundary conditions. We apply the criteria of positivity, L^∞-contractivity, and domination of semigroups, stated in an abstract setting in the previous chapters, to concrete situations of differential operators. The operators under consideration here may have real- or complex-valued measurable coefficients. The aim is to describe precisely how the above properties of the semigroup depend on the boundary conditions and on the coefficients of the operator.

4.1 EXAMPLES OF BOUNDARY CONDITIONS

Let Ω be an open subset of $\mathbb{R}^d$ $(d \geq 1)$, endowed with the Lebesgue measure dx. All the integrals over Ω are taken with respect to dx. We write $L^p(\Omega, \mathbb{K}) := L^p(\Omega, dx, \mathbb{K})$, where $\mathbb{K} = \mathbb{R}$ or $\mathbb{C}$. We will often consider the case $\mathbb{K} = \mathbb{C}$. The scalar product of $L^2(\Omega, \mathbb{K})$ is denoted by $(.; .)$. We denote by $W^{s,p}(\Omega), H^1(\Omega) = W^{1,2}(\Omega)$ and $H_0^1(\Omega)$ the classical Sobolev spaces. Recall that $H_0^1(\Omega)$ is the closure in $H^1(\Omega)$ of $C_c^\infty(\Omega)$, the space of C^∞-functions which are compactly supported in Ω. The space of distributions on Ω will be denoted by $(C_c^\infty(\Omega))'$.

We consider measurable functions a_{jk}, b_k, c_k, and a_0 $(1 \leq j, k \leq d)$ on Ω. We assume that the following uniform ellipticity condition holds.

(U.Ell) The functions a_{kj}, b_k, c_k, a_0 are bounded on Ω, i.e.,

$$a_{kj}, b_k, c_k, a_0 \in L^\infty(\Omega, \mathbb{C}) \text{ for all } 1 \leq j, k \leq d \qquad (4.1)$$

and the principal part is elliptic; i.e., there exists a constant $\eta > 0$ such that

$$\Re \sum_{j,k=1}^d a_{kj}(x) \xi_j \overline{\xi_k} \geq \eta |\xi|^2 \text{ for all } \xi \in \mathbb{C}^d, \text{ a.e. } x \in \Omega. \qquad (4.2)$$

The maximal possible η in (4.2) is called the ellipticity constant of the matrix $(a_{jk})_{1 \leq j, k \leq d}$.

Let $D_j = \frac{\partial}{\partial x_j}$ and define on $L^2(\Omega, \mathbb{C})$ the sesquilinear form

$$\mathfrak{a}_V(u, v) = \int_\Omega \left[\sum_{k,j=1}^d a_{kj} D_k u \overline{D_j v} + \sum_{k=1}^d (b_k \overline{v} D_k u + c_k u \overline{D_k v}) + a_0 u \overline{v} \right] dx,$$

with domain

$$D(\mathfrak{a}_V) = V,$$

where V is a closed subspace of $H^1(\Omega)$ with $H_0^1(\Omega) \subseteq V \subseteq H^1(\Omega)$. If $w \in \mathbb{R}$ is a constant, we denote by $\mathfrak{a}_V + w$ the form given by

$$(\mathfrak{a}_V + w)(u, v) := \mathfrak{a}_V(u, v) + w(u; v), \ u, v \in D(\mathfrak{a}_V).$$

Under the assumption (U.Ell), there exists a constant w such that

$$\Re \mathfrak{a}_V(u, u) + w\|u\|_2^2 \geq \frac{\eta}{2}\|u\|_{H^1(\Omega)}^2 \text{ for all } u \in V \tag{4.3}$$

and the form $\mathfrak{a}_V + w$ is continuous and closed. Indeed, Let $u \in V$ and set $M_k := \| |\Re(b_k + c_k)| + |\Im(b_k - c_k)| \|_\infty$. We have

$$\Re \mathfrak{a}_V(u, u) = \Re \sum_{k,j} \int_\Omega a_{kj} D_k u \overline{D_j u} dx + \sum_k \int_\Omega \Re(b_k + c_k) \Re(\overline{u} D_k u) dx$$

$$+ \sum_k \int_\Omega \Im(c_k - b_k) \Im(\overline{u} D_k u) dx + \int_\Omega \Re a_0 |u|^2 dx$$

$$\geq \eta \sum_k \int_\Omega |D_k u|^2 dx - \sum_k M_k \int_\Omega |\overline{u} D_k u| - \int_\Omega (\Re a_0)^- |u|^2 dx.$$

By the Cauchy-Schwarz inequality, we have for every $\varepsilon > 0$

$$\Re \mathfrak{a}_V(u, u) \geq \eta \sum_k \int_\Omega |D_k u|^2 dx - \varepsilon \sum_k \int_\Omega |D_k u|^2 dx$$

$$- \left(\varepsilon^{-1} \sum_k M_k^2 + \|(\Re a_0)^-\|_\infty \right) \int_\Omega |u|^2 dx,$$

which implies (4.3).

The proof of the continuity of $\mathfrak{a}_V + w$ follows from (4.3) and the inequality

$$|\mathfrak{a}_V(u, v)| \leq C \left[\sum_k \|D_k u\|_2^2 + \|u\|_2^2 \right]^{1/2} \left[\sum_k \|D_k v\|_2^2 + \|v\|_2^2 \right]^{1/2}$$

which holds for some positive constant C and all $u, v \in V$. The latter inequality follows easily from the Cauchy-Schwarz inequality and the boundedness of a_{kj}, b_k, c_k, and a_0.

Finally, the fact that the form $\mathfrak{a}_V + w$ is closed follows from the assumption that V is a closed subspace of $H^1(\Omega)$. Note that the norm of $H^1(\Omega)$ and $\sqrt{\mathfrak{a}_V(.,.) + w(.;.)}$ are equivalent on V. Therefore, the form $\mathfrak{a}_V + w$ satisfies the standard assumptions $(1.2)-(1.5)$. One can then associate with $\mathfrak{a}_V + w$ an operator on $L^2(\Omega, \mathbb{C})$. This means that we can associate with $\mathfrak{a}_V$ an operator A_V. Formally, A_V is given by the expression

$$ A_V u = - \sum_{k,j=1}^{d} D_j(a_{kj} D_k u) + \sum_{k=1}^{d} (b_k D_k u - D_k(c_k u)) + a_0 u. \quad (4.4) $$

The role of the space V is to impose boundary conditions. Although the dependence on V is not visible in the formal expression of A_V, it must be understood that for two different spaces V and W, the operators A_V and A_W are different as they are subject to two different types of boundary conditions.

We give now some examples of boundary conditions. We will say that we have:

(i) **Dirichlet boundary conditions** if $V = H_0^1(\Omega)$;

(ii) **Neumann boundary conditions** if $V = H^1(\Omega)$;

(iii) **mixed boundary conditions** if

$$ V = \overline{\{u_{|\Omega}, u \in C_c^\infty(\mathbb{R}^d \setminus \Gamma)\}}^{H^1(\Omega)}, \quad (4.5) $$

where Γ is a closed subset of $\partial\Omega$, $u_{|\Omega}$ denotes the restriction of u to Ω, and $\overline{\{\ldots\}}^{H^1(\Omega)}$ denotes the closure in $H^1(\Omega)$.

Roughly speaking, (i) corresponds to the condition $u = 0$ on the boundary $\partial\Omega$ of Ω, whereas (ii) corresponds to the condition

$$ \sum_{j=1}^{d} \left(\sum_{k=1}^{d} a_{kj} D_k u + c_j u \right) n_j = 0 \text{ on } \partial\Omega, $$

where $\vec{n} = (n_1, \ldots, n_d)$ denotes the outer unit normal on the boundary $\partial\Omega$. Mixed boundary conditions correspond to Dirichlet condition on Γ and the Neumann one on the rest of the boundary. All of this can be done precisely by applying Green's formula if both $\partial\Omega$ and the coefficients a_{kj}, c_k are smooth enough.

If Γ is empty, then (4.5) becomes

$$ V = \widetilde{H^1(\Omega)} := \overline{\{u_{|\Omega}, u \in C_c^\infty(\mathbb{R}^d)\}}^{H^1(\Omega)}. \quad (4.6) $$

The boundary condition given by such V is similar to the Neumann one. We call this **the good Neumann boundary conditions**. It is well-known that for domains with smooth boundary (Lipschitz boundary, for instance) one has $H^1(\Omega) = \widetilde{H^1(\Omega)}$. See, for instance, Adams [Ada75] or Brezis [Bre92].

The following is another type of boundary conditions.

(iv) **Robin boundary conditions.** Assume that Ω is bounded and has smooth boundary (e.g., Lipschitz boundary). In this case, every $u \in H^1(\Omega)$ has a trace on $\partial\Omega$. One can then consider perturbations of the previous form $\mathfrak{a}_V$ by integral terms on the boundary. More precisely, consider the sesquilinear form

$$\mathfrak{b}_{V,\alpha}(u,v) := \mathfrak{a}_V(u,v) + \int_{\partial\Omega} \alpha(x)u\bar{v}d\sigma, \quad D(\mathfrak{b}_{V,\alpha}) = V, \qquad (4.7)$$

where α is a non-negative and bounded function on $\partial\Omega$ and $d\sigma$ is the surface measure. The operator associated with this form has the same formal expression (4.4) but it is subject to different boundary conditions, called Robin boundary conditions. In the case where $V = H^1(\Omega)$, they correspond to the condition

$$\sum_{j=1}^{d}\left(\sum_{k=1}^{d}a_{kj}D_ku + c_ju\right)n_j + \alpha u = 0 \text{ on } \partial\Omega.$$

For V as in (4.5), the condition looks like the previous one on $\partial\Omega \setminus \Gamma$ and $u = 0$ on Γ.

The boundary conditions in (i)–(iv) are only examples of boundary conditions which fit into the framework of the present chapter. Other examples can also be considered.

Consider now a general space V. It follows from Theorem 1.52 that the operator $-A_V$ is the generator of a holomorphic semigroup $(e^{-tA_V})_{t\geq 0}$ on $L^2(\Omega, \mathbb{C})$. In particular, the parabolic equation (or the Cauchy problem)

$$(CP)\begin{cases} \frac{\partial}{\partial t}u(t,x) = \sum_{k,j=1}^{n}\frac{\partial}{\partial x_j}\left(a_{kj}(x)\frac{\partial}{\partial x_k}u(t,x)\right) - \sum_{k=1}^{n}b_k(x)\frac{\partial u(t,x)}{\partial x_k} \\ \qquad + \sum_{k=1}^{n}\frac{\partial}{\partial x_k}(c_k(x)u(t,x)) - a_0u(t,x) \\ \qquad = -A_Vu(t,x), \ t > 0, \\ u(0,.) = f \in L^2(\Omega, \mathbb{C}) \end{cases}$$

has a unique solution $u(t,.) = e^{-tA_V}f \in D(A_V)$ for all $t > 0$. The solution $u(t,.)$ satisfies the boundary conditions imposed by V. Now, as explained in Chapter 2, if the semigroup $(e^{-tA_V})_{t\geq 0}$ is L^{∞}-contractive, then it induces a strongly continuous semigroup on $L^p(\Omega, \mathbb{C})$ for each $p \in [2, \infty)$ and we obtain existence and uniqueness of the solution of the analog of (CP) in $L^p(\Omega, \mathbb{C})$. The same holds for $1 \leq p \leq 2$ if the adjoint semigroup $(e^{-tA_V^*})_{t\geq 0}$ is L^{∞}-contractive. For the parabolic equation (CP), the L^{∞}-contractivity of the semigroup $(e^{-tA_V})_{t\geq 0}$ can be rephrased as:

$$|f| \leq 1 \Rightarrow |u(t,.)| \leq 1 \text{ for all } t \geq 0.$$

Thus, it is a maximum principle for (CP). Note that the positivity of the semigroup $(e^{-tA_V})_{t\geq 0}$ can also be interpreted as a maximum principle for (CP).

In the next sections we study positivity, irreducibility, domination, and L^{∞}-contractivity of semigroups $(e^{-tA_V})_{t\geq 0}$, associated with different V's.

4.2 POSITIVITY AND IRREDUCIBILITY

We keep the framework and notation of the previous section. We assume the uniform ellipticity condition (U.Ell).

An application of Proposition 2.5 shows that the semigroup $(e^{-tA_V})_{t\geq 0}$ is real if and only if $\Re u \in V$ for all $u \in V$ and $\mathfrak{a}_V(u, v) \in \mathbb{R}$ for all real $u, v \in V$. Clearly, this is again equivalent to the fact that $\Re u \in V$ for all $u \in V$ and that for every $u, v \in V$

$$\mathfrak{a}_V(u, v) = \int_\Omega \left[\sum_{j,k=1}^d \Re(a_{kj}) D_k u \overline{D_j v} + \sum_{k=1}^d (\Re(b_k) \overline{v} D_k u + \Re(c_k) u \overline{D_k v}) \right. $$
$$\left. + \Re(a_0) u \overline{v} \right] dx. \quad (4.8)$$

This means that the form $\mathfrak{a}_V$ (and hence the operator A_V) is given by real-valued coefficients.

PROPOSITION 4.1 *The semigroup $(e^{-tA_V})_{t\geq 0}$ is real if and only if $\Re u \in V$ for all $u \in V$ and $\mathfrak{a}_V$ has real-valued coefficients.*

The following result characterizes the positivity of the semigroup.

THEOREM 4.2 *The semigroup $(e^{-tA_V})_{t\geq 0}$ is positive if and only if the following two conditions hold:*
1) $u \in V \Rightarrow (\Re u)^+ \in V$.
2) The form $\mathfrak{a}_V$ is given by real coefficients (i.e., $\mathfrak{a}_V$ satisfies (4.8) for all $u, v \in V$).

Proof. The fact that the conditions 1) and 2) are necessary follows immediately from the previous proposition and Theorem 2.6.

Assume now that 1) and 2) are satisfied. Applying 1) to $-u$ shows that $\Re u = (\Re u)^+ - (\Re u)^- \in V$ for all $u \in V$. It follows from the previous proposition that the semigroup is real. Since $D_j(\Re u)^+ = D_j(\Re u)\chi_{\{\Re u>0\}}$ (see Proposition 4.4 below), one has $\mathfrak{a}_V((\Re u)^+, (\Re u)^-) = 0$. Theorem 2.6 implies the positivity of the semigroup.   □

COROLLARY 4.3 *Assume that $\mathfrak{a}_V$ is given by real coefficients. Then for Dirichlet, Neumann, good Neumann, or mixed boundary conditions, the corresponding semigroup $(e^{-tA_V})_{t\geq 0}$ is positive.*

This result follows from Theorem 4.2 and the following proposition.

PROPOSITION 4.4 *Let V be one of the spaces $H_0^1(\Omega)$, $H^1(\Omega)$, $\widetilde{H^1(\Omega)}$, or $\overline{\{u_{|\Omega}, u \in C_c^\infty(\mathbb{R}^d \setminus \Gamma)\}}^{H^1(\Omega)}$ (where Γ is a closed subset of $\partial\Omega$). Then for every $u \in V$, we have $(\Re u)^+, |u| \in V$, and*

$$D_j(\Re u)^+ = D_j(\Re u)\chi_{\{\Re u>0\}} \text{ and } D_j|u| = \Re(\text{sign}(\overline{u})D_j u),$$

where $\chi_{\{\Re u>0\}}$ denotes the characteristic function of the set where $\Re u$ is > 0.

Proof. Assume first that $V = H^1(\Omega)$. Consider on $\mathbb{C}$, the function

$$f_\varepsilon(z) := \sqrt{|z|^2 + \varepsilon^2} - \varepsilon.$$

Clearly f_ε has partial derivatives $\frac{\partial}{\partial t}f_\varepsilon$ and $\frac{\partial}{\partial s}f_\varepsilon$ which are continuous and bounded on $\mathbb{C}$ (here $t = \Re z$ and $s = \Im z$) and $f_\varepsilon(0) = 0$. Thus by a classical result $f_\varepsilon(u) \in H^1(\Omega)$ for all $u \in H^1(\Omega)$.[1] Now

$$D_j f_\varepsilon(u) = \frac{\partial}{\partial t}f_\varepsilon(\Re u, \Im u)D_j(\Re u) + \frac{\partial}{\partial s}f_\varepsilon(\Re u, \Im u)D_j(\Im u)$$

$$= \frac{1}{\sqrt{|u|^2 + \varepsilon^2}}[\Re u D_j(\Re u) + \Im u D_j(\Im u)]$$

$$= \Re\left[D_j u \frac{\overline{u}}{\sqrt{|u|^2 + \varepsilon^2}}\right].$$

Let $\phi \in C_c^\infty(\Omega)$. We have for all $\varepsilon > 0$,

$$\int_\Omega f_\varepsilon(u)D_j\phi dx = -\int_\Omega \Re\left[D_j u \frac{\overline{u}}{\sqrt{|u|^2 + \varepsilon^2}}\right]\phi dx.$$

---

[1]This fact can also be deduced from Theorem 2.25.

Taking the limit when $\varepsilon \to 0$, it follows that

$$\int_\Omega |u| D_j \phi \, dx = -\int_\Omega \Re[\text{sign}(\overline{u}) D_j u] \phi \, dx.$$

This implies that $|u| \in H^1(\Omega)$ and $D_j |u| = \Re[\text{sign}(\overline{u}) D_j u]$. Applying this to $\Re u$ and using the fact that $(\Re u)^+ = \frac{1}{2}(|\Re u| + \Re u)$, one obtains $(\Re u)^+ \in H^1(\Omega)$ and $D_j (\Re u)^+ = D_j (\Re u) \chi_{\{\Re u > 0\}}$. For $V = H_0^1(\Omega)$, the assertions follow from the facts that $u^+ \in H^1(\Omega)$ for all real-valued $u \in C_c^\infty(\Omega)$ and that functions in $H^1(\Omega)$ which have compact supports in Ω are in $H_0^1(\Omega)$.

We prove the proposition in the case of mixed boundary conditions, i.e., we assume that $V = \overline{\{u_{|\Omega}, u \in C_c^\infty(\mathbb{R}^d \setminus \Gamma)\}}^{H^1(\Omega)}$ (the case where Γ is empty gives the good Neumann boundary conditions). First, it is clear that $\Re u \in V$ for all $u \in V$. Fix now $u = \Re u \in V$ and let $u_n \in C_c^\infty(\mathbb{R}^d \setminus \Gamma)$ be a sequence of real-valued functions which converges to u in $H^1(\Omega)$. It follows that (u_n^+) converges to u^+ in $H^1(\Omega)$. Let $\rho_n \in C_c^\infty(\mathbb{R}^d)$ be a regularizing sequence and denote by $*$ the convolution. Clearly, for each fixed n, the sequence $(\rho_m * u_n^+)_m$ converges in $H^1(\mathbb{R}^d)$ to u_n^+ and $\rho_m * u_n^+ \in C_c^\infty(\mathbb{R}^d \setminus \Gamma)$ for large m. This gives $u_n^+ \in V$ and hence $u^+ \in V$. A similar argument shows that $|u| \in V$ for all $u \in V$. $\square$

Example 4.2.1   *Let $V = \{u \in H^1(0,1), u(0) = iu(1)\}$. Define the form*

$$\mathfrak{a}_V(u,v) = \int_0^1 u'\overline{v'} dx, \ D(\mathfrak{a}_V) = V.$$

This form has real coefficients, but the semigroup is not real because V is not stable under the real part $\Re$. In particular, the semigroup is not positive. If we choose now

$$V = \{u \in H^1(0,1), u(0) = -u(1)\},$$

then the semigroup is real but not positive (the positive part u^+ does not operate on V).

We study now the irreducibility of the semigroup $(e^{-tA_V})_{t \geq 0}$. Since an irreducible semigroup is in particular positive, it follows that the conditions 1) and 2) of Theorem 4.2 are necessary for irreducibility of $(e^{-tA_V})_{t \geq 0}$.

THEOREM 4.5 *Assume that $(\Re u)^+ \in V$ for all $u \in V$ and that the form $\mathfrak{a}_V$ satisfies (4.8). Consider the following assertions:*
1) The semigroup $(e^{-tA_V})_{t \geq 0}$ is irreducible.
2) The open subset Ω is connected.
Then 2) implies 1). The converse holds if A_V is subject to Dirichlet, Neumann, mixed, or good Neumann boundary conditions.

Proof. Firstly, note that the semigroup $(e^{-tA_V})_{t\geq 0}$ is positive (cf. Corollary 4.3).

By Corollary 2.11, the irreducibility of $(e^{-tA_V})_{t\geq 0}$ is equivalent to the following: if Ω_1 is a subset of Ω such that

$$u \in V \Rightarrow \chi_{\Omega_1} u \in V, \tag{4.9}$$

then either $\lambda(\Omega_1) = 0$ or $\lambda(\Omega \setminus \Omega_1) = 0$, where λ denotes the Lebesgue measure on $\mathbb{R}^d$.

Assume that Ω is connected. Suppose for a contradiction that Ω_1 satisfies (4.9) and $\lambda(\Omega_1) > 0$, $\lambda(\Omega \setminus \Omega_1) > 0$. We have in particular,

$$u \in C_c^\infty(\Omega) \Rightarrow \chi_{\Omega_1} u \in H^1(\Omega). \tag{4.10}$$

The operator D_k satisfies $\chi_{\{v=0\}} D_k v = 0$ for all $v \in H^1(\Omega)$. This implies that $D_k(\chi_{\Omega_1} u) = \chi_{\Omega_1} D_k u$. Hence, $\chi_{\Omega_1} u \in W_0^{1,p}(\mathcal{O})$ for all $p \in [1, \infty]$, where $\mathcal{O}$ is an open subset of Ω with smooth boundary which contains the support of a fixed $u \in C_c^\infty(\Omega)$. (Note that there always exists an increasing sequence of open subsets U_n of Ω with smooth boundaries, such that $\cup_n U_n = \Omega$. Using the fact that the support of u is compact, one obtains from this that such $\mathcal{O}$ exists.) Choosing p large enough, we deduce from Sobolev embedding theorems that $\chi_{\Omega_1} u = v$ a.e. on $\mathcal{O}$, with v being a continuous function on $\mathcal{O}$.

Assume for a moment that there exists $x_0 \in \Omega$ such that for every $\eta > 0$

$$\lambda(B(x_0, \eta) \cap \Omega_1) > 0 \text{ and } \lambda(B(x_0, \eta) \cap \Omega_2) > 0, \tag{4.11}$$

where $\Omega_2 = \Omega \setminus \Omega_1$ and $B(x_0, \eta)$ denotes the open euclidean ball with center x_0 and radius η. We take $\eta > 0$ small enough such that $B(x_0, 2\eta) \subseteq \Omega$ and consider $u \in C_c^\infty(\Omega)$ such that $u(x) = 1$ for all $x \in B(x_0, \eta)$. We have for a.e. $x \in B(x_0, \eta) \cap \Omega_1$ and a.e. $y \in B(x_0, \eta) \cap \Omega_2$

$$1 = |\chi_{\Omega_1} u(x) - \chi_{\Omega_1} u(y)| = |v(x) - v(y)|.$$

From the fact that v is continuous, we see that this equality cannot hold.

Now we prove the existence of x_0 satisfying (4.11). Assume that for every $x \in \Omega$, there exists $\eta > 0$ such that either $\lambda(B(x, \eta) \cap \Omega_1) = 0$ or $\lambda(B(x, \eta) \cap \Omega_2) = 0$. Define $\mathcal{O}_1$ (respectively $\mathcal{O}_2$) as the union of all balls $B(x, \eta)$, where x and η are such that $\lambda(B(x, \eta) \cap \Omega_1) = 0$ (respectively $\lambda(B(x, \eta) \cap \Omega_2) = 0$). One checks easily that $\mathcal{O}_1$ and $\mathcal{O}_2$ are disjoint open subsets such that $\Omega \subseteq \mathcal{O}_1 \cup \mathcal{O}_2$. In addition, if $\Omega \subseteq \mathcal{O}_i$, then $\lambda(\Omega_i) = 0$. Since we have assumed that $\lambda(\Omega_i) > 0$ for $i = 1, 2$, we obtain a contradiction with the fact that Ω is connected. This proves the existence of x_0.

Assume now that we have one of the boundary conditions listed in the theorem. That is, respectively, $V = H_0^1(\Omega)$, $V = H^1(\Omega)$, V is as in (4.5), or V is as in (4.6). If Ω is not connected then $\Omega = \Omega_1 \cup \Omega_2$, where Ω_1 and Ω_2 are two disjoint open sets. It is not hard to see that Ω_1 satisfies (4.9) in each of the four cases above and this implies that semigroup cannot be irreducible. $\qquad\Box$

In the above theorem, we can assert that 1) implies 2) for several other boundary conditions. Those listed there are particular cases for which the implication 1) $\Rightarrow$ 2) holds. However, this implication is not true for all boundary conditions, as the following example shows.

Example 4.2.2 *Let $\Omega = (0,1) \cup (2,3)$. Define the form*

$$\mathfrak{a}_V(u,v) = \int_0^1 u'\overline{v'}dx, \ D(\mathfrak{a}_V) = V = \{u \in H^1(\Omega), u(0) = u(3)\}.$$

This corresponds to periodic boundary conditions at 0 and 3 and the Neumann conditions at 1 and 2. This form is well defined, since $H^1(\Omega) \subseteq C(\overline{\Omega})$ by Sobolev embedding. The semigroup $(e^{-tA_V})_{t\geq 0}$ associated with this form is irreducible. Indeed, by Theorem 4.2 and the fact that $u^+ \in H^1(\Omega)$ for every real $u \in H^1(\Omega)$ and $(u^+)' = u'\chi_{\{u>0\}}$, it follows that the semigroup is positive. Now if Ω_1 satisfies (4.10) and has nonzero measure, then one deduces easily that either $\Omega_1 = (0,1)$, $\Omega_1 = (2,3)$, or $\Omega_1 = \Omega$. Assume that $\Omega_1 = (0,1)$. Hence $(\chi_{(0,1)}u)(0) = (\chi_{(0,1)}u)(3)$ for all $u \in V$. We deduce that $u(0) = 0$ for all $u \in V$, which is not the case. The same conclusion holds if $\Omega = (2,3)$. Thus, $\Omega_1 = \Omega$ and we conclude by Theorem 2.10 or Theorem 4.5 that the semigroup is irreducible.

4.3 L^∞-CONTRACTIVITY

This section is devoted to the L^∞-contractivity property of the semigroup $(e^{-tA_V})_{t\geq 0}$. Our aim is to describe precisely in terms of the coefficients and also in terms of the boundary conditions imposed by V, whether or not L^∞-contractivity holds. In order to do this, we apply the criteria given in Chapter 2. Recall that by Theorem 2.15, the following condition is necessary for L^∞-contractivity:

$$u \in V \Rightarrow (1 \wedge |u|)\text{sign } u \in V. \qquad (4.12)$$

This restricts, independently of the coefficients of the form $\mathfrak{a}_V$, the range of boundary conditions for which L^∞-contractivity of $(e^{-tA_V})_{t\geq 0}$ may hold.

We first state and prove the following result.

THEOREM 4.6 *The semigroup $(e^{-tA_V})_{t\geq 0}$ is L^∞-contractive if and only if the following two conditions are satisfied:*

i) V satisfies (4.12).

ii) For all $u \in V$ such that $r\,\varphi_k\varphi_j \in L^1(\Omega, \mathbb{C})$ and $\varphi_k D_j r \in L^1(\Omega, \mathbb{C})$ for every $j, k = 1, \ldots, d$, where $r = |u|$ and $\varphi_j = \varphi_j(u) := \frac{\Im(\text{sign}(\overline{u})D_j u)}{r}\chi_{\{u \neq 0\}}$, we have

$$\int_\Omega \left[\sum_{j,k=1}^d \Re(a_{kj})\varphi_k\varphi_j r - \sum_{j,k=1}^d \Im(a_{kj})\varphi_k D_j r + \sum_{j=1}^d \Im(c_j - b_j)\varphi_j r \right.$$
$$\left. + \sum_{j=1}^d (\Re c_j)D_j r + \Re(a_0)r \right] dx \geq 0.$$

Proof. Let $u \in V$ and put $v = (r - 1)^+ \text{sign } u$. We have $D_j|u| = \Re(\text{sign}(\overline{u})D_j u)$ for every $u \in H^1(\Omega)$ (cf. Proposition 4.4), and by Proposition 4.11 below, we have $D_k v = (D_k r + i(r - 1)\varphi_k) \text{sign } u\, \chi_{\{r>1\}}$ and $D_k((r \wedge 1)\text{sign } u) = i\, \varphi_k \text{sign } u\, \chi_{\{r>1\}} + D_k u\, \chi_{\{r \leq 1\}}$. Hence

$$\Re a_V((r \wedge 1)\text{sign } u, (r - 1)^+ \text{sign } u)$$
$$= \int_\Omega \left[\sum_{j,k} \Re(a_{kj})\varphi_k\varphi_j (r - 1)^+ - \sum_{j,k} \Im(a_{kj})\varphi_k D_j r\, \chi_{\{r>1\}} \right.$$
$$+ \sum_{j=1} \Im(c_j - b_j)\varphi_j (r - 1)^+$$
$$\left. + \sum_j (\Re c_j)D_j r\, \chi_{\{r>1\}} + \Re(a_0)(r - 1)^+ \right] dx. \quad (4.13)$$

Assume now that conditions i) and ii) are satisfied. We have $v := (r - 1)^+ \text{sign } u \in V$ and $|v| = (r - 1)^+$, $\varphi_j(v) = \varphi_j(u)\,\chi_{\{r>1\}}$. It follows that $\varphi_k\varphi_j|v|$ and $\varphi_k D_j|v| \in L^1(\Omega, \mathbb{C})$. We apply ii) for v and obtain

$$\Re a_V((|u| \wedge 1)\text{sign } u, (|u| - 1)^+ \text{sign } u) \geq 0.$$

Theorem 2.15 implies that $(e^{-tA_V})_{t \geq 0}$ is L^∞-contractive.

Conversely, assume that $(e^{-tA_V})_{t \geq 0}$ is L^∞-contractive. As mentioned above, (4.12) holds. Again Theorem 2.15 implied to $\frac{u}{k}$ for $k > 0$ gives (4.13) with $(r - k)^+$ in place of $(r - 1)^+$ and $\chi_{\{r>k\}}$ in place of $\chi_{\{r>1\}}$. Letting $k \to 0$, we obtain with the help of the dominated convergence theorem the assertion ii). □

In the case of Dirichlet boundary conditions, i.e., $V = H_0^1(\Omega)$, a more precise result holds.

THEOREM 4.7 *The semigroup $(e^{-tA_{H_0^1}})_{t \geq 0}$ is L^∞-contractive if and only if the following four conditions are satisfied:*

i) $\Im(a_{kj} + a_{jk}) = 0$ *for all* $j, k \in \{1, \dots, d\}$,

ii) $f_0 = \Re a_0 - \sum\limits_{j=1}^{d} D_j(\Re c_j)$ *is a positive Radon measure on* Ω,

iii) $f_k = \sum\limits_{j=1}^{d} D_j \Im(a_{kj}) \in L^1_{\text{loc}}(\Omega)$, $k \in \{1, \dots, d\}$,

iv) $\sum\limits_{k,j=1}^{d} \Re(a_{kj}) \xi_k \xi_j + \sum\limits_{j=1}^{d} (\Im(c_j - b_j) + f_j) \xi_j + f_{0,r} \geq 0$ *a.e. on* Ω *for all* $\xi \in$

$\mathbb{R}^d$, *where* $f_{0,r}$ *is the regular part of the measure* f_0 *(i.e., the absolutely continuous part).*

Proof. Note that $H^1_0(\Omega)$ satisfies condition (4.12) (see Proposition 4.11). It follows that the semigroup $(e^{-tA_{H^1_0}})_{t \geq 0}$ is L^∞-contractive if and only if for every $u \in H^1_0(\Omega)$ such that $\varphi_k \varphi_j r, \varphi_k D_j r \in L^1(\Omega, \mathbb{C})$, we have

$$\int_\Omega \left[\sum_{k,j} \Re(a_{kj}) \varphi_k \varphi_j + \sum_j \Im(c_j - b_j) \varphi_j + \Re a_0 \right] r\, dx$$

$$\geq \int_\Omega \left[\sum_{k,j} \Im(a_{kj}) \varphi_k - \sum_j (\Re c_j) \right] D_j r\, dx, \qquad (4.14)$$

where $r = |u|$ and $\varphi_j = \frac{1}{r} \Im(D_j u \operatorname{sign} \bar{u}) \chi_{\{u \neq 0\}}$ as above. Assume that $(e^{-tA_{H^1_0}})_{t \geq 0}$ is L^∞-contractive. We apply (4.14) with $u = re^{i\varphi}$, where $r \in C^\infty_c(\Omega, \mathbb{R}^+)$ and $\varphi \in C^\infty(\mathbb{R}^d, \mathbb{R})$. Since $\varphi_j = D_j \varphi\, \chi_{\{r>0\}}$, we obtain

$$\sum_{k,j} \Re(a_{kj}) D_k \varphi\, D_j \varphi + \sum_j \Im(c_j - b_j) D_j \varphi + f_0$$

$$+ \sum_j f_j\, D_j \varphi + \sum_{k,j} \Im(a_{kj}) D_j D_k \varphi \geq 0 \text{ in } (C^\infty_c(\Omega))' \qquad (4.15)$$

for all $\varphi \in C^\infty(\mathbb{R}^d, \mathbb{R})$, where f_0, f_j are the distributions defined in ii) and iii), respectively.

We apply (4.15) with $\varphi = 0$ and obtain $f_0 \geq 0$ in $(C^\infty_c(\Omega))'$, that is, ii). On the other hand, it follows from (4.15) that $f_0 + \sum_j f_j D_j \varphi$ is a Radon measure on Ω whose singular part is non-negative for all $\varphi \in C^\infty(\mathbb{R}^d, \mathbb{R})$. For $k = 1, \dots, d$, choose $\varphi(x) = \lambda x_k$, $\lambda \in \mathbb{R}$, we deduce that f_k is a Radon measure with a trivial singular part and hence $f_k \in L^1_{\text{loc}}(\Omega)$, that is, iii).

Now let $x_0 \in \Omega$ be a Lebesgue point for $a_{kj}, c_j, b_j, f_j, f_{0,r}$ and apply (4.15) with $\varphi(x) = \frac{\lambda}{2}((x - x_0) . \xi)^2$, $\xi \in \mathbb{R}^d$, and $\lambda \in \mathbb{R}$ (where . denotes the classical inner product of $\mathbb{R}^d$). We obtain

$$\lambda^2 \sum_{k,j} \Re(a_{kj})\xi_k\,\xi_j((x-x_0).\xi)^2 + \lambda \sum_j (\Im(c_j - b_j) + f_j)((x-x_0).\xi)\xi_j$$

$$+ f_{0,r} + \lambda \sum_{k,j} \Im(a_{kj})\xi_k\,\xi_j \geq 0 \quad (\text{a.e. } x \in \Omega).$$

For $x = x_0$, this gives

$$f_{0,r}(x_0) + \lambda \sum_{k,j} \Im a_{kj}(x_0)\,\xi_k\,\xi_j \geq 0 \text{ for all } \lambda \in \mathbb{R}.$$

Since λ is arbitrary in $\mathbb{R}$, we obtain $\sum_{k,j} \Im a_{kj}\,\xi_k\,\xi_j = 0$ a.e. on Ω, for all $\xi \in \mathbb{R}^d$, that is, i).

Finally, we apply (4.15) with $\varphi(x) = \xi.x$ to obtain iv).

Conversely, we assume that the four conditions i)–iv) are satisfied. We show that $(e^{-tA_{H_0^1}})_{t \geq 0}$ is L^∞-contractive.

Using (4.13), Theorem 2.15 and the fact that $C_c^\infty(\Omega)$ is dense in $H_0^1(\Omega)$, it is enough to check that for every $u \in C_c^\infty(\Omega)$,

$$0 \leq \int_\Omega \left[\sum_{k,j} \Re(a_{kj})\varphi_k\,\varphi_j(r-1)^+ - \sum_{k,j} \Im(a_{kj})\varphi_k\,D_j(r-1)^+ \right.$$

$$+ \sum_j \Im(c_j - b_j)\varphi_j(r-1)^+ + \sum_j (\Re c_j)D_j(r-1)^+$$

$$\left. + \Re a_0(r-1)^+ \right] dx, \quad (4.16)$$

where r and φ_j are as above. Since $(r-1)^+ \in W^{1,\infty}(\Omega)$ and has compact support and $\sum_j D_j(\Re c_j) = \Re a_0 - f_0$ is a Radon measure on Ω (condition ii)), we have

$$\int_\Omega \sum_j \Re c_j\,D_j(r-1)^+ = \int_\Omega (f_0 - \Re a_0)(r-1)^+.$$

On the other hand, $\varphi_k \in C^\infty(\{r > 0\})$ and $\frac{\partial \varphi_k}{\partial x_j} = \frac{\partial \varphi_j}{\partial x_k}$. Indeed, on the open set $\{r > 0\}$, $\varphi_k = -i \text{ sign } \bar{u}\, D_k(\text{sign } u)$, hence $D_j\varphi_k = -i(-\varphi_j\varphi_k + (\text{sign } \bar{u})D_kD_j(\text{sign } u))$, which is symmetric with respect to (k, j).

Hence using iii) and i), we have

$$\sum_{k,j} D_j \Im(a_{kj}\varphi_k) = \sum_k \left(\sum_j D_j \Im a_{kj}\right)\varphi_k + \sum_{k,j} \Im a_{kj}\, D_j\varphi_k$$

$$= \sum_k f_k\varphi_k \quad \text{in } (C_c^\infty(\{r > 0\}))'.$$

Since $(r - 1)^+$ has compact support in $\{r > 0\}$

$$\int_\Omega -\sum_{k,j}(\Im a_{kj})\varphi_k\, D_j(r - 1)^+ = \int_\Omega \sum_k f_k\, \varphi_k(r - 1)^+.$$

Hence the right-hand side term of (4.16) is given by

$$\int_\Omega \left[\sum \Re a_{kj}\varphi_k\varphi_j + \sum(\Im(c_j - b_j) + f_j)\varphi_j + f_{0,r}\right](r - 1)^+ dx$$

$$+ \int_\Omega (r - 1)^+\, d(f_{0,s}),$$

where $f_{0,s} = f_0 - f_{0,r}$ is the singular part of f_0. From iv) and ii) we deduce that this quantity is nonnegative. $\qquad\square$

COROLLARY 4.8 *Assume that $\Re c_j$ and $\Im a_{kj} = -\Im a_{jk} \in W^{1,\infty}(\Omega)$ for $k, j \in \{1, \ldots, d\}$. Let $w \in \mathbb{R}$ be any constant such that*

$$w + f_0 - \frac{1}{4\eta}\sum_{j=1}^d |\Im(c_j - b_j) + f_j|^2 \geq 0 \text{ on } \Omega,$$

where η is the ellipticity constant. Then, the semigroup $(e^{-t(A_{H_0^1} + w)})_{t \geq 0}$ is L^∞-contractive.

Proof. The following trivial inequality holds

$$\sum_{j=1}^d (\Im(c_j - b_j) + f_j)\xi_j \geq -(4\eta)^{-1}\sum_{j=1}^d |\Im(c_j - b_j) + f_j|^2 - \eta\sum_{j=1}^d \xi_j^2. \quad (4.17)$$

This together with the ellipticity property (4.2) show that conditions i)−iv) of the Theorem 4.7 hold with $a_0 + w$ in place of a_0. $\qquad\square$

Note that the conditions i)−iv) in Theorem 4.7 are necessary conditions for the L^∞-contractivity of $(e^{-tA_V})_{t \geq 0}$ for every V satisfying (4.12). Indeed, by Theorem 2.15, the L^∞-contractivity of $(e^{-tA_V})_{t \geq 0}$ implies (4.12) and

$$\Re a_V((|u| \wedge 1)\, \text{sign}\, u, (|u| - 1)^+\, \text{sign}\, u) \geq 0$$

for all $u \in V$. Since $H_0^1(\Omega) \subseteq V$, this inequality holds for $u \in H_0^1(\Omega)$. In particular, the L^∞-contractivity of $(e^{-tA_V})_{t\geq 0}$ implies the L^∞-contractivity of $(e^{-tA_{H_0^1}})_{t\geq 0}$. We conclude by applying Theorem 4.7.

4.3.1 Validity of L^∞-contractivity: real principal part

As previously, we assume the uniform ellipticity condition (U.Ell). Recall again that η denotes the ellipticity constant in (4.2).

THEOREM 4.9 *Assume that for all $j, k \in \{1, \ldots, d\}$, the coefficients a_{kj} are real-valued and that $\Re c_j = 0$. Then the following assertions are equivalent:*
i) V satisfies (4.12).
ii) There exists a constant $w \in \mathbb{R}$ such that the semigroup $(e^{-t(A_V+w)})_{t\geq 0}$ is L^∞-contractive.
In this case, ii) holds with any constant w such that

$$w + \Re a_0 - \frac{1}{4\eta} \sum_{j=1}^{d} |\Im(c_j - b_j)|^2 \geq 0 \ on \ \Omega. \tag{4.18}$$

In particular, $(e^{-tA_V})_{t\geq 0}$ is L^∞-contractive if $\Re a_0 - \frac{1}{4\eta}\sum_{j=1}^{d}|\Im(c_j - b_j)|^2 \geq 0$ on Ω.

Proof. Since $(e^{-t(A_V+w)})_{t\geq 0}$ is the semigroup associated with the form $\mathfrak{a}_V(.,.) + w(.;.)$, with domain V, it follows from Theorem 2.15 that ii) implies i).

Assume now that i) holds. We apply Theorem 4.6. Let φ_j be as in that theorem. It is enough to prove that

$$\sum_{j,k=1}^{d} \Re(a_{kj})\varphi_k\varphi_j + \sum_{j=1}^{d} \Im(c_j - b_j)\varphi_j + \Re a_0 + w \geq 0 \ on \ \Omega.$$

Using the ellipticity assumption (U.Ell), we see that this inequality holds for every w satisfying (4.18). □

In the next result, we assume that the coefficients satisfy the hypothesis of the previous theorem.

COROLLARY 4.10 *If A_V is subject to Dirichlet, Neumann, good Neumann, or mixed boundary conditions, then $(e^{-t(A_V+w)})_{t\geq 0}$ is L^∞-contractive for every w satisfying (4.18).*

This result follows from the previous theorem and the next proposition. Note that for Dirichlet boundary conditions, we had a better result in Corollary 4.8. Note also that by combining Corollary 4.10 with Corollary 4.3, one obtains that, if all the coefficients are real-valued, then $(e^{-t(A_V+w)})_{t\geq 0}$ is sub-Markovian for every $w \in \mathbb{R}$ such that $w + a_0 \geq 0$.

PROPOSITION 4.11 *The spaces* $H_0^1(\Omega)$, $H^1(\Omega)$, $\widetilde{H^1(\Omega)}$, *and*
$\overline{\{u_{|\Omega}, u \in C_c^\infty(\mathbb{R}^d \setminus \Gamma)\}}^{H^1(\Omega)}$ *satisfy (4.12). Moreover,*

$$D_j((1 \wedge |u|)\text{sign } u) = i\frac{\Im(\text{sign}(\overline{u})D_j u)}{|u|}\text{sign } u\chi_{\{|u|>1\}} + \chi_{\{|u|\leq 1\}}D_j u$$

for all $u \in H^1(\Omega)$ *and all* $j \in \{1, \ldots, d\}$.

Proof. We first consider the case $V = H^1(\Omega)$. Since the idea of proof is similar to that of Proposition 4.4, we only sketch the proof. Let

$$f_\varepsilon(t) := \begin{cases} \sqrt{(t-1)^2 + \varepsilon^2} - \varepsilon & \text{if} \quad t > 1, \\ 0 & \text{if} \quad t \leq 1. \end{cases}$$

As in the proof of Proposition 4.4, since f_ε has bounded derivative on $\mathbb{R}$, $f_\varepsilon(u) \in H^1(\Omega)$ for all real-valued $u \in H^1(\Omega)$. Again, by Proposition 4.4, we conclude that $f_\varepsilon(|u|) \in H^1(\Omega)$ for all $u \in H^1(\Omega)$. Letting $\varepsilon \to 0$ in the expression of $D_k f_\varepsilon(|u|)$ one obtains $(|u| - 1)^+ \in H^1(\Omega)$ with

$$D_k(|u| - 1)^+ = \chi_{\{|u|>1\}}D_k|u| = \chi_{\{|u|>1\}}\Re(\text{sign}(\overline{u})D_k u).$$

From this, it follows that $\frac{u}{\sqrt{|u|^2+\varepsilon}}(|u| - 1)^+ \in H^1(\Omega)$ for all $\varepsilon > 0$ and $u \in H^1(\Omega)$. Letting $\varepsilon \to 0$ in the expression of $D_k\left[\frac{u}{\sqrt{|u|^2+\varepsilon}}(|u| - 1)^+\right]$ and arguing as in the proof of Proposition 4.4, it follows that $(|u| - 1)^+\text{sign } u \in H^1(\Omega)$. The equality $u - (1 \wedge |u|)\text{sign } u = (|u| - 1)^+\text{sign } u$ gives the desired assertion.

The corresponding result for $V = H_0^1(\Omega)$ follows from the fact that $(1 \wedge |u|)\text{sign } u \in H_0^1(\Omega)$ for all $u \in C_c^\infty(\Omega)$ (since it is in $H^1(\Omega)$ and has compact support in Ω).

Finally, if $V = \overline{\{u_{|\Omega}, u \in C_c^\infty(\mathbb{R}^d \setminus \Gamma)\}}^{H^1(\Omega)}$, the result follows as in Proposition 4.4 by considering $(1 \wedge |u|)\text{sign } u$ in place of u^+. $\qquad\square$

If no regularity assumption is imposed on $\Re c_j$, the conclusions of Theorem 4.9 and Corollary 4.10 do not hold. This is shown in the first example below. The same example shows also that the situation for the adjoint semigroup $(e^{-tA_V^*})_{t\geq 0}$ is different, even when $V = H_0^1(\Omega)$.

Example 4.3.1   *Let $\Omega = (0,1)$ and $b,c \in L^\infty(\Omega)$. Consider the form*

$$\mathfrak{a}_V(u,v) = \int_0^1 [u'\overline{v'} + b(x)u'\overline{v} + c(x)u\overline{v'}]dx, \ D(\mathfrak{a}_V) = V.$$

1) Let $c(x) = \sqrt{x}$ and $V = H_0^1(\Omega)$. If there exists a constant w such that $(e^{-t(A_V+w)})_{t\geq 0}$ is L^∞-contractive, then by ii) of Theorem 4.7, we must have $w - c'$ is non-negative on $(0,1)$. This is not the case for any w. If we take $b(x) = \sqrt{x}$ and $c = 0$, then there exists no w such that $(e^{-t(A_V^*+w)})_{t\geq 0}$ is L^∞-contractive.

2) Let $V = H^1(\Omega)$, $b(x) = 1$ and $c(x) = 0$. By Corollary 4.10, the semigroup $(e^{-tA_V})_{t\geq 0}$ is L^∞-contractive. However, there exists no $w \in \mathbb{R}$ such that $(e^{-t(A_V^*+w)})_{t\geq 0}$ is L^∞-contractive (despite the fact the coefficients are in $W^{1,\infty}(\Omega)$). Indeed, if such w exists, then for every $0 \leq u \in V$,

$$\mathfrak{a}_V((u-1)^+, 1\wedge u) + w((u-1)^+; 1\wedge u) \geq 0. \tag{4.19}$$

This gives

$$\int_0^1 \chi_{\{u>1\}}u'dx + w\int_0^1 (u-1)^+dx \geq 0.$$

Applying this for ku in place of u and letting $k \to \infty$, we obtain

$$\int_0^1 u'dx + w\int_0^1 udx \geq 0.$$

In other words, for every non-negative C^1-function u on Ω,

$$w\int_0^1 udx \geq u(0) - u(1).$$

Applying this for the sequence $u_n = (1-x)^n$ yields $w \geq n+1$ for every $n \in \mathbb{N}$ which is not possible. Hence (4.19) cannot hold for any w.

3) Let now $V = \{u \in H^1(0,1), u(0) = \alpha u(1)\}$, where $\alpha \in \mathbb{C}$. Assume that $b = c = 0$. It is easy to check that V satisfies (4.12) if and only if $|\alpha| = 1$. Thus, the semigroup $(e^{-tA_V})_{t\geq 0}$ is L^∞-contractive if and only if $|\alpha| = 1$. In particular, for $\alpha = i$ one obtains a semigroup which is L^∞-contractive but not positive (it is not even real) and for $\alpha = 2$, the semigroup is positive but not L^∞-contractive.

4.3.2 Absence of L^∞-contractivity: complex principal part

We consider in this section the form $\mathfrak{a}_V$ given by the principal part, that is,

$$\mathfrak{a}_V(u,v) = \sum_{k,j=1}^d \int_\Omega a_{kj} D_k u \overline{D_j v} dx, \; D(\mathfrak{a}_V) = V. \tag{4.20}$$

We assume as previously that the uniform ellipticity (U.Ell) holds.

We have seen in the previous subsection that if V satisfies (4.12) and if the coefficients a_{kj} are real, then the semigroup $(e^{-t A_V})_{t \geq 0}$ is L^∞-contractive. This paragraph is devoted to the necessity of having the a_{kj} to be real-valued. More precisely, we study the question whether the L^∞-contractivity of $(e^{-t A_V})_{t \geq 0}$ implies that

$$\mathfrak{a}_V(u,v) = \sum_{k,j=1}^d \int_\Omega \Re(a_{kj}) D_k u \overline{D_j v} dx \text{ for all } u,v \in V, \tag{4.21}$$

which entails that the form $\mathfrak{a}_V$ (and hence the operator A_V) has real-valued coefficients.

For Dirichlet boundary conditions, the answer follows from Theorem 4.7.

COROLLARY 4.12 *Suppose that the form $\mathfrak{a}_V$ is given by (4.20). The following assertions are equivalent:*
i) $(e^{-t A_{H_0^1}})_{t \geq 0}$ *is L^∞-contractive.*
ii) $(e^{-t A_{H_0^1}})_{t \geq 0}$ *is real.*
iii) $\mathfrak{a}_{H_0^1(\Omega)}(u,v) = \sum_{k,j} \int_\Omega \Re(a_{kj}) D_k u D_j \overline{v}, \text{for all } u,v \in H_0^1(\Omega).$
iv) $\Im(a_{kj} + a_{jk}) = 0, 1 \leq k, j \leq d$ *and* $\sum_j D_j(\Im a_{kj}) = 0$ *in* $(C_C^\infty(\Omega))'$
for $1 \leq k \leq d.$

Proof. The equivalence i) $\Longleftrightarrow$ iv) follows from Theorem 4.7. The implication iv) $\Longrightarrow$ iii) follows from the equality

$$\int_\Omega \sum_{k,j} (\Im a_{kj}) D_k u D_j \bar{x}$$

$$= -\int_\Omega \bar{v} \left[\sum_{k,j} D_j(\Im a_{kj}) D_k u + \frac{1}{2} \sum_{k,j} \Im(a_{kj} + a_{jk}) D_k D_j u \right] dx$$

for all $u,v \in C_c^\infty(\Omega)$. iii) $\Longrightarrow$ i) follows from Theorem 4.7. Finally, ii) $\Longleftrightarrow$ iii) follows from Proposition 4.1. □

We consider now more general boundary conditions. Recall that condition (4.12) is necessary for L^∞-contractivity of $(e^{-t A_V})_{t \geq 0}$. Thus, we consider only spaces V that satisfy (4.12).

PROPOSITION 4.13 *Suppose that the form $\mathfrak{a}_V$ is given by (4.20) and suppose that $a_{kj} = a_{jk}$ for $j, k \in \{1, \ldots, d\}$. If $(e^{-tA_V})_{t\geq0}$ is L^∞-contractive, then a_{kj} are real-valued for every j and k. In particular, the form $\mathfrak{a}_V$ satisfies (4.21).*

Proof. As explained after the proof of Corollary 4.8, if $(e^{-tA_V})_{t\geq0}$ is L^∞-contractive, then the same holds for $(e^{-tA_{H_0^1}})_{t\geq0}$. Therefore, condition iv) of Corollary 4.12 and the symmetry assumption $a_{kj} = a_{jk}$ give the proposition.   □

This result answers the above question for general boundary conditions V. However, we have assumed a rather restrictive condition on the coefficients. For general coefficients, we will see later that the result does not hold for general V.

We assume that V satisfies the following property:

$$u \in V \Rightarrow (\Re u)^+, (1 \wedge |u|)\mathrm{sign}\, u \in V. \tag{4.22}$$

We have

THEOREM 4.14 *Assume that the form $\mathfrak{a}_V$ is given by (4.20) and assume that V satisfies (4.22). If the semigroup $(e^{-tA_V})_{t\geq0}$ is L^∞-contractive, then $\mathfrak{a}_V$ satisfies (4.21).*

Proof. We show that the L^∞-contractivity of $(e^{-tA_V})_{t\geq0}$ implies that

$$\sum_{k,j=1}^{d} \int_\Omega (\Im a_{kj}) D_k u D_j \bar{v}\, dx = 0 \text{ for all } u, v \in V. \tag{4.23}$$

It follows from (4.22) that $\Re u$, $\Im u \in V$ for all $u \in V$. Hence, it is enough to prove (4.23) for all real $u, v \in V$.

Since $u^+, u^- \in V$ for all real $u \in V$, it is enough to prove (4.23) for $u, v \in V$ such that $u \geq 0$ and $v \geq 0$.

Assume for a moment that we have established (4.23) for non-negative $u, v \in V \cap L^\infty(\Omega, \mathbb{R})$. Let $0 \leq u, v \in V$. It follows from (4.22) that $t \wedge u$ and $s \wedge v \in V$ for all $t, s \in \mathbb{R}^+$. Thus we have

$$\sum_{k,j=1}^{d} \int_\Omega (\Im a_{kj}) D_k(t \wedge u) D_j(s \wedge v) dx = 0.$$

In other words,

$$\sum_{k,j=1}^{d} \int_\Omega (\Im a_{kj}) D_k u\, \chi_{\{u<t\}} D_j v chi_{\{v<s\}} dx = 0.$$

Letting $t \to +\infty$ and $s \to +\infty$, we obtain (4.23) for u and v. This shows that it is enough to prove (4.23) for non-negative $u, v \in V \cap L^\infty(\Omega, \mathbb{R})$.
It follows from (4.22) and Theorems 4.2 and 4.9 that the semigroup associated with the symmetric form

$$\mathfrak{b}(u, v) = \sum_{k=1}^{d} \int_\Omega D_k u \overline{D_k v} dx, \quad D(\mathfrak{b}) = V$$

is sub-Markovian. On the other hand, the function $p(s) = \frac{1}{2} (e^{is} - 1)$ is a normal contraction on $\mathbb{R}$. Theorems 2.25 and 2.28 assert that $p(V) \subseteq V$ and $V \cap L^\infty(\Omega, \mathbb{C})$ is an algebra for the standard multiplication of functions. Thus, for $0 \le u, v \in V \cap L^\infty(\Omega, \mathbb{R})$, we write $ue^{iv} = 2up(v) + u$ and deduce that $ue^{iv} \in V$.
Now we apply Theorem 2.15 with $ue^{iv} \in V$ in place of u. We obtain

$$\Re \sum_{k,j=1}^{d} \int_\Omega a_{kj} \, D_k((1 \wedge u)e^{iv}) \, D_j((u-1)^+ e^{-iv}) dx \ge 0.$$

Since $D_k(1 \wedge u) = D_k u \chi_{\{u<1\}}$ and $D_k(u-1)^+ = D_k u \chi_{\{u>1\}}$, we obtain

$$\sum_{k,j=1}^{d} \int_\Omega \Re(a_{kj}) D_k v D_j v (u-1)^+ dx$$

$$\ge \sum_{k,j=1}^{d} \int_\Omega \Im(a_{kj}) D_j u D_k v \chi_{\{u>1\}} dx.$$

We apply the same arguments to ue^{-iv} to obtain

$$\sum_{k,j=1}^{d} \int_\Omega \Re(a_{kj}) D_k v D_j v (u-1)^+ dx$$

$$\ge \left| \sum_{k,j=1}^{d} \int_\Omega \Im(a_{kj}) D_j u D_k v \chi_{\{u>1\}} dx \right|.$$

Applying this inequality with $\frac{u}{\varepsilon}$ (with $\varepsilon > 0$) instead of u and letting $\varepsilon \to 0$, we obtain by the help of the dominated convergence theorem

$$\sum_{k,j=1}^{d} \int_\Omega \Re(a_{kj}) D_k v D_j v dx \ge \left| \sum_{k,j=1}^{d} \int_\Omega \Im(a_{kj}) D_j u D_k v dx \right|. \tag{4.24}$$

Replacing v by λv in (4.24) and letting $\lambda \to 0$ yields

$$\sum_{k,j=1}^{d} \int_{\Omega} (\Im a_{kj}) D_j u D_k v \, dx = 0,$$

which is the previous assertion.   □

The next example shows that the assumption (4.22) cannot be omitted in the previous theorem.

Example 4.3.2   *Let $\Omega = (0,1) \times (0,1)$ and*

$$V_0 = \{ v \in H^1(\Omega, \mathbb{C}), \ v(x,0) = v(x,1) = 0 \text{ and } v(0,y) = v(1,y) \}.$$

Let $\varphi \in C^\infty(\mathbb{R}^2, \mathbb{R})$ such that $\frac{\partial \varphi}{\partial y}(0,y) < \frac{\partial \varphi}{\partial y}(1,y)$ for all $y \in \mathbb{R}$. Set

$$V := e^{i\varphi} V_0.$$

Fix $0 < \theta < 1$ and consider the sesquilinear form defined on V by

$$a_V(u,v) = \iint_{\Omega} \left(\frac{\partial u}{\partial x} \frac{\partial \overline{v}}{\partial x} + \frac{\partial u}{\partial y} \frac{\partial \overline{v}}{\partial y} \right) dx dy$$

$$+ i\theta \iint_{\Omega} \left(\frac{\partial u}{\partial x} \frac{\partial \overline{v}}{\partial y} - \frac{\partial u}{\partial y} \frac{\partial \overline{v}}{\partial x} \right) dx dy.$$

The coefficients of a_V satisfy the uniform ellipticity assumption (U.Ell) and we have

PROPOSITION 4.15   *The semigroup $(e^{-t A_V})_{t \geq 0}$ associated with the form a_V is L^∞-contractive but the form a_V does not satisfy (4.21).*

Proof. Consider for $u, v \in V$

$$I(u,v) := \iint_{\Omega} \left(\frac{\partial u}{\partial x} \frac{\partial \overline{v}}{\partial y} - \frac{\partial u}{\partial y} \frac{\partial \overline{v}}{\partial x} \right) dx dy.$$

For $u, v \in V \cap C^2(\overline{\Omega}, \mathbb{C})$, we have by the Green-Riemann formula and the fact that $u(x,0) = u(x,1) = 0$,

$$I(u,v) = \iint_{\Omega} \left(\frac{\partial}{\partial x} \left(u \frac{\partial \overline{v}}{\partial y} \right) - \frac{\partial}{\partial y} \left(u \frac{\partial \overline{v}}{\partial x} \right) \right) dx dy$$

$$= \int_{\partial \Omega} u \frac{\partial \overline{v}}{\partial x} dx + u \frac{\partial \overline{v}}{\partial y} dy$$

$$= \int_0^1 \left(u(1,y) \frac{\partial \overline{v}}{\partial y}(1,y) - u(0,y) \frac{\partial \overline{v}}{\partial y}(0,y) \right) dy.$$

Since $ue^{-i\varphi}, ve^{-i\varphi} \in V_0$ it follows that

$$u(1,y)\, e^{-i\varphi(1,y)} = u(0,y)\, e^{-i\varphi(0,y)},$$

$$\frac{d}{dy}\left(v(1,y)\, e^{-i\varphi(1,y)}\right) = \frac{d}{dy}\left(v(0,y)\, e^{-i\varphi(0,y)}\right).$$

Hence

$$u(1,y)\,\frac{\partial\overline{v}}{\partial y}(1,y) - u(0,y)\,\frac{\partial\overline{v}}{\partial y}(0,y)$$

$$= i\, u(0,y)\,\overline{v}(0,u)\left(\frac{\partial\varphi}{\partial y}(0,y) - \frac{\partial\varphi}{\partial y}(1,y)\right).$$

Thus,

$$I(u,v) = i\int_0^1 u(0,y)\,\overline{v}(0,y)\left(\frac{\partial\varphi}{\partial y}(0,y) - \frac{\partial\varphi}{\partial y}(1,y)\right)\, dy. \qquad (4.25)$$

This extends to all $u, v \in V$ by a density argument. From this expression and the assumption $\frac{\partial\varphi}{\partial y}(0,y) - \frac{\partial\varphi}{\partial y}(1,y) < 0$, we deduce that $I(u,u) \neq 0$ for all u such that $u(0,.)$ is not the zero function on $]0,1[$. This proves that the form a_V cannot satisfy (4.21).

Now we show that $(e^{-tA_V})_{t\geq 0}$ is L^∞-contractive. For this, we apply Theorem 2.13.

Firstly, it is easy to check that $(|u| - 1)^+ \operatorname{sign} u \in V$ for all $u \in V$. Thus, we only have to check that

$$\Re a_V((|u| \wedge 1) \operatorname{sign} u, (|u| - 1)^+ \operatorname{sign} u) \geq 0.$$

Clearly, it is enough to prove that

$$\Re(iI\left((|u| \wedge 1) \operatorname{sign} u,\ (|u| - 1)^+ \operatorname{sign} u\right)) \geq 0.$$

But this follows from (4.25) since

$$iI\left((|u| \wedge 1) \operatorname{sign} u,\ (|u| - 1)^+ \operatorname{sign} u\right)$$

$$= \int_0^1 \left(|u(0,y)| - 1\right)^+ \left(\frac{\partial\varphi}{\partial y}(1,y) - \frac{\partial\varphi}{\partial y}(0,y)\right)\, dy$$

$$\geq 0.$$

This finishes the proof of the proposition.   □

We point out that the conclusion of Theorem 4.14 does not hold if we merely assume that there exists a positive constant w such that the semigroup $(e^{-t(A_V+w)})_{t\geq 0}$ is L^∞-contractive. This can be seen from Corollary 4.8.

In the two-dimension case, we have the following result.

PROPOSITION 4.16 *Assume that $d = 2$ and let $\mathfrak{a}_V$ be as in (4.20). Then there exists a non-negative constant w such that $(e^{-t(A_{H_0^1}+w)})_{t\geq 0}$ is L^∞-contractive if and only if $\Im a_{11} = \Im a_{22} = 0$ and $\Im a_{12} = -\Im a_{21} \in W^{1,\infty}(\Omega)$.*

Proof. It follows from Theorem 4.7 that the semigroup $(e^{-t(A_{H_0^1}+w)})_{t\geq 0}$ is L^∞-contractive if and only if $\Im a_{11} = \Im a_{22} = 0$, $\Im(a_{12} + a_{21}) = 0$, $\frac{\partial}{\partial x_2}\Im a_{12}, \frac{\partial}{\partial x_1}\Im a_{21} \in L^1_{\text{loc}}(\Omega)$ and

$$\sum_{k,j=1}^{2} \Re a_{kj}\xi_k\xi_j + \frac{\partial}{\partial x_2}(\Im a_{12})\,\xi_1 + \frac{\partial}{\partial x_1}(\Im a_{21})\,\xi_2 + w \geq 0 \text{ for all } \xi_1, \xi_2 \in \mathbb{R}.$$

(4.26)

Set $h = \Im a_{12} = -\Im a_{21}$ and let

$$C := \max_{\xi_1^2+\xi_2^2=1} \sum_{k,j=1}^{2} \Re a_{kj}\,\xi_k\,\xi_j, \text{ and } c := \min_{\xi_1^2+\xi_2^2=1} \sum_{k,j=1}^{2} \Re a_{kj}\,\xi_k\,\xi_j.$$

Clearly, (4.26) implies that

$$|\nabla h(x)|^2 \leq 4wC \text{ a.e. } x \in \Omega$$

and (4.26) holds if

$$|\nabla h(x)|^2 \leq 4wc \text{ a.e. } x \in \Omega.$$

This proves the proposition. $\square$

4.4 THE CONSERVATION PROPERTY

In this section we discuss the conservation property

$$e^{-tA_V}1 = 1 \text{ for all } t \geq 0.$$

(4.27)

Here 1 denotes the constant function with value 1. Clearly, this property cannot hold for all boundary conditions. To see this, one can consider the case of a bounded domain with smooth boundary and A_V the Laplacian with Dirichlet boundary conditions. In this case, $e^{-tA_V}1 \in H_0^1(\Omega)$ for all $t > 0$ and cannot coincide with 1.

For the same reason, the conservation property cannot hold for mixed boundary conditions (if Γ in (4.5) is not empty). We will see that it does hold for Neumann and good Neumann boundary conditions.

The notation and assumptions here are the same as in Section 4.1, that is, the form $\mathfrak{a}_V$ is given by

$$\mathfrak{a}_V(u,v) = \int_\Omega \left[\sum_{k,j=1}^d a_{kj} D_k u \overline{D_j v} + \sum_{k=1}^d (b_k \overline{v} D_k u + c_k u \overline{D_k v}) + a_0 u \overline{v} \right] dx$$

with domain

$$D(\mathfrak{a}_V) = V \text{ with } V = H^1(\Omega) \text{ or } = \widetilde{H^1(\Omega)}. \tag{4.28}$$

We assume again that the coefficients a_{kj}, b_k, c_k, and a_0 are bounded measurable (possibly complex-valued) functions and satisfy the uniform ellipticity condition (U.Ell). Let Ω be an open subset of $\mathbb{R}^d$.

THEOREM 4.17 *Let V be either $H^1(\Omega)$ or $\widetilde{H^1(\Omega)}$. Assume that*

$$\int_\Omega \left(\sum_j b_j D_j u + a_0 u \right) dx = 0 \text{ for all } u \in V \cap W^{1,1}(\Omega). \tag{4.29}$$

Assume in addition that the semigroup $(e^{-tA_V})_{t \geq 0}$ extends from $L^2(\Omega, \mathbb{C}) \cap L^1(\Omega, \mathbb{C})$ to a strongly continuous semigroup on $L^1(\Omega, \mathbb{C})$. Then

$$\int_\Omega e^{-tA_V} u \, dx = \int_\Omega u \, dx \text{ for all } t \geq 0, u \in L^2(\Omega, \mathbb{C}) \cap L^1(\Omega, \mathbb{C}). \tag{4.30}$$

In other words, if we also denote by $(e^{-tA_V})_{t \geq 0}$ the strongly continuous semigroup in $L^1(\Omega, \mathbb{C})$, then the adjoint $(e^{-tA_V^*})_{t \geq 0}$ satisfies the conservation property (4.27).

If there exists a constant w such that the semigroup $(e^{-t(A_V^* + w)})_{t \geq 0}$ is L^∞-contractive, then the semigroup $(e^{-tA_V})_{t \geq 0}$ extends from $L^2(\Omega, \mathbb{C}) \cap L^1(\Omega, \mathbb{C})$ to a strongly continuous semigroup on $L^1(\Omega, \mathbb{C})$ (see Chapter 2). In this case, one has only to check condition (4.29) to obtain (4.27) for the adjoint semigroup.

The assumptions of the previous theorem hold in the particular case where a_{kj} are real-valued functions and $b_k = c_k = a_0 = 0$ for $1 \leq k, j \leq d$.

In order to prove Theorem 4.17, we will need the following lemma.

LEMMA 4.18 *Suppose that $V = H^1(\Omega)$ or $V = \widetilde{H^1(\Omega)}$. Then there exist two constants C_1 and ω_1 such that for every $R > 0$, every $u \in L^2(\Omega, \mathbb{C})$ with support contained in $\{|x| \leq R\}$ and every $r > 0$, we have*

$$\left(\int_{\Omega \cap \{|x| \geq R+r\}} (|e^{-tA_V} u|^2 + \sum_j |D_j e^{-tA_V} u|^2) dx \right)^{1/2}$$

$$\leq C_1 (1 + t^{-1})^{1/2} e^{-r + \omega_1 t} \|u\|_2, \tag{4.31}$$

for all $t > 0$.

Proof. Assume first that $V = H^1(\Omega)$.

Set $\psi(x) := (|x| - R)^+ \wedge r$ and let $S(t) := e^\psi e^{-tAv} e^{-\psi}$. We have $\psi \in W^{1,\infty}(\mathbb{R}^d)$, $\psi(x) = r$ on the set $\{|x| \geq R + r\}$, $|D_j\psi| \leq 1$ a.e. on $\mathbb{R}^d$ for $j \in \{1, \ldots, d\}$, and $u = ue^{-\psi}$ on Ω. Hence

$$e^{-tAv} u = e^{-\psi} S(t) u$$
$$D_j(e^{-tAv} u) = e^{-\psi}(D_j S(t)u - S(t)u D_j\psi),$$

and the term in the left-hand side of (4.31) is bounded by

$$e^{-r}\left((1 + 2d)\|S(t)u\|_2^2 + 2\sum_j \|D_j S(t)u\|_2^2\right)^{1/2}. \qquad (4.32)$$

Now we estimate (4.32). Note that $S(t) = e^{-tB_V}$, where B_V is the operator associated with the form

$$\mathfrak{b}_V(f, g) = \mathfrak{a}_V(e^{-\psi} f, e^\psi g) \text{ for all } f, g \in V = H^1(\Omega).$$

The form $\mathfrak{b}_V$ has the same principal part as $\mathfrak{a}_V$. Hence there exist ω_1 and C such that

$$\|S(t)\|_{\mathcal{L}(L^2)} \leq e^{\omega_1 t}, \quad \|B_V S(t)\|_{\mathcal{L}(L^2)} \leq \frac{Ce^{\omega_1 t}}{t},$$

$$\frac{\eta}{2}\sum_j \|D_j f\|_2^2 \leq \Re \int B_V f \bar{f} dx + \omega_1 \|f\|_2^2 \text{ for all } f \in D(B_V).$$

From this and the bound (4.32) we obtain the lemma for the case $V = H^1(\Omega)$.

If $V = \widetilde{H^1(\Omega)}$, the proof is similar. It only remains to check that the form $\mathfrak{b}_V$ is well defined. That is, we need that

$$f \in \widetilde{H^1(\Omega)} \Rightarrow e^{-\psi} f, e^\psi f \in \widetilde{H^1(\Omega)}.$$

We show that for every $\phi \in W^{1,\infty}(\mathbb{R}^d)$ we have

$$f \in \widetilde{H^1(\Omega)} \Rightarrow \phi f \in \widetilde{H^1(\Omega)}. \qquad (4.33)$$

Let $f_n \in C_c^\infty(\mathbb{R}^d)$ be such that $f_{n|\Omega}$ converges to f in $H^1(\Omega)$. Let ρ_n be a regularizing sequence. Fix n. For each m, we have $(\phi * \rho_m)f_n \in C_c^\infty(\mathbb{R}^d)$ and this sequence converges in $H^1(\mathbb{R}^d)$ to ϕf_n. Since ϕf_n converges to ϕf in $H^1(\Omega)$ we deduce that $\phi f \in \widetilde{H^1(\Omega)}$. $\qquad \square$

Proof of Theorem 4.17. By a density argument, it is enough to prove (4.30) for $u \in L^2(\Omega, \mathbb{C})$ with compact support. Let $R > 0$ be such that the support of u is contained in $\{|x| < R\}$. Let $\rho \in C_c^\infty(\mathbb{R}^d)$ with $0 \le \rho \le 1$, $\rho = 1$ on $\{|x| \le 1\}$. Set $\rho_n(x) := \rho(\frac{x}{n})$ and $u(t) := e^{-tA_V} u$. For every $t > 0$, we have

$$I_n(t) := \int_\Omega (A_V u(t)) \rho_n \, dx = \mathfrak{a}_V(u(t), \rho_n)$$

$$= \int_\Omega \sum_j \left(\sum_k a_{kj} D_k u(t) + (c_j - b_j) u(t) \right) D_j \, \rho_n dx,$$

where we have used the fact that $\rho_n \in V$ and the following consequence of (4.29):

$$\int_\Omega \left(\sum_j b_j \rho_n D_j u(t) + a_0 u(t) \rho_n \right) dx = - \int_\Omega \sum_j b_j u(t) D_j \rho_n dx.$$

Since $D_j \rho_n(x) = \frac{1}{n} D_j \rho(\frac{x}{n}) = 0$ for $|x| \le n$, we have

$$|I_n(t)|$$

$$\le M \sum_j \|D_j \rho_n\|_2 \left(\int_{\Omega \cap \{|x| \ge n\}} \left(|u(t)|^2 + \sum_k |D_k u(t)|^2 \right) dx \right)^{1/2},$$

where $M = \max(\|a_{kj}\|_\infty, \|c_j - b_j\|_\infty)$.

Now using Lemma 4.18, we see that for $n > R$

$$|I_n(t)| \le M \, C_1 \sum_j \|D_j \rho\|_2 \, n^{\frac{d}{2}-1} e^{-n} (1 + t^{-1})^{\frac{1}{2}} e^{R + \omega_1 t} \|u\|_2.$$

But $I_n(t) = -\frac{d}{dt} \int_\Omega u(t) \rho_n dx$ and $u(t)$ is continuous from $[0, \infty[$ into $L^1(\Omega, \mathbb{C})$ by our assumption on the semigroup. Hence for $0 < s < t$, we have

$$\int_\Omega u(t) dx - \int_\Omega u(s) dx = \lim_{n \to \infty} \left(\int_\Omega u(t) \rho_n dx - \int_\Omega u(s) \rho_n dx \right)$$

$$= \lim_{n \to \infty} \int_s^t I_n(\tau) d\tau = 0.$$

Letting $s \to 0$, we obtain (4.30). $\square$

As an application of Theorem 4.17, we have the following stronger version of Theorem 4.14 which in turn is limited to Neumann and good Neumann boundary conditions.

THEOREM 4.19 *Let $\mathfrak{a}_V$ be as above and assume that V is either $H^1(\Omega)$ or $\widetilde{H^1(\Omega)}$. Assume that*

$$\int_\Omega \left(\sum_j c_j D_j u + a_0 u \right) dx = 0 \ for \ all \ u \in V \cap W^{1,1}(\Omega). \qquad (4.34)$$

If the semigroup $(e^{-tA_V})_{t \geq 0}$ is L^∞-contractive, then

$$\mathfrak{a}_V(u, v) = \int_\Omega \sum_{k,j} \Re(a_{kj}) D_k u \overline{D_j v} dx + \sum_j \Re(b_j - c_j) \overline{v} D_j u \, dx \qquad (4.35)$$

for all $u, v \in V$.

We first show the following elementary lemma.

LEMMA 4.20 *Let $T \in \mathcal{L}(L^\infty(\Omega, \mathbb{C}))$ be a contraction operator such that $T1 = 1$. Then T is real; that is, $Tu \in L^\infty(\Omega, \mathbb{R})$ for all $u \in L^\infty(\Omega, \mathbb{R})$.*

Proof. Let $u \in L^\infty(\Omega, \mathbb{R})$ with $\|u\|_\infty \leq 1$. Fix $\lambda \in \mathbb{R}$. Using the assumptions on T, we have

$$\begin{aligned}
(1 - \lambda \Im Tu)^2 + \lambda^2 (\Re Tu)^2 &= |1 + i\lambda \, Tu|^2 \\
&= |T(1 + i\lambda u)|^2 \\
&\leq \|1 + i\lambda u\|_\infty^2 \leq 1 + \lambda^2.
\end{aligned}$$

Hence

$$\lambda^2 (1 - |Tu|^2) + 2\lambda \Im Tu \geq 0 \text{ for all } \lambda \in \mathbb{R}.$$

This implies that $\Im Tu = 0$. $\qquad\qquad\qquad\qquad\qquad\qquad\qquad\qquad \square$

Proof of Theorem 4.19. If the semigroup $(e^{-tA_V})_{t \geq 0}$ is L^∞-contractive, then the adjoint semigroup $(e^{-tA_V^*})_{t \geq 0}$ extends to a strongly continuous contraction semigroup on $L^1(\Omega, \mathbb{C})$ (see Chapter 2). Using (4.34) we can apply Theorem 4.17 and obtain that $e^{-tA_V} 1 = 1$ for all $t > 0$. The previous lemma implies then that $(e^{-tA_V})_{t \geq 0}$ is real. By Proposition 4.1 it follows that for all $u, v \in V$

$$\begin{aligned}
\mathfrak{a}_V(u, v) = \int_\Omega \Bigg[\sum_{k,j=1}^d \Re(a_{kj}) D_k u \overline{D_j v} + \sum_{k=1}^d (\Re(b_k) \overline{v} D_k u + \Re(c_k) u \overline{D_k v}) \\
+ \Re(a_0) u \overline{v} \Bigg] dx.
\end{aligned}$$

But if $u, v \in V \cap L^\infty(\Omega, \mathbb{R})$ we have $u.v \in V$ and it follows from (4.34) that for all $u, v \in V \cap L^\infty(\Omega, \mathbb{R})$

$$\int_\Omega \left[\sum_j \Re c_j(uD_jv + vD_ju) + \Re a_0 uv \right] dx$$

$$= \Re \int_\Omega \left[\sum_j c_j D_j(uv) + a_0 uv \right] dx$$

$$= 0.$$

This implies (4.35) for $u, v \in V \cap L^\infty(\Omega, \mathbb{R})$. As in the proof of Theorem 4.14, by applying this to $u \wedge t$ and $u \wedge s$ and letting $t \to \infty$ and $s \to \infty$, we extend this to all real $u, v \in V$. Now since $\Re u \in V$ for all $u \in V$, we obtain (4.35) for all $u, v \in V$. $\qquad\square$

4.5 DOMINATION

The aim of this section is to study the domination property for semigroups generated by uniformly elliptic operators. For general semigroups associated with sesquilinear forms, criteria for the domination property are given in Chapter 2. We apply some of those criteria to the semigroups $(e^{-tA_V})_{t \geq 0}$. As previously, we define the form $\mathfrak{a}_V$ by

$$\mathfrak{a}_V(u, v) = \int_\Omega \left[\sum_{k,j=1}^d a_{kj} D_k u \overline{D_j v} + \sum_{k=1}^d (b_k \overline{v} D_k u + c_k u \overline{D_k v}) + a_0 u \overline{v} \right] dx.$$

The domain of $\mathfrak{a}_V$ is V. Here Ω is any open subset of $\mathbb{R}^d$ and we assume again that the coefficients satisfy the uniform ellipticity condition (U.Ell).

THEOREM 4.21 *Assume that a_{kj}, b_k, c_k, and a_0 are real-valued for all $j, k \in \{1, \ldots, d\}$. The following assertions are equivalent:*
1) $u \in V$ implies $(\Re u)^+ \in V$.
2) The semigroup $(e^{-tA_{H_0^1}})_{t \geq 0}$ is dominated by the semigroup $(e^{-tA_V})_{t \geq 0}$. That is, the following inequality holds for all $t > 0$ and all $f \in L^2(\Omega, \mathbb{C})$:

$$|e^{-tA_{H_0^1}} f| \leq e^{-tA_V} |f|. \tag{4.36}$$

Now let $V = \overline{\{u_{|\Omega}, u \in C_c^\infty(\mathbb{R}^d \setminus \Gamma)\}}^{H^1(\Omega)}$, where Γ is a closed subset of $\partial\Omega$. Then $(e^{-tA_V})_{t \geq 0}$ is dominated by the semigroups $(e^{-tA_{H^1(\Omega)}})_{t \geq 0}$ and $(e^{-tA_{\widetilde{H^1(\Omega)}}})_{t \geq 0}$.

Proof. By Theorem 4.2, the positivity of the semigroup $(e^{-tA_V})_{t\geq 0}$ implies assertion 1). In particular, 2) $\Longrightarrow$ 1).

Assume now that assertion 1) is satisfied. Again, Theorem 4.2 shows that the semigroup $(e^{-tA_V})_{t\geq 0}$ is positive. For the same reason, the semigroup $(e^{-tA_{H_0^1}})_{t\geq 0}$ is also positive. To establish the domination, we show that $H_0^1(\Omega)$ is an ideal of V and apply Corollary 2.22.

By Proposition 2.23, it suffices to show that if $0 \leq v \leq u, u \in H_0^1(\Omega)$ and $v \in V$, then $v \in H_0^1(\Omega)$.

Let $(u_n)_n \in C_c^\infty(\Omega)$ be such that $(u_n)_n$ converges to u in $H^1(\Omega)$. Set $v_n = \inf(|u_n|, v)$. For each n, v_n has compact support in Ω and hence $v_n \in H_0^1(\Omega)$. But $(v_n)_n$ converges to v in $H^1(\Omega)$ (this follows from the continuity of the absolute value in $H^1(\Omega)$). Thus, $v \in H_0^1(\Omega)$. This shows that $H_0^1(\Omega)$ is an ideal of V and assertion 2) holds.

In order to show the last claim of the theorem we have to show, as above, that V is an ideal of $H^1(\Omega)$ and of $\widetilde{H^1(\Omega)}$. Since $\widetilde{H^1(\Omega)} \subseteq H^1(\Omega)$, it is enough to prove that V is an ideal of $H^1(\Omega)$. Let $0 \leq v \leq u$ with $u \in V$ and $v \in H^1(\Omega)$. Let $u_n \in C_c^\infty(\mathbb{R}^d \setminus \Gamma)$ be a sequence which converges to u in $H^1(\Omega)$. Let $v_n = \inf(u_n, v)$. As mentioned above, v_n converges to v in $H^1(\Omega)$ and thus it suffices to prove that $v_n \in V$ for each fixed n. Let $\rho_n \in C_c^\infty(\mathbb{R}^d)$ be a regularizing sequence. The sequence $\rho_m * v_n$ converges in $H^1(\Omega)$ to v_n as $m \to \infty$. Since v_n has compact support contained in $\mathbb{R}^d \setminus \Gamma$, it follows that for m large, $\rho_m * v_n$ is in $C_c^\infty(\mathbb{R}^d \setminus \Gamma)$ and this proves that $v_n \in V$. $\square$

It follows from the above result that the semigroup associated with the Dirichlet boundary conditions is the smallest semigroup (for the domination property). However, the semigroup associated with the Neumann boundary conditions does not dominate all the others. Indeed, it is easy to construct subspaces V which are not ideals of $H^1(\Omega)$. For example

$$V = \{u \in H^1(0,1), u(0) = u(1)\}$$

is not an ideal of $H^1(0,1)$. To see this, pick a function $v \in H^1(0,1)$ with $0 \leq v \leq 1$ and $v(0) \neq u(1)$. Then $v \notin V$ but $0 \leq v \leq 1 \in V$.

Now let V be a closed subspace of $H^1(\Omega)$ which contains $H_0^1(\Omega)$, and let $\phi \in W^{1,\infty}(\Omega)$. Define

$$W = e^{i\phi}V.$$

We have

PROPOSITION 4.22 *Assume that a_{kj}, b_k, c_k, and a_0 are real-valued for all $j, k \in \{1, \ldots, d\}$. Assume also that $(\Re u)^+ \in V$ for all $u \in V$. Then, the semigroup $(e^{-tA_W})_{t\geq 0}$ is dominated by $(e^{-tA_V})_{t\geq 0}$.*

Proof. By Corollary 2.22 (or Theorem 2.21) it is enough to prove that W is an ideal of V. So let $u \in W$ and $v \in V$ such that $|v| \leq |u|$. We need to show that $v\text{sign}\, u \in W$. Let now $u_0 \in V$ such that $u = e^{i\phi}u_0$. Thus $|v| \leq |u_0|$. The assumption on V implies that the semigroup $(e^{-tA_V})_{t \geq 0}$ is positive and hence V is an ideal of itself (cf. Proposition 2.20). This implies that $v\text{sign}\, u_0 \in V$. Now $v\text{sign}\, u = v\text{sign}\, u_0 e^{i\phi} \in W$. $\square$

The next result shows that for Dirichlet boundary conditions, the semigroup increases as the domain increases. To be more precise, let $\Omega_1 \subseteq \Omega_2$, where Ω_i are open sets of $\mathbb{R}^d$. Define now the form

$$\mathfrak{a}_{H_0^1(\Omega_2)}(u, v) = \int_{\Omega_2} \sum_{k,j=1}^{d} a_{kj}D_k u \overline{D_j v} + \sum_{k=1}^{d}(b_k \overline{v} D_k u + c_k u \overline{D_k v}) + a_0 u \overline{v}\, dx,$$

where the coefficients satisfy (U.Ell) in Ω_2. We can now define the form $\mathfrak{a}_{H_0^1(\Omega_1)}$ by the same expression by taking the restrictions of the coefficients a_{kj}, b_k, c_k, a_0 to Ω_1. Then we have

PROPOSITION 4.23 *Assume that $a_{kj}, b_k, c_k,$ and a_0 are real-valued for all $j, k \in \{1, \ldots, d\}$. Then the semigroup $(e^{-tA_{H_0^1(\Omega_1)}})_{t \geq 0}$ is dominated by $(e^{-tA_{H_0^1(\Omega_2)}})_{t \geq 0}$.*

Proof. Since $\Omega_1 \subseteq \Omega_2$, the form $\mathfrak{a}_{H_0^1(\Omega_1)}$ is seen as a non-densely defined form on $L^2(\Omega_2, \mathbb{C})$. Applying the results of Sections 2.3 and 2.6 (for non-dense forms), it is enough to prove that $H_0^1(\Omega_1)$ is an ideal of $H_0^1(\Omega_2)$. In order to do this, consider $0 \leq v \leq u$, where $u \in H_0^1(\Omega_1)$ and $v \in H_0^1(\Omega_2)$. Let $(u_n) \in C_c^\infty(\Omega_1)$ which converges to u in $H^1(\Omega_1)$ and $(v_n) \in C_c^\infty(\Omega_2)$ which converges to v in $H^1(\Omega_2)$. The sequence $\inf(u_n, v_n)$ converges to v in $H^1(\Omega_2)$, and for each n, $\inf(u_n, v_n)$ has a compact support contained in Ω_1. This implies that $\inf(u_n, v_n) \in H_0^1(\Omega_1)$ and hence $v \in H_0^1(\Omega_1)$. $\square$

We consider now elliptic operators with Robin boundary conditions. Assume that Ω is a bounded smooth domain of $\mathbb{R}^d$ and let $\mathfrak{b}_{V,\alpha}$ be the form defined in (4.7). We denote by $B_{V,\alpha}$ the operator associated with the form $\mathfrak{b}_{V,\alpha}$ and set $A_V := B_{V,0}$ (i.e., the operator corresponding to the case $\alpha = 0$ on $\partial\Omega$).

PROPOSITION 4.24 *Assume that $a_{kj}, b_k, c_k,$ and a_0 are real-valued for all $j, k \in \{1, \ldots, d\}$. Assume also that $(\Re u)^+ \in V$ for all $u \in V$ and that $\alpha \geq 0$ on $\partial\Omega$. Then the semigroup $(e^{-tB_{V,\alpha}})_{t \geq 0}$ is dominated by $(e^{-tA_V})_{t \geq 0}$.*

Proof. By assumption on V, the semigroups $(e^{-tA_V})_{t \geq 0}$ and $(e^{-tB_{V,\alpha}})_{t \geq 0}$ are positive (see Theorem 4.2; the proof of positivity of $(e^{-tB_{V,\alpha}})_{t \geq 0}$ is similar to that of $(e^{-tA_V})_{t\geq 0})$. It follows from Proposition 2.20 that V is an ideal of itself. Thus we only have to check that

$$\mathfrak{b}_{V,\alpha}(u,v) \geq \mathfrak{a}_V(u,v)$$

for all $0 \leq u, v \in V$ and apply Theorem 2.24. This inequality is satisfied since α is non-negative on $\partial\Omega$. $\square$

An immediate consequence of the previous proposition and Theorem 4.9 is that if V satisfies in addition (4.12) and $c_k = 0$ for all $k \in \{1, \dots, d\}$, then the semigroup $(e^{-t(B_{V,\alpha}+w)})_{t\geq 0}$ is L^∞-contractive for some constant w.

We have seen above that many results on domination and contractivity hold for elliptic operators with real-valued coefficients. In order to consider operators with complex-valued coefficients, we may ask whether one can dominate the associated semigroup when the coefficients are complex-valued by the semigroup of a similar operator having only real-valued coefficients. We show that it is possible to prove such domination under an appropriate condition on the imaginary parts of the principal coefficients a_{kj}.

We first consider the case of Dirichlet boundary conditions and then show how the domination result extends to other boundary conditions.

We suppose that

$$\Im(a_{kj} + a_{jk}) = 0, \quad f_k := \sum_{j=1}^{d} D_j(\Im a_{kj}) \in L^1_{loc}(\Omega) \tag{4.37}$$

for all $k, j \in \{1, \dots, d\}$, where $D_j \Im a_{kj}$ is taken in the distributional sense.

Let η be the ellipticity constant of (a_{kj}) and set

$$m(x) := \frac{\displaystyle\sum_{k=1}^{d}[f_k + \Im(c_k - b_k)]^2}{4\eta}. \tag{4.38}$$

Define the form

$$\mathfrak{b}_{H_0^1}(u,v) = \int_\Omega \left[\sum_{k,j=1}^{d} \Re(a_{kj})D_k u \overline{D_j v} + \sum_{k=1}^{d}(\Re(b_k)\overline{v}D_k u + \Re(c_k)u\overline{D_k v}) \right.$$

$$\left. + \Re(a_0)u\overline{v} \right] dx - \int_\Omega mu\overline{v}dx$$

$$=: \mathcal{E}_{H_0^1}(u,v) - \int_\Omega mu\overline{v}dx.$$

We assume that the potential function m is form-bounded with respect to the form $\mathcal{E}_{H_0^1}$, with relative bound < 1, that is, there exist $\beta \in \mathbb{R}$ and $0 \leq \alpha < 1$ such that

$$\int_\Omega m(x)|u|^2 dx \leq \beta \int_\Omega |u|^2 dx + \alpha \Re \mathcal{E}_{H_0^1}(u, u) \text{ for all } u \in H_0^1(\Omega). \quad (4.39)$$

It follows from Theorem 1.19 that the form $\mathfrak{b}_{H_0^1}$, with domain $D(\mathfrak{b}_{H_0^1}) = H_0^1(\Omega)$, is well defined and there exists a constant $w \in \mathbb{R}$ such $\mathfrak{b} + w$ is accretive, continuous, and closed. Let us denote by $B_{H_0^1}$ the operator associated with the form $\mathfrak{b}_{H_0^1}$. Recall that $A_{H_0^1}$ is the operator associated with the form

$$\mathfrak{a}_{H_0^1}(u, v) = \int_\Omega \left[\sum_{k,j=1}^d a_{kj} D_k u \overline{D_j v} + \sum_{k=1}^d (b_k \overline{v} D_k u + c_k u \overline{D_k v}) + a_0 u \overline{v} \right] dx,$$

where the coefficients a_{kj}, b_k, c_k, a_0 are complex-valued and satisfy (U.Ell).

We have

THEOREM 4.25 *Assume that (4.37) and (4.39) hold. Then, for every $t \geq 0$ and every $f \in L^2(\Omega)$*

$$|e^{-tA_{H_0^1}} f| \leq e^{-tB_{H_0^1}} |f|.$$

We first prove the following lemma.

LEMMA 4.26 *Let $u, v \in H^1(\Omega)$ be such that $u(x)\overline{v(x)} \geq 0$ (for* a.e. *$x \in \Omega$). We have for each $k \in \{1, \ldots, d\}$*
1) $\Im(\overline{v} D_k u) = |v| \Im(\text{sign}(\overline{u}) D_k u)$.
2) $|v| \Im(\text{sign}(\overline{u}) D_k u) = |u| \Im(\text{sign}(\overline{v}) D_k v)$.

Proof. Let u and v be as in the lemma. Since $\chi_{\{u=0\}} D_k u = 0$, we have

$$\overline{v} D_k u = \overline{v} D_k u \frac{v\overline{u}}{|u||v|} \chi_{\{u \neq 0\}} \chi_{\{v \neq 0\}} = |v| D_k u \frac{\overline{u}}{|u|} \chi_{\{u \neq 0\}}.$$

Assertion 1) follows then by taking the imaginary parts.

In order to prove assertion 2), we write

$$|v|u = |v|u \frac{v\overline{u}}{|u||v|} \chi_{\{u \neq 0\}} \chi_{\{v \neq 0\}} = |u|v.$$

Hence,

$$u D_k |v| + |v| D_k u = v D_k |u| + |u| D_k v.$$

We multiply each term by $\mathrm{sign}\overline{u} = \frac{\overline{u}}{|u|}\chi_{\{u\neq 0\}}$ and take the imaginary parts to obtain

$$|v|\Im(\mathrm{sign}(\overline{u})D_k u) = \Im(\overline{u}\chi_{\{u\neq 0\}}D_k v) = \Im(\overline{u}D_k v).$$

This together with assertion 1) (with u in place of v and vice-versa) gives 2). □

Proof of Theorem 4.25. Since $H_0^1(\Omega)$ is an ideal of itself (see, e.g. the proof of Theorem 4.21), it suffices to prove that

$$\Re\mathfrak{a}_{H_0^1}(u,v) \geq \mathfrak{b}_{H_0^1}(u,v) \text{ for } u,v \in H_0^1(\Omega) \text{ with } u\overline{v} \geq 0 \qquad (4.40)$$

and then apply Theorem 2.21.

Let $u,v \in H_0^1(\Omega)$ be such that $u\overline{v} \geq 0$. We have

$$D_k u \overline{D_j v} = D_k u \frac{\overline{u}}{|u|}\chi_{\{u\neq 0\}}\overline{D_j v}\frac{v}{|v|}\chi_{\{v\neq 0\}}.$$

Hence

$$I_1 := \Re \sum_{k,j=1}^{d} \int_\Omega \Re(a_{kj})D_k u\overline{D_j v}dx$$

$$= \sum_{k,j=1}^{d} \int_\Omega \Re(a_{kj})\Re(\mathrm{sign}(\overline{u})D_k u)\Re(\mathrm{sign}(\overline{v})D_j v)dx$$

$$+ \sum_{k,j=1}^{d} \int_\Omega \Re(a_{kj})\Im(\mathrm{sign}(\overline{u})D_k u)\Im(\mathrm{sign}(\overline{v})D_j v)dx$$

$$= \sum_{k,j=1}^{d} \int_\Omega \Re(a_{kj})\Re(\mathrm{sign}(\overline{u})D_k u)\Re(\mathrm{sign}(\overline{v})D_j v)dx$$

$$+ \sum_{k,j=1}^{d} \int_\Omega \Re(a_{kj})\Im(\mathrm{sign}(\overline{u})D_k u)\Im(\mathrm{sign}(\overline{u})D_j u)\frac{|v|}{|u|}\chi_{\{u\neq 0\}}dx,$$

where we have used Lemma 4.26 in order to write the last equality. By Proposition 4.4,

$$D_k|u| = \Re(\mathrm{sign}(\overline{u})D_k u) \text{ for all } u \in H^1(\Omega).$$

This gives

$$I_1 = \sum_{k,j=1}^{d} \int_{\Omega} \Re(a_{kj}) D_k |u| D_j |v| dx \qquad (4.41)$$

$$+ \sum_{k,j=1}^{d} \int_{\Omega} \Re(a_{kj}) \Im(\text{sign}(\overline{u}) D_k u) \Im(\text{sign}(\overline{u}) D_j u) \frac{|v|}{|u|} \chi_{\{u \neq 0\}} dx.$$

We have now to handle the imaginary part. For arbitrary $u, v \in C_c^{\infty}(\Omega)$,

$$I_2 := \Re \sum_{k,j=1}^{d} \int_{\Omega} i \Im(a_{kj}) D_k u \overline{D_j v} dx$$

$$= -\Im \sum_{k,j=1}^{d} \int_{\Omega} \Im(a_{kj}) D_k u \overline{D_j v} dx$$

$$= \Im \sum_{k,j=1}^{d} \int_{\Omega} D_j \Im(a_{kj}) \overline{v} D_k u dx + \Im \sum_{k,j=1}^{d} \int_{\Omega} \Im(a_{kj}) \overline{v} D_k D_j u dx.$$

By assumptions $\Im(a_{kj} + a_{jk}) = 0$, hence

$$I_2 = \sum_{k=1}^{d} \int_{\Omega} f_k \Im(\overline{v} D_k u) dx. \qquad (4.42)$$

Using the Cauchy-Schwarz inequality and assumption (4.39), we see that (4.42) extends to all $u, v \in H_0^1(\Omega)$.

Assume again that $u\overline{v} \geq 0$. By Lemma 4.26, (4.42) becomes

$$I_2 = \sum_{k=1}^{d} \int_{\Omega} f_k \Im(\text{sign}(\overline{u}) D_k u) |v| dx. \qquad (4.43)$$

We come now to the terms of order 1. We have for $u, v \in H_0^1(\Omega)$ with $u\overline{v} \geq 0$,

$$I_3 = \Re \sum_{k=1}^{d} \int_{\Omega} b_k \overline{v} D_k u + c_k u \overline{D_k v} dx$$

$$= \sum_{k=1}^{d} \int_{\Omega} \Big[\Re(b_k) \Re(\overline{v} D_k u) - \Im(b_k) \Im(\overline{v} D_k u) + \Re(c_k) \Re(\overline{u} D_k v)$$

$$+ \Im(c_k) \Im(\overline{u} D_k v) \Big] dx.$$

As in the proof of Lemma 4.26, we have

$$\bar{v}D_k u = D_k u \frac{v\bar{u}}{|u||v|}\bar{v}\chi_{\{u\neq0\}}\chi_{\{v\neq0\}} = \bar{u}D_k u \frac{|v|}{|u|}\chi_{\{u\neq0\}}\chi_{\{v\neq0\}}\,dx$$

and thus,

$$\Re(\bar{v}D_k u) = \Re(\text{sign}(\bar{u})D_k u)|v| = |v|D_k|u|.$$

Using this and Lemma 4.26, we can rewrite I_3 as

$$I_3 = \sum_{k=1}^{d}\int_\Omega \Re(b_k)|v|D_k|u| + \Re(c_k)|u|D_k|v|\,dx$$

$$+ \sum_{k=1}^{d}\int_\Omega (\Im c_k - \Im b_k)\Im(\text{sign}(\bar{u})D_k u)|v|\,dx. \qquad (4.44)$$

Concerning the term a_0, we have

$$I_4 := \Re\int_\Omega a_0 u\bar{v}\,dx = \int_\Omega \Re a_0|u||v|\,dx \qquad (4.45)$$

for all $u, v \in H_0^1(\Omega)$ such that $u\bar{v} \geq 0$.

Since $\Re\mathfrak{a}_{H_0^1}(u,v) = I_1 + I_2 + I_3 + I_4$, we obtain from (4.41), (4.43), (4.44), and (4.45),

$$\Re\mathfrak{a}_{H_0^1}(u,v) = \int_\Omega \left[\sum_{k,j=1}^{d}\Re(a_{kj})D_k|u|D_j|v| + \sum_{k=1}^{d}(\Re(b_k)D_k|u||v| \right.$$

$$\left. + \Re(c_k)|u|D_k|v|) + \Re(a_0)|u||v| \right]dx$$

$$+ \sum_{k,j=1}^{d}\int_\Omega \Re(a_{kj})\Im(\text{sign}(\bar{u})D_k u)\Im(\text{sign}(\bar{u})D_j u)\frac{|v|}{|u|}\chi_{\{u\neq0\}}\,dx$$

$$+ \sum_{k=1}^{d}\int_\Omega \left[f_k + \Im(c_k - b_k) \right]\Im(\text{sign}(\bar{u})D_k u)|v|\,dx.$$

It follows from the ellipticity assumption (4.2) that

$$\sum_{k,j=1}^{d} \int_{\Omega} \Re(a_{kj}) \Im(\text{sign}(\overline{u}) D_k u) \Im(\text{sign}(\overline{u}) D_j u) \frac{|v|}{|u|} \chi_{\{u \neq 0\}}$$

$$+ \sum_{k=1}^{d} \int_{\Omega} \left[f_k + \Im(c_k - b_k) \right] \Im(\text{sign}(\overline{u}) D_k u) |v| dx$$

$$\geq - \int_{\Omega} m(x) |u| |v| dx,$$

and this gives (4.40). □

It is clear from this proof that we may replace $\Re a_0$ in the last theorem by $(\Re a_0)^-$.

We want now to consider other boundary conditions in the previous theorem. Let V, $\mathfrak{a}_V$, and A_V be as above. Let $\mathcal{E}_V$ be the form given by the same expression as $\mathcal{E}_{H_0^1}$, but with domain $D(\mathcal{E}_V) = V$. Assume that (4.39) holds with $\mathcal{E}_{H_0^1}$ replaced by $\mathcal{E}_V$ and $H_0^1(\Omega)$ is replaced by V, that is,

$$\int_{\Omega} m(x) |u|^2 dx \leq \beta \int_{\Omega} |u|^2 dx + \alpha \Re \mathcal{E}_V(u, u) \text{ for all } u \in V, \qquad (4.46)$$

with some constants $\beta \in \mathbb{R}$ and $\alpha < 1$. Define

$$\mathfrak{b}_V(u, v) := \mathcal{E}_V(u, v) - \int_{\Omega} m(x) u \overline{v} dx \text{ for all } u, v \in D(\mathfrak{b}_V) = V.$$

Denote by B_V the operator associated with the form $\mathfrak{b}_V$.
The above theorem can be extended to the boundary conditions given by V if we assume that

$$\sum_{k,j=1}^{d} \int_{\Omega} \Im(a_{kj}) \phi \overline{D_j v} dx$$

$$= - \sum_{k,j=1}^{d} \int_{\Omega} D_j \Im(a_{kj}) \phi \overline{v} dx - \sum_{k,j=1}^{d} \int_{\Omega} \Im(a_{kj}) \overline{v} D_j \phi dx \qquad (4.47)$$

for all $v \in V, \phi \in C^{\infty}(\Omega) \cap H^1(\Omega)$ (this means that we assume that $D_j \Im a_{kj}$ exist as functions). Roughly speaking, this assumption means that $\Im a_{kj}$ are smooth and $= 0$ on parts of the boundary of Ω, where functions in V do not necessarily vanish.
We have

THEOREM 4.27 *Assume that V satisfies*

$$u \in V \Rightarrow (\Re u)^+ \in V$$

and assume that (4.37), (4.46), and (4.47) hold. Then, for every $t \geq 0$ and every $f \in L^2(\Omega)$

$$|e^{-tA_V} f| \leq e^{-tB_V} |f|.$$

Proof. The assumption on V together with $\mathfrak{b}_V(u^+, u^-) \leq 0$ implies the positivity of the semigroup $(e^{-tB_V})_{t \geq 0}$.

Using again Theorem 2.21, it suffices to prove that

$$\Re a_V(u, v) \geq \mathfrak{b}_V(|u|, |v|) \text{ for every } u, v \in V \text{ such that } u\bar{v} \geq 0. \quad (4.48)$$

The proof is the same as that of Theorem 4.25. The only place where we used the fact that $V = H_0^1(\Omega)$ is in the proof of (4.42). Now, in order to prove (4.42) for $u, v \in V$ we proceed as above by taking first $u \in C^\infty(\Omega) \cap H^1(\Omega)$ and $v \in V$, then use (4.47) to integrate by parts. This gives (4.42) for u and v as above. The Meyers-Serrin theorem ([Ada75], p. 52) and the assumption that m is $\mathcal{E}_V$-bounded show that (4.42) holds for all $u \in H^1(\Omega)$ and $v \in V$. $\qquad \square$

4.6 L^p-CONTRACTIVITY FOR $1 < p < \infty$

We consider again the form a_V defined by

$$a_V(u, v) = \int_\Omega \left[\sum_{k,j=1}^d a_{kj} D_k u \overline{D_j v} + \sum_{k=1}^d (b_k \bar{v} D_k u + c_k u \overline{D_k v}) + a_0 u \bar{v} \right] dx,$$

where the coefficients a_{kj}, b_k, c_k, a_0 are complex-valued and satisfy (U.Ell). We have seen in Section 4.3 that the L^∞-contractivity of $(e^{-t(A_V + w)})_{t \geq 0}$ does not necessarily hold with some constant w, even when $V = H_0^1(\Omega)$ and all the coefficients are real. Some regularity of the coefficients $\Re c_k$ is necessary (see Theorem 4.7, ii)).

In this section, we prove that for each $p \in (1, \infty)$, there exists a constant w_p, such that the semigroup $(e^{-t(A_V + w_p)})_{t \geq 0}$ is contractive on $L^p(\Omega)$. This holds without any further assumption on $\Re b_k$ and $\Re c_k$.

Note first that the semigroup $(e^{-tA_V})_{t \geq 0}$ is defined on $L^2(\Omega)$ and satisfies the estimate

$$\|e^{-t(A_V + w)}\|_{\mathcal{L}(L^2)} \leq 1 \text{ for all } t \geq 0 \quad (4.49)$$

where w may be chosen as follows

$$w = \frac{1}{4\eta} \sum_{k=1}^{d} \| |\Re(b_k + c_k)| + |\Im(b_k - c_k)| \|_\infty^2 + \|(\Re a_0)^-\|_\infty. \qquad (4.50)$$

This follows from the inequality $\Re \mathfrak{a}_V(u, u) + w(u; u) \geq 0$ (for all $u \in V$), which can be shown in the same way as (4.3).

The next result shows that this estimate extends to $L^p(\Omega)$. More precisely,

THEOREM 4.28 *Assume that (U.Ell) holds and that the coefficients a_{kj} are real-valued functions for $1 \leq k, j \leq d$. Suppose that V satisfies (4.12). Then for every $p \in (1, +\infty)$, the semigroup $(e^{-tA_V})_{t \geq 0}$ extends boundedly to $L^p(\Omega)$. In addition,*

$$\|e^{-tA_V}\|_{\mathcal{L}(L^p(\Omega))} \leq e^{w_p t} \ for \ all \ t \geq 0, \qquad (4.51)$$

where

$$w_p = \|(\Re a_0)^-\|_\infty + \frac{1}{\eta}\left(\frac{1}{p} + \frac{1}{2}\right) \sum_{k=1}^{d} \|b_k - c_k\|_\infty^2 + \frac{p}{\eta} \sum_{k=1}^{d} \|\Re c_k\|_\infty^2$$

for $p \in [2, \infty)$; and

$$w_p = \|(\Re a_0)^-\|_\infty + \frac{1}{\eta}\frac{3p-2}{2p} \sum_{k=1}^{d} \|b_k - c_k\|_\infty^2 + \frac{p}{\eta(p-1)} \sum_{k=1}^{d} \|\Re b_k\|_\infty^2$$

for $p \in (1, 2]$.

Proof. Define for each $z \in \mathbb{C}$, the form $\mathfrak{a}_V(z)$ by

$$\mathfrak{a}_V(z)(u, v) := \frac{1}{2} \sum_{k,j=1}^{d} \int_\Omega a_{kj} D_k u \overline{D_j v} dx$$

$$+ z \sum_{k=1}^{d} \int_\Omega \left[(\Re c_k)\overline{v} D_k u + (\Re c_k)u\overline{D_k v} \right] dx,$$

with domain $D(\mathfrak{a}_V(z)) = V$. We denote by $(T_z(t))_{t \geq 0}$ the semigroup generated by (minus) the operator associated with $\mathfrak{a}_V(z)$. By Theorem 4.9, the semigroup $(T_{is}(t))_{t \geq 0}$ and its adjoint are both L^∞-contractive for every $s \in \mathbb{R}$.

For $z = 1 + is$ ($s \in \mathbb{R}$), the estimate (4.49) (applied to the semigroup associated with the form $\mathfrak{a}_V(1 + is)$) gives

$$\|T_{1+is}(t)\|_{\mathcal{L}(L^2(\Omega))} \leq e^{w_2' t},$$

where $w_2' = \frac{2}{\eta} \sum_{k=1}^{d} \|\Re c_k\|_\infty^2$.

On the other hand, it follows from Kato [Kat80], Theorems VII-4.2 and IX-2.6 that $T_z(t)$ depends analytically on z (for each $t \geq 0$). The Stein interpolation theorem allows to interpolate between the previous L^2- and L^∞-estimates. Thus, for every $p \in [2, \infty]$

$$\|T_{2/p}(t)\|_{\mathcal{L}(L^p(\Omega))} \leq e^{\frac{2}{p} w_2' t}.$$

Applying this estimate to the semigroup associated with the form where c_k is changed into $\frac{p}{2} c_k$ yields

$$\|T_1(t)\|_{\mathcal{L}(L^p(\Omega))} \leq e^{\frac{p}{2} w_2' t} = e^{\frac{p}{\eta} \sum_{k=1}^{d} \|\Re c_k\|_\infty^2 t}. \tag{4.52}$$

Define now the form

$$\mathfrak{b}_V(u, v) := \frac{1}{2} \sum_{k,j=1}^{d} \int_\Omega a_{kj} D_k u \overline{D_j v}\, dx$$

$$+ \sum_{k=1}^{d} \int_\Omega \left[(b_k - \Re c_k) \overline{v} D_k u + i \Im c_k u \overline{D_k v} \right] dx + \int_\Omega a_0 u \overline{v}\, dx$$

(with domain $D(\mathfrak{b}_V) = V$) and denote by $(S(t))_{t \geq 0}$ the semigroup generated by (minus) its associated operator. By Theorem 4.9, $(e^{-w_0 t} S(t))_{t \geq 0}$ is L^∞-contractive for every w_0 such that

$$w_0 + \Re a_0 - \frac{1}{2\eta} \sum_{k=1}^{d} |\Im(c_k - b_k)|^2 \geq 0 \text{ on } \Omega.$$

In particular, this holds for

$$w_0 = \left\| (\Re a_0)^- + \frac{1}{2\eta} \sum_{k=1}^{d} |\Im(c_k - b_k)|^2 \right\|_\infty.$$

This L^∞-estimate and (4.49) (applied to the semigroup associated with $\mathfrak{b}_V$) imply that for every $p \in [2, +\infty]$

$$\|S(t)\|_{\mathcal{L}(L^p(\Omega))} \leq e^{[\frac{2}{p} w_2'' + w_0(1 - \frac{2}{p})]t}, \tag{4.53}$$

where $w_2'' = \frac{1}{2\eta} \sum_{k=1}^{d} \||\Re(b_k - c_k)| + |\Im(b_k - c_k)|\|_\infty^2 + \|(\Re a_0)^-\|_\infty$.

Since the form $\mathfrak{a}_V$ is the sum

$$\mathfrak{a}_V = \mathfrak{a}_V(1) + \mathfrak{b}_V$$

it follows from the Trotter-Kato product formula (cf. [Kat78]) that

$$e^{-tA_V} f = \lim_{n \to +\infty} (T_1(t/n)S(t/n))^n f \qquad (4.54)$$

for every $f \in L^2(\Omega)$. This together with (4.52) and (4.53) gives

$$\|e^{-tA_V}\|_{\mathcal{L}(L^p(\Omega))} \le e^{[\frac{2}{p}w_2'' + w_0(1-\frac{2}{p}) + \frac{p}{2}w_2']t}. \qquad (4.55)$$

This shows the desired estimate on $L^p(\Omega)$ for $p \in [2, \infty)$. The estimate on $L^p(\Omega)$ for $p \in (1, 2]$ is obtained by applying the previous one to the adjoint semigroup $(e^{-tA_V^*})_{t \ge 0}$ and arguing by duality. $\square$

For complex-valued coefficients a_{kj}, we have the following

THEOREM 4.29 *Assume that* $f_k := \sum_{k=1}^d D_j \Im a_{kj} \in L^\infty(\Omega)$ *for* $1 \le k \le d$ *and let* $m(x)$ *be as in (4.38). Assume that* $\Im(a_{kj} + a_{jk}) = 0$ *for all* $1 \le k, j \le d$ *(respectively, that the hypotheses of Theorem 4.27 are satisfied if* $V \ne H_0^1(\Omega)$). *Then the conclusion of the above theorem holds for the semigroup* $(e^{-tA_{H_0^1}})_{t \ge 0}$ *(respectively, for* $(e^{-tA_V})_{t \ge 0}$) *with* $\|(\Re a_0)^-\|_\infty$ *replaced by* $\|(\Re a_0 - m)^-\|_\infty$ *in the expression of* w_p.

Proof. Apply Theorem 4.25 (respectively, Theorem 4.27) and the previous result. $\square$

4.7 OPERATORS WITH UNBOUNDED COEFFICIENTS

In this section we show that some of the results presented in the foregoing sections can be extended to operators having unbounded coefficients and to Schrödinger type operators.

Let Ω be an open set of $\mathbb{R}^d$. Assume that

$$a_{kj} = a_{jk} \in L_{\text{loc}}^1(\Omega, \mathbb{R}), \; 1 \le k, j \le d, \; \Re \sum_{j,k=1}^d a_{kj}\xi_k\overline{\xi_j} \ge \eta|\xi|^2 \quad (4.56)$$

for all $\xi \in \mathbb{C}^d$, where $\eta > 0$ is a constant. Let $0 \le m \in L_{\text{loc}}^1(\Omega)$ and define the symmetric form

$$\mathfrak{b}_V(u, v) := \int_\Omega \sum_{k,j=1}^d a_{kj} D_k u \overline{D_j v} \, dx + \int_\Omega m u \overline{v} \, dx.$$

The domain of $\mathfrak{b}_V$ is given by

$$D(\mathfrak{b}_V) = \left\{ u \in V, \int_\Omega \sum_{k,j=1}^d a_{kj} D_k u \overline{D_j u} dx < \infty \text{ and } \int_\Omega m|u|^2 dx < \infty \right\},$$

where V is a closed subspace of $H^1(\Omega)$ and such that $H_0^1(\Omega) \subseteq V \subseteq H^1(\Omega)$.

PROPOSITION 4.30 *The form $\mathfrak{b}_V$ is densely defined and closed.*

Proof. The form $\mathfrak{b}_V$ is densely defined since $C_c^\infty(\Omega) \subseteq D(\mathfrak{b}_V)$. We show that $\mathfrak{b}_V$ is closed. Let $(u_n)_n$ be a Cauchy sequence in $D(\mathfrak{b}_V)$. It follows from the ellipticity assumption (4.56) and the fact that m is non-negative that $(u_n)_n$ is a Cauchy sequence in V and hence it converges in V. Let u be its limit. Taking a subsequence (u_{n_k}) such that $D_j u_{n_k}(x)$ converges to $D_j u(x)$ for a.e. $x \in \Omega$ and all $j \in \{1, \dots, d\}$, one obtains by Fatou's lemma that $u \in D(\mathfrak{b}_V)$. (To see that such a subsequence exists, one can argue as follows: $D_1 u_n$ converges to $D_1 u$ in $L^2(\Omega, \mathbb{C})$ and hence there exists a subsequence $D_1 u_{\phi(n)}$ which converges a.e. to $D_1 u$. Now $D_2 u_{\phi(n)}$ converges in $L^2(\Omega, \mathbb{C})$ and hence we can extract a subsequence of $D_2 u_{\phi(n)}$ which converges a.e. to $D_2 u$. Iterating this d times one obtains the subsequence (u_{n_k}) with the desired property.)

Apply again the a.e. convergence of the subsequence u_{n_k} and Fatou's lemma to obtain

$$\mathfrak{b}_V(u_n - u, u_n - u)$$
$$= \int_\Omega \lim_k \left[\sum_{k,j} a_{kj} D_k(u_n - u_{n_k}) \overline{D_j(u_n - u_{n_k})} + m|u_n - u_{n_k}|^2 \right] dx$$
$$\leq \liminf_k \int_\Omega \left[\sum_{k,j} a_{kj} D_k(u_n - u_{n_k}) \overline{D_j(u_n - u_{n_k})} + m|u_n - u_{n_k}|^2 \right] dx.$$

From this it follows that $\mathfrak{b}_V(u_n - u, u_n - u) \to 0$ as $n \to \infty$. Thus, the form $\mathfrak{b}_V$ is closed. □

Now let $b_k, c_k, a_0 \in L^\infty(\Omega, \mathbb{C})$ for $1 \leq k \leq d$ and define the form

$$\mathfrak{a}_V(u, v) = \mathfrak{b}_V(u, v) + \sum_{k=1}^d \int_\Omega \left[b_k \overline{v} D_k u + c_k u \overline{D_k v} \right] dx + \int_\Omega a_0 u \overline{v} dx,$$
$$D(\mathfrak{a}_V) = D(\mathfrak{b}_V).$$

Using the fact that b_k, c_k, a_0 are bounded, one can check that there exists a constant w' such that the form $\mathfrak{a}_V + w'$ is non-negative, continuous, and closed. Denote by A_V the operator associated with $\mathfrak{a}_V$. Several results given in previous sections for operators with bounded coefficients can be extended to the present situation. We summarize the relevant ones in the following

THEOREM 4.31 *Assume that (4.56) holds and that $b_k, c_k, a_0 \in L^\infty(\Omega, \mathbb{C})$ for $1 \le k \le d$. We have:*
1) Assume that for each k, the coefficients b_k, c_k and a_0 are real. If $(\Re u)^+ \in V$ for all $u \in V$, then the semigroup $(e^{-tA_V})_{t\ge 0}$ is positive.
2) Assume that $(1 \wedge |u|)\text{sign } u \in V$ for all $u \in V$ and that $\Re c_j = 0$ for all $j = 1, \dots, d$. Then for every constant w such that

$$w + m + \Re a_0 - \frac{1}{4\eta} \sum_{j=1}^d |\Im(c_j - b_j)|^2 \ge 0 \text{ on } \Omega,$$

the semigroup $(e^{-t(A_V + w)})_{t\ge 0}$ is L^∞-contractive.
3) Assume now that all the coefficients b_k, c_k, and a_0 are real.
i) If $(\Re u)^+ \in V$ for all $u \in V$, then the semigroup $(e^{-tA_{H_0^1(\Omega)}})_{t\ge 0}$ is dominated by $(e^{-tA_V})_{t\ge 0}$.
ii) Let $V = \overline{\{u_{|\Omega}, u \in C_c^\infty(\mathbb{R}^d \setminus \Gamma)\}}^{H^1(\Omega)}$, where Γ is a closed subset of $\partial\Omega$. Then $(e^{-tA_V})_{t\ge 0}$ is dominated by the two semigroups $(e^{-tA_{H^1(\Omega)}})_{t\ge 0}$ and $(e^{-tA_{\widetilde{H^1(\Omega)}}})_{t\ge 0}$.

Proof. We only sketch the proof since there is no new difficulties.

Assertion 1) follows from Theorem 2.6 once we establish that $(\Re u)^+ \in D(\mathfrak{a}_V)$ for all $u \in D(\mathfrak{a}_V)$. Since this is true by assumption on the space V and since $D_j(\Re u)^+ = \chi_{\{\Re u > 0\}} D_j \Re u$, it follows that $(\Re u)^+ \in D(\mathfrak{a}_V)$.

Assertion 2) holds as in Theorem 4.9 once we have $(1 \wedge |u|)\text{sign } u \in D(\mathfrak{a}_V)$ for all $u \in D(\mathfrak{a}_V)$. This follows from the assumption on V and the expression of $D_j[(1 \wedge |u|)\text{sign } u]$. Finally, as in the previous section we obtain assertion 3) once we establish the ideal property. It follows easily from the definition of $D(\mathfrak{a}_V)$ and the fact that $H_0^1(\Omega)$ is an ideal of every V (as in i)) that $D(\mathfrak{a}_{H_0^1(\Omega)})$ is an ideal of $D(\mathfrak{a}_V)$. Again, since $\overline{\{u_{|\Omega}, u \in C_c^\infty(\mathbb{R}^d \setminus \Gamma)\}}^{H^1(\Omega)}$ is an ideal of $H^1(\Omega)$ and of $\widetilde{H^1(\Omega)}$, we obtain that $D(\mathfrak{a}_V)$ is an ideal of $D(\mathfrak{a}_{H^1(\Omega)})$ and of $D(\mathfrak{a}_{\widetilde{H^1(\Omega)}})$. $\square$

It should be noted that the boundedness assumption of the coefficients b_k, c_k, and a_0 does not play any role in the proof. We have assumed that these coefficients are bounded only for simplicity and also to ensure the continuity and closability of the form $\mathfrak{a}_V + w'$. We could, however, include more general coefficients. For example, if we assume that $|b_k|^2$, $|c_k|^2$ and

a_0 are form bounded with respect to the form

$$\mathfrak{c}_V(u,v) := \sum_k \int_\Omega D_k u \overline{D_k v}\,dx$$

with form bound sufficiently small, then continuity and closability of $\mathfrak{a}_V + w'$ for some w' hold. In order to see this, we write

$$\left|\Re \int_\Omega b_k \overline{u} D_k u\,dx\right| \leq \frac{1}{2\varepsilon}\int_\Omega |b_k|^2 |u|^2 dx + \frac{\varepsilon}{2}\int_\Omega |D_k u|^2 dx$$

which is valid for all $\varepsilon > 0$. The fact that $|b_k|^2$ is $\mathfrak{c}_V$-bounded means that for some constants $\alpha_k, \beta_k \geq 0$,

$$\int_\Omega |b_k|^2 |u|^2 dx \leq \alpha_k \int_\Omega |D_k u|^2 dx + \beta_k \int_\Omega |u|^2 dx.$$

Using this for each b_k, c_k, a_0 and using the ellipticity assumption (4.56), we obtain the desired conclusion if the constants α_k are small enough.

Consider on $L^2(\Omega, \mathbb{R})$ the symmetric form

$$\mathfrak{a}(u,v) = \int_\Omega \sum_{k,j=1}^d a_{kj} D_k u D_j v\,dx + \int_\Omega muv\,dx, \quad D(\mathfrak{a}) = C_c^\infty(\Omega),$$

where $0 \leq m \in L^1_{\text{loc}}(\Omega)$, the coefficients $a_{kj} \in L^1_{\text{loc}}(\Omega)$ and satisfy (4.56). We have seen that the form $\mathfrak{b}_{H_0^1}$, defined above, is closed. Hence, $\mathfrak{a}$ is closable. We denote again by $\mathfrak{a}$ its closure and by A its associated operator. We have

PROPOSITION 4.32 *The semigroup $(e^{-tA})_{t\geq 0}$ is sub-Markovian.*

Proof. Assume for a moment that $|u| \in D(\mathfrak{a})$ for all $u \in C_c^\infty(\Omega)$. By Theorem 4.31, the semigroup associated with the form $\mathfrak{b}_{H_0^1}$ is sub-Markovian and hence Theorem 2.6 (or Corollary 2.18) implies

$$\mathfrak{a}(|u|,|u|) = \mathfrak{b}_{H_0^1}(|u|,|u|) \leq \mathfrak{b}_{H_0^1}(u,u) = \mathfrak{a}(u,u). \tag{4.57}$$

This implies that $\mathfrak{a}(u^+, u^-) \leq 0$ and we conclude again by Theorem 2.6 that the semigroup $(e^{-tA})_{t\geq 0}$ is positive.

It remains to prove that $|u| \in D(\mathfrak{a})$ for all $u \in C_c^\infty(\Omega)$. Fix $u \in C_c^\infty(\Omega)$ and let (ρ_n) be a regularizing sequence. For n large enough, $\rho_n * |u| \in C_c^\infty(\Omega)$. We have for each $j \in \{1,\dots,d\}$

$$|D_j(\rho_n * |u|) - D_j|u|| = |\rho_n * D_j|u|) - D_j|u|| \leq 2\|D_j u\|_\infty.$$

For n large enough, $\rho_n * |u| - |u|$ has support contained in a compact set which is independent of n. Thus, using the fact that $m, a_{kj} \in L^1_{\text{loc}}(\Omega)$, we obtain by the dominated convergence theorem that

$$\int_\Omega \sum_{k,j=1}^d a_{kj} D_k(\rho_n * |u| - |u|) D_j(\rho_n * |u| - |u|) dx + \int_\Omega m|\rho_n * |u| - |u||^2 dx$$

converges to 0 as $n \to \infty$. This shows that $|u| \in D(\mathfrak{a})$.

The proof of the L^∞-contractivity is similar. We just replace $|u|$ in the above arguments by $1 \wedge u$ for non-negative u and apply Corollary 2.18. □

Notes

Treatment of second-order elliptic operators can be found in several books. The sub-Markovian property for second-order symmetric operators with real-valued coefficients is studied in Davies [Dav89], [Dav95d], Fukushima [Fuk80], Fukushima, Oshima, and Takeda [FOT94]. Ma and Röckner [MaRö92] consider nonsymmetric operators with real-valued coefficients.

The novelty in the present chapter is that we are able to treat operators with complex-valued coefficients and describe precisely in terms of boundary conditions and of the coefficients when positivity, L^∞-contractivity, or domination properties hold.

Section 4.2. The presentation of this section follows Ouhabaz [Ouh92b], [Ouh96], and [Ouh02]. Proposition 4.1 and Theorem 4.2 are taken from [Ouh92b] and [Ouh96]. As mentioned above, results of the same type are contained in [Dav89], [Dav95d], [Fuk80], [FOT94], and [MaRö92]. In the symmetric case $a_{kj} = a_{jk}$ and $b_k = c_k = 0$ for all $1 \leq k, j \leq d$, the fact that 2) $\Longrightarrow$ 1) in Theorem 4.5 is shown by Arendt [Are01] by using the maximum principle and by Davies [Dav89] (cf. Theorem 3.3.5) as a consequence of lower bounds for the heat kernel.

Section 4.3. The results in this section are mainly taken from Ouhabaz [Ouh92b], [Ouh96] and Auscher et al. [ABBO00]. Theorems 4.6 and 4.7 and Corollary 4.8 are proved in [ABBO00]. Theorem 4.9 and Corollary 4.10 are taken from [Ouh92b] and [Ouh96]. Similar results for symmetric real-valued coefficients can be found in [Dav89], [Fuk80], [FOT94] and in [MaRö92] for nonsymmetric operators (with real-valued coefficients). Related results to Proposition 4.11 can be found in Dautray-Lions [DaLi88], Section 7 and Adams [Ada75], Lemma 8.31. The same proposition is proved in Arendt and ter Elst [ArEl97]. Corollary 4.12 and Theorem 4.14 are proved in Auscher et al. [ABBO00]. Related results to Corollary 4.12 for weakly coupled systems are proved in the case of smooth coefficients (and also smooth open set Ω) by Kresin and Maz'ya [KrMa94] and Langer [Lan99].

Section 4.4. This section is taken from Auscher et al. [ABBO00] in which Theorem 4.17 is proved. General criteria for the conservation property with applications to second-order symmetric operators with real-valued coefficients are given in Oshima [Osh92] and Davies [Dav85]. Similar results to Lemma 4.18 can be found in Auscher, Coulhon, and Tchamitchian [ACT96] and Davies [Dav95c]. Lemma 4.20 is taken from Clément et al. [CHADP87].

Section 4.5. Theorem 4.21 and Proposition 4.23 are taken from Ouhabaz [Ouh96]. These results show the role of the ideal property in the domination of semigroups. We mention that a description of closed ideals of regular Dirichlet spaces is given in Stollmann [Sto93]. See also the recent work by Arendt and Warma [ArWa03] for related results to Theorem 4.21. Theorems 4.25 and 4.27 are shown in [Ouh02]. The idea behind the last two results is the well-known diamagnetic inequality for Schrödinger operators with magnetic fields.

Section 4.6. Theorem 4.28 is proved in Ouhabaz [Ouh02]. It was proved for elliptic operators with real-valued coefficients (but with less precise constant w_p) by Robinson [Rob91], Daners [Dan00], and Karrmann [Kar01]. Note that Theorem 4.29 cannot hold for operators with arbitrary complex-valued a_{kj}. For operators $A = -\sum_{k,j} D_j(a_{kj}D_k)$ with complex-valued coefficients (considered on $L^2(\mathbb{R}^d, dx)$ with $d \geq 3$), one obtains by using the Sobolev embedding that $(e^{-tA})_{t\geq 0}$ extends to a strongly continuous semigroup on $L^p(\mathbb{R}^d)$ for $p \in [\frac{2d}{d+2}, \frac{2d}{d-2}]$; see Davies [Dav95c]. It is proved in [Dav97a] that this result is sharp in the sense that for every $p \notin [\frac{2d}{d+2}, \frac{2d}{d-2}]$, there exists an operator A for which e^{-tA} cannot be extended from $L^2 \cap L^p$ to a bounded operator on $L^p(\mathbb{R}^d)$ for any $t > 0$.

Section 4.7. Similar results to those in assertions 1) and 2) of Theorem 4.31 can be found in Ma and Röckner [MaRö92]. More can be said on elliptic operators with unbounded coefficients; see recent works by Liskevich [Lis96], Liskevich, Sobol and Vogt [LSV02], Sobol and Vogt [SoVo02], Metafune et al. [MPRS02].

Chapter Five

DEGENERATE-ELLIPTIC OPERATORS

We have studied in the last chapter contractivity properties of semigroups associated with second-order uniformly elliptic operators. In particular, we have seen that it is possible in several cases to extend the semigroup initially defined on $L^2(\Omega, \mathbb{C})$ to $L^p(\Omega, \mathbb{C})$ for $p \neq 2$. In the present chapter, we study similar questions for second-order degenerate-elliptic operators. More precisely, we consider operators of the type

$$Au = -\sum_{k,j=1}^{d} D_j(a_{kj}D_k u) + \sum_{k=1}^{d} b_k D_k u + a_0 u, \qquad (5.1)$$

where the matrix $(a_{kj})_{k,j}$ satisfies the following weaker assumption than (4.2):

$$\sum_{j,k=1}^{d} a_{kj}(x)\xi_j\xi_k \geq 0 \text{ for all } \xi \in \mathbb{R}^d, \text{ a.e. } x \in \Omega. \qquad (5.2)$$

In this case, any realization of the operator A is called a degenerate-elliptic operator. For such operators, several difficulties which we did not meet previously for uniformly elliptic operators occur now. The first one is the L^2-theory, that is, the construction of a realization of A which generates a strongly continuous semigroup on $L^2(\Omega, \mathbb{R})$. If one applies the sesquilinear form method, then because of the absence of the ellipticity assumption (4.2), it becomes difficult to check continuity and closability of the form. In the symmetric case, one can still apply this technique. We shall assume a smoothness condition of the coefficients which guarantees the closability of the symmetric form. One advantadge of the sesquilinear form method is that it allows one to treat operators with unbounded coefficients. The nonsymmetric case will be treated by a perturbation method, based on Theorem 1.50. In both cases, we obtain a strongly continuous semigroup on L^2, which can be extended to other L^p-spaces. This extension to L^p will be achieved by applying the results of the previous chapter together with an approximation argument.

5.1 SYMMETRIC DEGENERATE-ELLIPTIC OPERATORS

Let Ω be an open subset of $\mathbb{R}^d$. Assume that a_{jk} ($1 \le k, j \le d$) and a_0 are measurable functions on Ω, such that

$$a_{kj} = a_{jk} \in H^1_{\text{loc}}(\Omega, \mathbb{R}), \ 1 \le k, j \le d \text{ and } 0 \le a_0 \in L^1_{\text{loc}}(\Omega). \quad (5.3)$$

Here, $u \in H^1_{\text{loc}}(\Omega, \mathbb{R})$ means that u is a real-valued function such that $\phi u \in H^1(\Omega)$ for every $\phi \in C^\infty_c(\Omega)$.

Define on $L^2(\Omega, \mathbb{R})$ the symmetric form

$$\mathfrak{a}(u, v) = \int_\Omega \left[\sum_{k,j=1}^d a_{kj}(x) D_k u D_j v + a_0(x) uv \right] dx, \ \ D(\mathfrak{a}) = C^\infty_c(\Omega).$$

Note that by assumptions (5.3), the operator

$$Au = - \sum_{k,j=1}^d D_j(a_{kj} D_k u) + a_0 u, \ u \in D(A) = C^\infty_c(\Omega)$$

is well defined. It is a symmetric and accretive operator. By Lemma 1.29, the form $\mathfrak{a}$ is closable. We denote by $\overline{\mathfrak{a}}$ its closure and keep the same notation A for the self-adjoint operator associated with $\overline{\mathfrak{a}}$ (i.e., the Friedrichs extension of the initial operator A with domain $C^\infty_c(\Omega)$). Again, $(e^{-tA})_{t \ge 0}$ denotes the semigroup generated by $-A$. We have

THEOREM 5.1 *Assume that (5.2) and (5.3) are satisfied. Then the semigroup $(e^{-tA})_{t \ge 0}$ is sub-Markovian.*

Proof. Define the following sequence of symmetric forms:

$$\mathfrak{a}_n(u, v) = \int_\Omega \left[\sum_{k,j=1}^d (a_{kj}(x) + \frac{1}{n} \delta_{kj}) D_k u D_j v + a_0(x) uv \right] dx,$$
$$D(\mathfrak{a}_n) = C^\infty_c(\Omega),$$

where $\delta_{kj} = 0$ if $k \ne j$ and $\delta_{kk} = 1$ for all $1 \le k, j \le d$.

Denote by $\overline{\mathfrak{a}_n}$ the closure of $\mathfrak{a}_n$ and by A_n the operator associated with $\overline{\mathfrak{a}_n}$. The sequence of symmetric forms $(\mathfrak{a}_n)_n$ is nonincreasing. This means that $\mathfrak{a}_{n+1}(u, u) \le \mathfrak{a}_n(u, u)$ for all $n \ge 0$. The same property holds then for the closures $(\overline{\mathfrak{a}_n})_n$. In addition, it is clear that $D(\overline{\mathfrak{a}_n}) \subseteq D(\overline{\mathfrak{a}})$ for all n. Thus, the monotone convergence theorem for forms (see, e.g., Kato [Kat80], Chap. VIII) implies that for each $t \ge 0$, the sequence $(e^{-tA_n})_n$ converges strongly in $L^2(\Omega, \mathbb{R})$ to e^{-tA}.

By Proposition 4.32, the semigroup $(e^{-tA_n})_{t \ge 0}$ is sub-Markovian. Letting $n \to \infty$, we obtain that $(e^{-tA})_{t \ge 0}$ is sub-Markovian. $\qquad \square$

5.2 OPERATORS WITH TERMS OF ORDER 1

5.2.1 The L^2-theory

In this section, we study degenerate-elliptic operators of the type (5.1). We consider here the L^2-theory. This will be achieved by using Theorem 1.50. We make the following assumption on the coefficients:

$$a_{kj} = a_{jk} \in W^{2,\infty}(\mathbb{R}^d, \mathbb{R}), b_k \in W^{1,\infty}(\mathbb{R}^d, \mathbb{R}), \ a_0 \in L^\infty(\mathbb{R}^d, \mathbb{R}). \quad (5.4)$$

Note that we assume here that all the coefficients are real-valued functions and that (5.4) holds for all $k, j = 1, ..., d$.

Under the assumption (5.4), the operator

$$Au = -\sum_{k,j=1}^{d} D_j(a_{kj} D_k u) + \sum_{k=1}^{d} b_k D_k u + a_0 u, \ u \in D(A) = C_c^\infty(\mathbb{R}^d)$$

is well defined as an operator on $L^2 = L^2(\mathbb{R}^d, \mathbb{R})$. Moreover, there exists $w \in \mathbb{R}$ such that $A + wI$ is accretive. Indeed, let $u \in C_c^\infty(\mathbb{R}^d)$ and apply (5.2) (with $\Omega = \mathbb{R}^d$ here) to obtain

$$\left(-\sum_{k,j=1}^{d} D_j(a_{kj} D_k u); u \right) = \sum_{k,j=1}^{d} \int_{\mathbb{R}^d} a_{kj} D_k u D_j u dx \geq 0.$$

For $k \in \{1, \ldots, d\}$,

$$(b_k D_k u; u) = -\int_{\mathbb{R}^d} [|u|^2 D_k b_k + b_k u D_k u] dx,$$

and thus,

$$(b_k D_k u; u) = -\frac{1}{2} \int_{\mathbb{R}^d} D_k b_k |u|^2 dx \geq -\frac{1}{2} \|D_k b_k\|_\infty (u; u).$$

Finally,

$$(a_0 u; u) \geq -\|a_0\|_\infty (u; u).$$

These inequalities show that for some $w \in \mathbb{R}$, $((A + w)u; u) \geq 0$ for all $u \in C_c^\infty(\mathbb{R}^d)$. This means that $A + wI$ is accretive. By Lemma 1.47, A is closable. Let $\overline{A}$ denote its closure. We have

THEOREM 5.2 *Assume that (5.2) (with $\Omega = \mathbb{R}^d$) and (5.4) are satisfied. Then there exists a constant $w \in \mathbb{R}$ such that $\overline{A} + wI$ is m-accretive. Moreover, the domain of $\overline{A}$ is the maximal domain in the sense that*

$$D(\overline{A}) = \left\{ u \in L^2, -\sum_{k,j=1}^{d} D_j(a_{kj}D_k u) + \sum_{k=1}^{d} b_k D_k u + a_0 u \right.$$

$$\left. \text{(as a distribution)} \in L^2 \right\}.$$

As a consequence of this result and Theorem 1.49, we have

COROLLARY 5.3 *Assume (5.2) and (5.4). Then $-\overline{A}$ generates a strongly continuous semigroup $(e^{-t\overline{A}})_{t \geq 0}$ on $L^2(\mathbb{R}^d, \mathbb{R})$, such that*

$$\|e^{-t\overline{A}}f\|_2 \leq e^{wt}\|f\|_2 \text{ for all } f \in L^2(\mathbb{R}^d, \mathbb{C}), \ t \geq 0.$$

We also mention the following corollary.

COROLLARY 5.4 *Assume (5.2), (5.4) and that $b_k = 0$ for $k = 1, \ldots, d$. Then $\overline{A}$ is self-adjoint. In other words, the operator A is essentially self-adjoint on $C_c^\infty(\mathbb{R}^d)$.*

Proof. Under the assumptions of the corollary, the operator A is symmetric and hence $\overline{A}$ is symmetric, too. This means that $(\overline{A})^*$ is an extension of $\overline{A}$. Recall the definition of the adjoint operator

$$D(A^*) = \{u \in L^2, \exists v \in L^2 : (A\phi, u) = (\phi, v) \ \forall \phi \in D(A)\}, \ A^* u = v.$$

It follows from this that

$$D(A^*) = \left\{ u \in L^2(\mathbb{R}^d, \mathbb{R}), -\sum_{k,j=1}^{d} D_j(a_{kj}D_k u) + a_0 u \right.$$

$$\left. \text{(as a distribution)} \in L^2(\mathbb{R}^d, \mathbb{R}) \right\}.$$

Theorem 5.2 implies that $D(A^*) = D(\overline{A})$. Since $D((\overline{A})^*) \subseteq D(A^*) = D(\overline{A})$, we conclude that $(\overline{A})^* = \overline{A}$. $\qquad \square$

In order to prove Theorem 5.2, we will need the following lemma. Denote by $\text{tr}(C)$ the trace of a given symmetric matrix C. We have

LEMMA 5.5 *(Oleinik inequality) Let $C(x) = (c_{kj}(x))_{1 \leq k,j \leq d}$ be a real symmetric matrix satisfying (5.2) for $x \in \mathbb{R}^d$. Assume that $c_{kj} \in W^{2,\infty}(\mathbb{R}^d)$*

*for all $1 \leq k, j \leq d$. Then, there exists a constant M depending only on d
and $\|D_h^2 c_{kj}\|_\infty$, such that for every real symmetric matrix $U = (u_{kj})_{1 \leq k,j \leq d}$,*

$$[\mathrm{tr}(D_l C(x) U)]^2 \leq M \mathrm{tr}(U C(x) U) \ for \ all \ x \in \mathbb{R}^d, l \in \{1, \ldots, d\}.$$

Proof. Fix $l \in \{1, \ldots, d\}$ and a non-negative function $f \in W^{2,\infty}(\mathbb{R}^d)$.
Consider a regularizing sequence $\rho_n \geq 0$. Since $\rho_n * f$ is C^2, one has by
Taylor's formula,

$$0 \leq (\rho_n * f)(x + he_l)$$
$$\leq (\rho_n * f)(x) + h(\rho_n * D_l f)(x) + \frac{h^2}{2} \|(\rho_n * D_l^2 f)\|_\infty$$
$$\leq (\rho_n * f)(x) + h(\rho_n * D_l f)(x) + \frac{h^2}{2} \|D_l^2 f\|_\infty$$

for all $x \in \mathbb{R}^d$ and all $h \in \mathbb{R}$. Here $(e_1, \ldots, e_d)$ is the standard basis of $\mathbb{R}^d$.
Letting $n \to \infty$ yields

$$0 \leq f(x) + h D_l f(x) + \frac{h^2}{2} \|D_l^2 f\|_\infty.$$

Since this inequality holds for all $h \in \mathbb{R}$, it follows that

$$|D_l f(x)|^2 \leq 2 f(x) \|D_l^2 f\|_\infty \ for \ all \ x \in \mathbb{R}^d. \tag{5.5}$$

Set

$$M_0 := \sup\{| < D_l^2 C(x)\xi, \xi >_{\mathbb{R}^d} |, 1 \leq l \leq d, x \in \mathbb{R}^d, \|\xi\|_{\mathbb{R}^d} = 1\},$$

where $\|.\|_{\mathbb{R}^d}$ denotes the Euclidean norm of $\mathbb{R}^d$ and $< ., . >_{\mathbb{R}^d}$ the Euclidean
scalar product.

Let U be a real symmetric matrix. Let P be an orthogonal matrix such
that $P^{-1} U P$ is diagonal. We use the notation V_{jk} to denote the coefficients
of a matrix V.

Since $(P^{-1} C(x) P)_{jj} \geq 0$ for all $x \in \mathbb{R}^d$ we can use (5.5) to obtain

$$|D_l(P^{-1}C(x)P)_{jj}|^2 \leq 2(P^{-1}C(x)P)_{jj} \|D_l^2(P^{-1}C(x)P)_{jj}\|_\infty$$
$$= 2(P^{-1}C(x)P)_{jj} \sup_x | < D_l^2 C(x) Pe_j, Pe_j > |$$
$$\leq 2M_0(P^{-1}C(x)P)_{jj}.$$

It follows that

$$
|\operatorname{tr}(D_l C(x).U)|^2 = |\operatorname{tr}(P^{-1} D_l C(x) P P^{-1} U P)|^2
$$

$$
= |\sum_{j=1}^{d} (P^{-1} D_l C(x) P)_{jj} (P^{-1} U P)_{jj}|^2
$$

$$
\leq d \sum_{j=1}^{d} |(P^{-1} D_l C(x) P)_{jj}|^2 |(P^{-1} U P)_{jj}|^2
$$

$$
\leq 2 M_0 d \sum_{j=1}^{d} (P^{-1} U P)_{jj} (P^{-1} C(x) P)_{jj} (P^{-1} U P)_{jj}
$$

$$
= 2 M_0 d \operatorname{tr}(UCU).
$$

This proves the lemma with $M = 2 d M_0$. $\square$

Note that the constant M in this lemma depends only on d and $\|D_h^2 c_{kj}\|_\infty$, and hence we do not need to assume that $c_{jk} \in W^{2,\infty}$. The lemma holds if the coefficients c_{jk} have only bounded second-order derivatives.

Proof of Theorem 5.2. Since $a_0 \in L^\infty(\mathbb{R}^d, \mathbb{R})$, then by bounded perturbation arguments, we may assume that $a_0 = 0$. Thus we consider that A is given by

$$
Au = -\sum_{k,j=1}^{d} D_j(a_{kj} D_k u) + \sum_{k=1}^{d} b_k D_k u, \ u \in D(A) = C_c^\infty(\mathbb{R}^d).
$$

We have seen that for some constant w, the operator $A + wI$ is accretive. This implies that $\overline{A} + wI$ is accretive, too. In order to show that $\overline{A} + wI$ is m-accretive, we apply Theorem 1.50 with $S = -\Delta = -\frac{\partial^2}{\partial x_1^2} - \cdots - \frac{\partial^2}{\partial x_d^2}$ ((minus) the Laplace operator).

Since the coefficients a_{kj}, b_k are bounded, there exist positive constants α_0 and α, such that

$$
\|Au\|_2 \leq \alpha_0 \|u\|_{H^2(\mathbb{R}^d)} \leq \alpha [\|u\|_2 + \|\Delta u\|_2] \text{ for all } u \in C_c^\infty(\mathbb{R}^d). \quad (5.6)
$$

Using this and the density of $C_c^\infty(\mathbb{R}^d)$ in $H^2(\mathbb{R}^d) = D(S)$, we obtain

$$
D(S) \subseteq D(\overline{A}).
$$

Now we show that there exists a constant $\beta \in \mathbb{R}$ such that

$$
(\overline{A}u; -\Delta u) \geq -\beta(u; -\Delta u) \text{ for all } u \in H^2(\mathbb{R}^d). \quad (5.7)
$$

Note that because of (5.6), it suffices to prove (5.7) for all $u \in C_c^\infty(\mathbb{R}^d)$. Fix $u \in C_c^\infty(\mathbb{R}^d)$ and let $U := (D_k D_j u)_{1 \le k, j \le d}$ be the Jacobian matrix of u. Integration by parts gives

$$\left(-\sum_{k,j=1}^{d} D_j(a_{kj} D_k u); -\Delta u \right)$$

$$= -\sum_{k,j,l=1}^{d} \int_{\mathbb{R}^d} \left[D_j(D_l a_{kj} D_k u) D_l u + D_j(a_{kj} D_l D_k u) D_l u \right] dx$$

$$= -\sum_{k,j,l=1}^{d} \int_{\mathbb{R}^d} D_j(D_l a_{kj} D_k u) D_l u\, dx + \sum_{k,j,l=1}^{d} \int_{\mathbb{R}^d} a_{kj} D_l D_k u D_j D_l u\, dx.$$

The second term satisfies

$$\sum_{k,j,l=1}^{d} \int_{\mathbb{R}^d} a_{kj} D_l D_k u D_j D_l u\, dx = \int_{\mathbb{R}^d} \mathrm{tr}(U(x)(a_{kj}(x))U(x))\, dx \quad (5.8)$$

and the first one can be rewritten as

$$-\sum_{k,j,l=1}^{d} \int_{\mathbb{R}^d} D_j D_l a_{kj} D_k u D_l u\, dx - \sum_{k,j,l=1}^{d} \int_{\mathbb{R}^d} D_l a_{kj} D_j D_k u D_l u\, dx$$

$$=: I + II.$$

Since all the terms $D_j D_l a_{kj}$ are bounded on $\mathbb{R}^d$, we have

$$I \ge -M_1 \sum_k \int_{\mathbb{R}^d} |D_k u|^2 dx = -M_1(u; -\Delta u) \qquad (5.9)$$

for some positive constant M_1. The term II satisfies

$$II = -\sum_l \int_{\mathbb{R}^d} \mathrm{tr}((D_l a_{kj})U) D_l u\, dx$$

$$\ge -\frac{\varepsilon}{2} \sum_l \int_{\mathbb{R}^d} [\mathrm{tr}((D_l a_{kj})U)]^2 dx - \frac{1}{2\varepsilon} \int_{\mathbb{R}^d} |D_l u|^2 dx,$$

where the last inequality is valid for every $\varepsilon > 0$. Applying now Lemma 5.5, we obtain

$$II \ge -\varepsilon M_2 \int_{\mathbb{R}^d} \mathrm{tr}(U(x)(a_{kj}(x))U(x))\, dx - \frac{1}{2\varepsilon}(u; -\Delta u), \qquad (5.10)$$

where M_2 is a constant. Choosing $\varepsilon > 0$ small enough and using (5.8), (5.9), (5.10), we see that for some constant $\beta_0 > 0$

$$\left(-\sum_{k,j=1}^{d} D_j(a_{kj}D_k u); -\Delta u \right) \geq -\beta_0(u; -\Delta u). \qquad (5.11)$$

We study now terms of order 1. We integrate again by parts

$$\left(\sum_{k=1}^{d} b_k D_k u; -\Delta u \right)$$

$$= \sum_{k,l=1}^{d} \int_{\mathbb{R}^d} D_l b_k D_k u D_l u dx + \sum_{k,l=1}^{d} \int_{\mathbb{R}^d} b_k D_l D_k u D_l u dx$$

$$= \sum_{k,l=1}^{d} \int_{\mathbb{R}^d} D_l b_k D_k u D_l u dx - \sum_{k,l=1}^{d} \int_{\mathbb{R}^d} D_k b_k |D_l u|^2 dx$$

$$- \sum_{k,l=1}^{d} \int_{\mathbb{R}^d} b_k D_l u D_k D_l u dx.$$

Using the assumption $b_k \in W^{1,\infty}(\mathbb{R}^d)$, we can find two constants M_3 and M_4 such that

$$\left(\sum_{k=1}^{d} b_k D_k u; -\Delta u \right) \geq -M_3 \sum_{k} \int |D_k u|^2 dx - \frac{1}{2} \sum_{k,l=1}^{d} \int_{\mathbb{R}^d} D_k b_k [D_l u]^2 dx$$

$$\geq -M_4 \sum_{k} \int |D_k u|^2 dx.$$

We have then shown that

$$\left(\sum_{k=1}^{d} b_k D_k u; -\Delta u \right) \geq -M_4(u; -\Delta u). \qquad (5.12)$$

Combining (5.12) and (5.11), one obtains (5.7).

Theorem 1.50 implies that $\overline{A} + wI$ is m-accretive for some constant w. Finally, we show that

$$D(\overline{A}) = \left\{ u \in L^2, -\sum_{k,j=1}^{d} D_j(a_{kj}D_k u) + \sum_{k=1}^{d} b_k D_k u \right.$$

$$\left. \text{(as a distribution)} \in L^2 \right\}.$$

Note first that

$$-\sum_{k,j=1}^{d} D_j(a_{kj}D_k u) + \sum_{k=1}^{d} b_k D_k u$$

$$= -\sum_{k,j} D_j D_k(a_{kj}u) + \sum_{j}\left[D_j\left(\sum_{k} D_k a_{kj}u + b_j u\right)\right] - \sum_{j} D_j b_j u.$$

Using the assumptions on the coefficients a_{kj} and b_k, we see that for each $u \in L^2(\mathbb{R}^d, \mathbb{R})$, the term on the right-hand side is a distribution.

Define now the operator $A_{\max}$ by the same expression as A but with domain

$$D(A_{\max}) = \left\{u \in L^2, -\sum_{k,j=1}^{d} D_j(a_{kj}D_k u) + \sum_{k=1}^{d} b_k D_k u \right.$$

$$\left. (\text{as a distribution}) \in L^2\right\}.$$

Our aim is to show that $\overline{A} = A_{\max}$. Obviously, $A_{\max}$ is an extension of $\overline{A}$.

Define the operator

$$Bu := -\sum_{k,j=1}^{d} D_j(a_{kj}D_k u) - \sum_{k=1}^{d} b_k D_k u - \sum_{k=1}^{d} D_k b_k u,$$

$$D(B) = C_c^{\infty}(\mathbb{R}^d).$$

Since the coefficients of the operator B satisfy the same assumptions as those of A, there exists $w \in \mathbb{R}$ such that the operator $wI + \overline{B}$ is m-accretive (here $\overline{B}$ denotes the closure of B).

Fix $\lambda > 0$ large enough and let $u \in D(A_{\max})$ be such that $\lambda u + A_{\max}u = 0$. Then for every $\phi \in C_c^{\infty}(\mathbb{R}^d)$,

$$0 = (\lambda u + A_{\max}u; \phi) = (u; \lambda\phi + B\phi).$$

This implies that $0 = (u; \lambda\phi + \overline{B}\phi)$ for every $\phi \in D(\overline{B})$. Thus, if λ is large enough, $u = 0$.

If $u \in D(A_{\max})$, then $\lambda u + A_{\max}u \in L^2(\mathbb{R}^d)$, and since $wI + \overline{A}$ is m-accretive, there exists $\psi \in D(\overline{A})$ such that

$$\lambda u + A_{\max}u = \lambda\psi + \overline{A}\psi.$$

Thus $\lambda(u-\psi) + A_{\max}(u-\psi) = 0$ and we have just proved that this implies $u = \psi$. This shows that $D(A_{\max}) = D(\overline{A})$. $\square$

5.2.2 Positivity and L^∞-contractivity

We assume that the assumptions of Theorem 5.2 hold. By Corollary 5.3, the operator $-\overline{A}$ generates a strongly continuous semigroup $(e^{-t\overline{A}})_{t\geq 0}$ on $L^2(\mathbb{R}^d, \mathbb{R})$. We can extend this semigroup to a strongly continuous semigroup on $L^p(\mathbb{R}^d, \mathbb{R})$ for all $p \in [1, \infty)$. More precisely,

THEOREM 5.6 *Assume that the assumptions of Theorem 5.2 are satisfied. Then the semigroup $(e^{-t\overline{A}})_{t\geq 0}$ is positive and there exists a constant w such that $(e^{-t(\overline{A}+w)})_{t\geq 0}$ is L^∞-contractive. If $a_0 \geq 0$, we may take $w = 0$. In particular, $(e^{-t\overline{A}})_{t\geq 0}$ extends to a strongly continuous semigroup on $L^p(\mathbb{R}^d, \mathbb{R})$ for all $p \in [1, \infty)$.*

Proof. Define

$$A_n = -\sum_{k,j=1}^{d} D_j \left((a_{kj}(x) + \frac{1}{n}\delta_{kj})D_k u \right) + \sum_{k=1}^{d} b_k D_k u + a_0 u,$$

$$D(A_n) = C_c^\infty(\mathbb{R}^d).$$

Applying Theorem 5.2, we obtain that A_n is closable and there exists a constant w' (independent of n) such that $\overline{A}_n + w'I$ is m-accretive. Corollaries 4.3 and 4.10 show that there exists w, such that for every n, the semigroup $(e^{-t(\overline{A}_n+w)})_{t\geq 0}$ is sub-Markovian. In addition, if $a_0 \geq 0$, then $(e^{-t\overline{A}_n})_{t\geq 0}$ is sub-Markovian. Theorem 5.7 below shows that the same properties hold for $(e^{-t\overline{A}})_{t\geq 0}$. As a consequence, $(e^{-t\overline{A}})_{t\geq 0}$ extends to $L^p(\mathbb{R}^d, \mathbb{R})$ (as a strongly continuous semigroup) for all $p \in [2, \infty)$. This also holds for $p \in [1, 2]$ by duality, since the coefficients of the adjoint A^* satisfy the same assumptions as those of A. The strong continuity on $L^1(\mathbb{R}^d, \mathbb{R})$ was shown in a general setting in Chapter 2. $\qquad\square$

We quote the following convergence theorem without proof (see, e.g., Pazy [Paz83], Chap. 3, Theorem 4.5)

THEOREM 5.7 *Assume that $-A_n$ and $-A$ are generators of strongly continuous semigroups $(e^{-tA_n})_{t\geq 0}$ and $(e^{-tA})_{t\geq 0}$ on a Banach space X. Assume that $(e^{-tA_n})_{t\geq 0}$ is uniformly bounded (with respect to n) in $\mathcal{L}(X)$ and assume that there exists a core D of A such that*
1) $D \subseteq D(A_n)$ for all n;
2) $A_n u$ converges to Au as $n \to \infty$, for every $u \in D$.
Then for every $t \geq 0$, e^{-tA_n} converges strongly to e^{-tA}.

Notes
Section 5.1. The smoothness condition (5.3) is assumed in order to guarantee the

closability of the symmetric form defined there. More general results on closability of such forms can be found in Ma and Röckner [MaRö92] and the references there. Theorem 5.1 is well-known and even more general results of this type can be found in Fukushima [Fuk80] and Fukushima, Oshima and Takeda [FOT94]. Degenerate-elliptic equations $\sum_{k,j=1}^{d} D_k(a_{kj}D_j)u + a(x)u = f(x)$ with boundary conditions have been studied by several authors. See Fichera [Fic56], Oleinik [Ole67], and Devinatz [Dev78]. More recent results dealing with regularity properties of solutions to degenerate-elliptic equations are proved by Fabes, Kenig, and Serapioni [FKS82], Franchi and Serapioni [FrSe87], Franchi, Serapioni, and Serra Cassano [FSSC98], and Taira, Favini, and Romanelli [TFR00].

Section 5.2. The idea of using Theorem 1.50 and the Oleinik inequality to prove Theorem 5.2 is mentioned in Wong-Dzung [Won81]. A more elaborate proof is used in Wong-Dzung [Won81] and [Won83] to prove an L^p-version of Theorem 5.2 under the condition $a_{kj} \in C^2$ with bounded second order derivatives and $b_k \in C^1$ with bounded first order derivatives. Related results can also be found in Stroock and Varadhan [StVa79], Chapter 3. Lemma 5.5 is proved in Oleinik [Ole67] and in Stroock and Varadhan [StVa79] (Lemma 3.2.3).

Chapter Six

GAUSSIAN UPPER BOUNDS FOR HEAT KERNELS

6.1 HEAT KERNEL BOUNDS, SOBOLEV, NASH, AND GAGLIARDO-NIRENBERG INEQUALITIES

Let (X, μ) be a σ-finite measure space and let $L^p := L^p(X, \mu)$ be the corresponding complex or real Lebesgue spaces. In this section we study $L^2 - L^\infty$ estimates of the type:

$$\|e^{-tA}\|_{\mathcal{L}(L^2, L^\infty)} \le ct^{-d/4} \text{ for all } t > 0. \tag{6.1}$$

Since the operators under consideration in this book are defined by sesquilinear forms, we shall focus on characterizations of (6.1) in terms of forms. We mainly concentrate on the case of symmetric forms and assume that the associated semigroup is L^∞-contractive. The latter property can be removed in the statements. It will be assumed only for simplicity.

We start with the following extrapolation result.

LEMMA 6.1 *Let $(T(t))_{t \ge 0}$ be a strongly continuous semigroup on L^2. Assume that $(T(t))_{t \ge 0}$ is L^∞-contractive. If there exists $r > 2$ such that $T(t)$ maps L^2 into L^r and*

$$\|T(t)\|_{\mathcal{L}(L^2, L^r)} \le ct^{-\alpha} \text{ for all } t > 0,$$

where c and α are positive constants, then $T(t)$ is bounded from L^2 into L^∞ and

$$\|T(t)\|_{\mathcal{L}(L^2, L^\infty)} \le c't^{-\frac{r\alpha}{r-2}} \text{ for all } t > 0,$$

where c' is a positive constant depending only on r, α, and c.

Proof. By the Riesz-Thorin interpolation theorem, we have for every $p \in [2, \infty)$

$$\|T(t)\|_{\mathcal{L}(L^p, L^{pr/2})} \le c^{2/p} t^{-2\alpha/p} \text{ for all } t > 0. \tag{6.2}$$

Let $t_k := \frac{r-1}{r} r^{-k}$ and $p_k := 2(\frac{r}{2})^k$ for $k \geq 0$. Thus,

$$\sum_{k=0}^{\infty} t_k = 1 \text{ and } \sum_{k=0}^{\infty} \frac{1}{p_k} = \frac{r}{2(r-2)}.$$

Applying (6.2) with $p = p_k$ yields

$$\|T(t)\|_{\mathcal{L}(L^2, L^\infty)} \leq \prod_{k=0}^{\infty} \|T(t t_k)\|_{\mathcal{L}(L^{p_k}, L^{p_{k+1}})}$$

$$\leq \prod_{k=0}^{\infty} c^{2/p_k} t^{-2\alpha/p_k} t_k^{-2\alpha/p_k}$$

$$= c' t^{-\frac{r\alpha}{r-2}},$$

which proves the lemma.   □

The conclusion of Lemma 6.1 holds without assuming that $(T(t))_{t\geq 0}$ is L^∞-contractive; it is enough to assume that the semigroup is uniformly bounded on L^∞. We refer the reader to Coulhon [Cou91] and Varopoulos et al. [VSC92].

Throughout this section, $\mathfrak{a}$ denotes a densely defined, symmetric, accretive, and closed form on L^2. We denote by A its associated self-adjoint operator and by $(e^{-tA})_{t\geq 0}$ the semigroup generated by $-A$. We will use the classical formula:

$$\mathfrak{a}(u, v) = (A^{1/2} u, A^{1/2} v) \text{ for all } u, v \in D(A^{1/2}) = D(\mathfrak{a}),$$

where $A^{1/2}$ is the square root of A.[1]

The following result shows that $L^2 - L^\infty$ polynomial decay of the semigroup $(e^{-tA})_{t\geq 0}$ is equivalent to a Gagliardo-Nirenberg type inequality. More precisely,

THEOREM 6.2 *Assume that the semigroup $(e^{-tA})_{t\geq 0}$ is L^∞-contractive. The following assertions are equivalent (here c, c', and d are positive constants):*

1) $\|e^{-tA}\|_{\mathcal{L}(L^2, L^\infty)} \leq c t^{-d/4}$ for all $t > 0$.

2) For every $q \in (2, \infty]$ such that $d \frac{q-2}{2q} < 1$, we have

$$\|u\|_q \leq c' \mathfrak{a}(u, u)^{d\frac{q-2}{4q}} \|u\|_2^{1 - d\frac{q-2}{2q}} \text{ for all } u \in D(\mathfrak{a}).$$

---

[1] $A^{1/2}$ is the self-adjoint accretive operator such that $(A^{1/2})^2 = A$. See, e.g., Kato [Kat80] or Chapter 8.

3) *There exists* $q \in (2, \infty]$, $d\frac{q-2}{2q} < 1$ *such that*

$$\|u\|_q \le c' \mathfrak{a}(u, u)^{d\frac{q-2}{4q}} \|u\|_2^{1-d\frac{q-2}{2q}} \quad for \ all \ u \in D(\mathfrak{a}).$$

Proof. Of course, 2) $\Longrightarrow$ 3). Assume now that 3) holds. For every $u \in D(\mathfrak{a})$ and $t > 0$,

$$\|e^{-tA}u\|_q \le c' \mathfrak{a}(e^{-tA}u, e^{-tA}u)^{d\frac{q-2}{4q}} \|e^{-tA}u\|_2^{1-d\frac{q-2}{2q}}$$

$$\le c' \|u\|_2^{1-d\frac{q-2}{2q}} \|Ae^{-tA}u\|_2^{d\frac{q-2}{4q}} \|e^{-tA}u\|_2^{d\frac{q-2}{4q}}$$

$$\le c'' t^{-d\frac{q-2}{4q}} \|u\|_2.$$

Assertion 1) holds now by applying the previous lemma.

It remains to prove that 1) $\Longrightarrow$ 2). By the Riesz-Thorin interpolation theorem, it follows from 1) that

$$\|e^{-tA}\|_{\mathcal{L}(L^2, L^q)} \le c^{\frac{q-2}{q}} t^{-d\frac{q-2}{4q}} \quad \text{for all } t > 0. \tag{6.3}$$

Let $u \in D(\mathfrak{a}) \setminus \{0\}$ and write $u = e^{-tA}u + \int_0^t Ae^{-sA}u \, ds$. Using (6.3) we have for some positive constants c_i,

$$\|u\|_q \le \|e^{-tA}u\|_q + \int_0^t \|e^{-s/2A}Ae^{-s/2A}u\|_q ds$$

$$\le c_1 \left[t^{-d\frac{q-2}{4q}} \|u\|_2 + \int_0^t s^{-d\frac{q-2}{4q}} \|A^{1/2}e^{-s/2A}A^{1/2}u\|_2 ds \right]$$

$$\le c_2 \left[t^{-d\frac{q-2}{4q}} \|u\|_2 + \int_0^t s^{-d\frac{q-2}{4q}-\frac{1}{2}} \|A^{1/2}u\|_2 ds \right]$$

$$\le c_3 [t^{-d\frac{q-2}{4q}} \|u\|_2 + t^{-d\frac{q-2}{4q}+\frac{1}{2}} \mathfrak{a}(u, u)^{1/2}].$$

Assertion 2) follows by choosing $t = \|u\|_2^2 \mathfrak{a}(u, u)^{-1}$ (observe that 1) implies that 0 cannot be an eigenvalue of A, and hence $\mathfrak{a}(u, u) \ne 0$). $\quad \square$

Let $(e^{-tB})_{t \ge 0}$ be a bounded holomorphic semigroup on L^p for some $p \in [1, \infty)$. Using a similar proof to that of Theorem 6.2, one shows that the estimate

$$\|e^{-tB}\|_{\mathcal{L}(L^p, L^\infty)} \le ct^{-\frac{d}{2p}} \quad \text{for all } t > 0$$

is equivalent to the inequality

$$\|u\|_\infty \le c' \|u\|_p^{1-\frac{d}{p\beta}} \|B^{\beta/2}u\|_p^{\frac{d}{p\beta}} \quad \text{for all } u \in D(B^{\beta/2}) \tag{6.4}$$

for every (or some) β such that $\beta > d/p$.[2]

The next result gives a characterization in terms of the Nash inequality.

THEOREM 6.3 *Assume that the symmetric semigroup* $(e^{-tA})_{t\geq 0}$ *is* L^∞-*contractive. The following assertions are equivalent (here* c, c', *and* d *are positive constants):*
1) $\|e^{-tA}\|_{\mathcal{L}(L^1, L^2)} \leq ct^{-d/4}$ *for all* $t > 0$.
2) $\|u\|_2^{2+4/d} \leq c' \mathfrak{a}(u, u).\|u\|_1^{4/d}$ *for all* $u \in D(\mathfrak{a}) \cap L^1$.

Proof. Assume that 1) holds and let $u \in D(\mathfrak{a}) \cap L^1$ be a nontrivial function. It follows from the equation $u = e^{-tA}u + \int_0^t A^{1/2}e^{-sA}A^{1/2}u\,ds$ that

$$\|u\|_2 \leq c_1 \left[t^{-d/4}\|u\|_1 + \int_0^t \frac{1}{\sqrt{s}}\|A^{1/2}u\|_2 ds \right]$$
$$\leq c_2 [t^{-d/4}\|u\|_1 + \sqrt{t}\mathfrak{a}(u, u)^{1/2}].$$

Assertion 2) follows by choosing $t = \mathfrak{a}(u, u)^{-\frac{2}{d+2}}\|u\|_1^{\frac{4}{d+2}}$.

Conversely, assume that 2) holds and let $f \in L^2 \cap L^1$ be a nontrivial function. Set $\phi(t) := \frac{1}{2}\|e^{-tA}f\|_2^2$. We have for $t > 0$

$$\frac{d}{dt}\phi(t) = -\mathfrak{a}(e^{-tA}f, e^{-tA}f)$$
$$\leq -\frac{1}{c'}\|e^{-tA}f\|_2^{2+4/d}\|e^{-tA}f\|_1^{-4/d}$$
$$\leq -c_1\phi(t)^{1+2/d}\|f\|_1^{-d/4}.$$

This means that $\frac{d}{dt}\phi(t)^{-2/d} \geq c_2\|f\|_1^{-4/d}$. Integration from 0 to t yields assertion 1). $\square$

We quote the following characterization by the Sobolev inequality.

THEOREM 6.4 *Let* $d > 2$ *be a constant. Assume that the symmetric semi-group* $(e^{-tA})_{t\geq 0}$ *is* L^∞-*contractive. The following assertions are equivalent:*
1) $\|e^{-tA}\|_{\mathcal{L}(L^2, L^\infty)} \leq ct^{-d/4}$ *for all* $t > 0$.
2) $\|u\|_{\frac{2d}{d-2}}^2 \leq c'\mathfrak{a}(u, u)$ *for all* $u \in D(\mathfrak{a})$.

Assertion 2) applied to $u = e^{-tA}f$ gives

$$\|e^{-tA}f\|_{\frac{2d}{d-2}} \leq c_1 t^{-1/2}\|f\|_2.$$

---

[2]For more information and details, see Coulhon [Cou92] or Varopoulos et al. [VSC92].

This and Lemma 6.1 give 1). The converse holds as in Davies [Dav89], p. 76, Varopoulos et al. [VSC92], p. 21, or Varopoulos [Var85] (there the positivity of the semigroup is not needed in the proof). Assertion 2) means that the operator $A^{-1/2}$ given by

$$A^{-1/2}u = \Gamma\left(\frac{1}{2}\right)^{-1}\int_0^\infty t^{-1/2}e^{-tA}u\,dt$$

is bounded from L^2 into $L^{\frac{2d}{d-2}}$.

Remark. i) As mentioned previously, the extrapolation result (Lemma 6.1) holds without the L^∞-contractivity assumption. Thus, the previous theorems hold as well without this assumption. One only assumes that the semigroup $(e^{-tA})_{t\geq 0}$ is uniformly bounded on L^∞.

ii) In the above theorems, 2) $\Longrightarrow$ 1) does not use the fact that $\mathfrak{a}$ is symmetric. This implication holds in the more general situation of nonsymmetric operators.

iii) If we apply the above theorems to $e^{-t}e^{-tA}$ instead of e^{-tA}, we obtain that the bound in 1) for $0 < t \leq 1$ is equivalent to 2), where $\mathfrak{a}(u,u)$ is now replaced by $\mathfrak{a}(u,u) + \|u\|_2^2$.

iv) The $L^2 - L^\infty$ estimate in 1) yields by duality the same estimate for the $L^1 - L^2$ norm. Therefore

$$\|e^{-tA}\|_{\mathcal{L}(L^1, L^\infty)} \leq Ct^{-d/2} \text{ for all } t > 0,$$

where C is a positive constant. This is equivalent to the fact that e^{-tA} is given by a kernel $p(t, x, y)$, that is, a measurable function on $X \times X$ such that

$$e^{-tA}f(x) = \int_X p(t, x, y)f(y)d\mu(y) \text{ a.e. } x \in X \text{ for all } t > 0, f \in L^2$$

and such that

$$|p(t, x, y)| \leq Ct^{-d/2} \text{ for all } t > 0. \tag{6.5}$$

We will call the kernel $p(t, x, y)$ the heat kernel of A.

The following lemma shows that $L^2 - L^\infty$ estimates of a strongly continuous semigroup can be improved by taking into account $L^2 - L^2$ estimates. More precisely,

LEMMA 6.5 *Let $(T(t))_{t\geq 0}$ be a strongly continuous semigroup on L^2 such that*

$$\|T(t)\|_{\mathcal{L}(L^2)} \leq Me^{-wt} \text{ for all } t \geq 0 \tag{6.6}$$

for some constants $M \geq 1$ and $w \in \mathbb{R}$. Assume that $T(t)$ is bounded from L^2 into L^∞ with norm satisfying

$$\|T(t)\|_{\mathcal{L}(L^2, L^\infty)} \leq Ct^{-d/4}e^{\alpha t} \ \text{for all} \ t > 0 \qquad (6.7)$$

where $C, d > 0$ and α are constants. Then

$$\|T(t)\|_{\mathcal{L}(L^2, L^\infty)} \leq CMet^{-d/4}e^{-wt}\left[1 + \max(\alpha + w, 0)t\right]^{d/4} \ \text{for all} \ t > 0. \qquad (6.8)$$

Proof. Of course, we may assume that $\beta := \alpha + w > 0$; otherwise the conclusion follows directly from (6.7).

Set $S(t) := e^{wt}T(t)$ for $t \geq 0$. By assumption,

$$\|S(t)\|_{\mathcal{L}(L^2)} \leq M \ \text{and} \ \|S(t)\|_{\mathcal{L}(L^2, L^\infty)} \leq Ct^{-d/4}e^{\beta t} \ \text{for all} \ t > 0.$$

Clearly, if $\beta t \leq 1$ then

$$\|S(t)\|_{\mathcal{L}(L^2, L^\infty)} \leq Cet^{-d/4}.$$

Assume now that $\beta t > 1$. We have

$$\begin{aligned}
\|S(t)\|_{\mathcal{L}(L^2, L^\infty)} &\leq \left\|S\left(\frac{1}{\beta}\right)\right\|_{\mathcal{L}(L^2, L^\infty)} \left\|S\left(t - \frac{1}{\beta}\right)\right\|_{\mathcal{L}(L^2)} \\
&\leq MCe\beta^{d/4} \\
&= MCet^{-d/4}(t\beta)^{d/4} \\
&\leq MCet^{-d/4}[1 + t\beta]^{d/4}.
\end{aligned}$$

Thus, we have proved

$$\|S(t)\|_{\mathcal{L}(L^2, L^\infty)} \leq MCet^{-d/4}[1 + t\beta]^{d/4}$$

for every $t > 0$. $\qquad \square$

6.2 HÖLDER-CONTINUITY ESTIMATES OF THE HEAT KERNEL

Theorem 6.2 (and the comments following its proof) shows that the estimate (6.5) is equivalent to Gagliardo-Nirenberg inequalities. We prove now that Hölder continuity of the heat kernel can also be characterized by inequalities of similar type. Before we give the precise statement, we need to fix some notation and assumptions.

Let $-A$ be the generator of a bounded holomorphic semigroup $(e^{-tA})_{t \geq 0}$ on $L^2(X, \mu)$ (here A is not necessarily self-adjoint). We assume that the semigroup $(e^{-tA})_{t\geq 0}$ is uniformly bounded on $L^p := L^p(X,\mu)$ for all $1 \leq p \leq \infty$ and it is given by a heat kernel $p(t,x,y)$. It follows that $(e^{-tA})_{t\geq 0}$ is bounded holomorphic on $L^p, 1 < p < \infty$ (cf. Section 3.3). Denote by $-A_p$ the corresponding generator on $L^p, 1 < p < \infty$ ($A_2 = A$). Let ϱ be a metric on X. Under these assumptions on the semigroup $(e^{-tA})_{t\geq 0}$, we have the following characterization of Hölder continuity of $p(t,x,y)$.

THEOREM 6.6 *Assume that $p(t,x,y)$ satisfies*

$$|p(t,x,y) - p(t,x',y)| \leq Ct^{-\frac{d}{2}-\frac{\eta}{2}}\varrho(x,x')^\eta, \qquad (6.9)$$

for all $t > 0, x, x' \in X$, and μ a.e. $y \in X$, where $\eta \in (0,1), C$, and d are positive constants. Then there exists a constant $C' > 0$ such that

$$\mathrm{Sup}_{x\neq x'} \frac{|f(x) - f(x')|}{\varrho(x,x')^{\frac{\eta}{p}}} \leq C'\|f\|_p^{1-\frac{1}{\beta}(\frac{\eta}{p}+\frac{d}{p})}\|A_p^{\frac{\beta}{2}}f\|_p^{\frac{1}{\beta}(\frac{\eta}{p}+\frac{d}{p})} \qquad (6.10)$$

for all $p \in (1,\infty)$ and β such that $\beta > \frac{\eta}{p} + \frac{d}{p}$ and all $f \in D(A_p^{\frac{\beta}{2}})$.

Conversely, assume that (6.5) and (6.10) hold for some β and $p > 1$ with $\beta > \frac{\eta}{p} + \frac{d}{p}$. Then, $p(t,x,y)$ satisfies (6.9) with $\frac{\eta}{p}$ in place of η.

The same conclusions hold with $p = 1$ if we assume that $(e^{-tA})_{t\geq 0}$ is bounded holomorphic on L^1.

Proof. Throughout this proof, we will denote by $C_1, C_2, \ldots$ all inessential constants.

Assume that (6.5) and (6.10) hold with some β and $p > 1$ ($\beta > \frac{\eta}{p} + \frac{d}{p}$). Using (6.5) and the fact that the semigroup $(e^{-tA})_{t\geq 0}$ is bounded holomorphic on L^p, it is easy to see that for $t > 0$, e^{-tA} maps L^1 into $D(A_p^m)$ for all $m \geq 0$. We apply now (6.10) for $e^{-tA}f$ where f is any function in L^1. We have

$$|e^{-tA}f(x) - e^{-tA}f(x')|$$
$$\leq C'\varrho(x,x')^{\frac{\eta}{p}}\|e^{-tA}f\|_p^{1-\frac{1}{\beta}(\frac{\eta}{p}+\frac{d}{p})}\|A_p^{\frac{\beta}{2}}e^{-tA}f\|_p^{\frac{1}{\beta}(\frac{\eta}{p}+\frac{d}{p})}.$$

From (6.5) we have

$$\|e^{-tA}f\|_p \leq C_1 t^{\frac{-d}{2}(1-1/p)}\|f\|_1.$$

By analyticity of the semigroup,

$$\|A_p^{\frac{\beta}{2}}e^{-tA}f\|_p \leq C_2 t^{\frac{-\beta}{2}}\|e^{-\frac{t}{2}A}f\|_p.$$

Hence

$$|e^{-tA}f(x) - e^{-tA}f(x')| \leq C_3 t^{-\frac{d}{2}-\frac{\eta}{2p}}\varrho(x,x')^{\frac{\eta}{p}}\|f\|_1.$$

This implies (6.9) with $\frac{\eta}{p}$ in place of η.

The same argument shows that if the semigroup $(e^{-tA})_{t\geq 0}$ is bounded holomorphic on L^1, then the conclusion holds with $p = 1$.

Conversely, assume that (6.9) holds. This implies that for every x and x'

$$\|p(t, x, .) - p(t, x', .)\|_\infty \leq Ct^{-\frac{d}{2}-\frac{\eta}{2}}\varrho(x, x')^\eta.$$

The fact that the semigroup is uniformly bounded on L^∞ implies that

$$\|p(t, x, .)\|_1 \leq C_4 \text{ for all } x \in X, t > 0.$$

Now the interpolation inequality implies that for all $p \in (1, \infty)$

$$\|p(t, x, .) - p(t, x', .)\|_p \leq C_5 t^{(1-\frac{1}{p})(-\frac{d}{2}-\frac{\eta}{2})}\varrho(x, x')^{\eta(1-\frac{1}{p})}. \tag{6.11}$$

Let now $f \in L^p$ and $t > 0$. By Hölder's inequality

$$|e^{-tA}f(x) - e^{-tA}f(x')| = \left|\int_X (p(t, x, y) - p(t, x', y))f(y)d\mu(y)\right|$$
$$\leq \|p(t, x, .) - p(t, x', .)\|_q \|f\|_p,$$

where $\frac{1}{p} + \frac{1}{q} = 1$. Using (6.11) we obtain

$$|e^{-tA}f(x) - e^{-tA}f(x')| \leq C_6\varrho(x, x')^{\frac{\eta}{p}}t^{-\frac{d}{2p}-\frac{\eta}{2p}}\|f\|_p. \tag{6.12}$$

We prove now (6.10). Suppose first that $\beta \leq 2$. Let $f \in D(A_p^{\frac{\beta}{2}})$ with $\beta > \frac{\eta}{p} + \frac{d}{p}$. Write

$$f = e^{-tA}f + \int_0^t e^{-\frac{s}{2}A}A_p^{1-\frac{\beta}{2}}e^{-\frac{s}{2}A}A_p^{\frac{\beta}{2}}f\,ds. \tag{6.13}$$

Using (6.12) it follows that

$$|e^{-\frac{s}{2}A}A_p^{1-\frac{\beta}{2}}e^{-\frac{s}{2}A}A_p^{\frac{\beta}{2}}f(x) - e^{-\frac{s}{2}A}A_p^{1-\frac{\beta}{2}}e^{-\frac{s}{2}A}A_p^{\frac{\beta}{2}}f(x')|$$
$$\leq C_6\varrho(x, x')^{\frac{\eta}{p}}s^{-\frac{d}{2p}-\frac{\eta}{2p}}\|A_p^{1-\frac{\beta}{2}}e^{-\frac{s}{2}A}A_p^{\frac{\beta}{2}}f\|_p,$$

and since the semigroup is bounded holomorphic on L^p,

$$\|A_p^{1-\frac{\beta}{2}}e^{-\frac{s}{2}A}A_p^{\frac{\beta}{2}}f\|_p \leq C_7 s^{\frac{\beta}{2}-1}\|A_p^{\frac{\beta}{2}}f\|_p.$$

From (6.12), (6.13), and these two estimates, it follows that for $x \neq x'$

$$\frac{|f(x) - f(x')|}{\varrho(x, x')^{\frac{\eta}{p}}} \leq C_6 t^{-\frac{\eta}{2p}-\frac{d}{2p}}\|f\|_p + C_8 \int_0^t s^{\frac{\beta}{2}-1-\frac{\eta}{2p}-\frac{d}{2p}}ds\|A_p^{\frac{\beta}{2}}f\|_p$$

$$\leq C_9[t^{-\frac{\eta}{2p}-\frac{d}{2p}}\|f\|_p + t^{\frac{\beta}{2}-\frac{\eta}{2p}-\frac{d}{2p}}\|A_p^{\frac{\beta}{2}}f\|_p].$$

By choosing $t = \|f\|_p^{\frac{2}{\beta}}\|A_p^{\frac{\beta}{2}}f\|_p^{\frac{-2}{\beta}}$ we obtain (6.10).

If $\beta > 2$, one iterates (6.13) n times and writes for $n \geq \frac{\beta}{2}$

$$f = e^{-tA}f + tA_pe^{-tA}f + \ldots + \frac{t^{n-1}}{(n-1)!}A_p^{n-1}e^{-tA}f$$

$$+ \frac{1}{(n-1)!}\int_0^t s^{n-1}e^{-\frac{s}{2}A}A_p^{n-\frac{\beta}{2}}e^{-\frac{s}{2}A}A^{\frac{\beta}{2}}f\,ds.$$

Arguing as previously, the desired conclusion follows.   □

Hölder continuity of the heat kernel can also be characterized by a Sobolev type inequality. This was proved recently by Coulhon [Cou03]. He has proved by interpolation methods that Hölder continuity of the heat kernel is equivalent to

$$\sup_{x \neq x'} \frac{|f(x) - f(x')|}{\varrho(x, x')^{\beta - \frac{d}{p}}} \leq C'\|A_p^{\beta/2}f\|_p \tag{6.14}$$

for all p and $\beta > \frac{d}{p}$.

Besides the fact that Hölder continuity of the heat kernel implies regularity of the solution to the corresponding evolution equation, it is a very useful tool in obtaining lower bounds for the heat kernel. We will come back to this in Chapter 7 and show how Gaussian lower bounds can be deduced from Gaussian upper bounds and Hölder continuity.

6.3 GAUSSIAN UPPER BOUNDS

We now consider uniformly elliptic operators as in Chapter 4. Let Ω be an open subset of $\mathbb{R}^d$, endowed with the Lebesgue measure dx. Define the sesquilinear form

$$\mathfrak{a}_V(u, v) = \int_\Omega \left[\sum_{k,j=1}^d a_{kj}D_ku\overline{D_jv} + \sum_{k=1}^d (b_k\overline{v}D_ku + c_ku\overline{D_kv}) + a_0u\overline{v} \right] dx$$

with domain $D(\mathfrak{a}_V) = V$, where V is a closed subspace of $H^1(\Omega)$ that contains $H_0^1(\Omega)$. We assume the uniform ellipticity condition (U.Ell), which we recall again:

(U.Ell) The functions a_{kj}, b_k, c_k, a_0 are bounded on Ω, that is,

$$a_{kj}, b_k, c_k, a_0 \in L^\infty(\Omega, \mathbb{C}) \text{ for all } 1 \leq j, k \leq d, \tag{6.15}$$

and there exists a constant $\eta > 0$ such that

$$\Re \sum_{j,k=1}^{d} a_{kj}(x)\xi_j\overline{\xi_k} \geq \eta|\xi|^2 \text{ for all } \xi \in \mathbb{C}^d \text{ a.e. } x \in \Omega. \qquad (6.16)$$

As in Chapter 4, we denote by A_V the operator associated with the form $\mathfrak{a}_V$. In this section we study Gaussian upper bounds for heat kernels of operators A_V. In order to prove such bounds, we need to show the boundedness from L^1 into L^∞ of e^{-tA_V}, with an estimate of the norm $\|e^{-tA_V}\|_{\mathcal{L}(L^1,L^\infty)}$ which depends in a good way on the coefficients of the operator A_V.

We have seen in the previous section that $L^1 - L^\infty$ estimates for the semigroup can be deduced from the Sobolev inequality. For this reason, we need to assume that

$$V \text{ is continuously embedded into } L^{2^*}(\Omega), \qquad (6.17)$$

where $2^* = \frac{2d}{d-2}$ if $d \geq 3$, $2^* = \infty$ if $d = 1$ and 2^* is any number in $(2, \infty)$ if $d = 2$.

If $V = H_0^1(\Omega)$, where Ω is an arbitrary domain of $\mathbb{R}^d$, this embedding holds and one has for $d \geq 3$

$$\int_\Omega |\nabla u|^2 \geq c\|u\|_{2^*}^2 \text{ for all } u \in H_0^1(\Omega). \qquad (6.18)$$

If $V = H^1(\Omega)$, then (6.17) holds provided Ω has smooth boundary. For example, if Ω has the extension property (i.e., there exists a bounded linear operator $P : H^1(\Omega) \to H^1(\mathbb{R}^d)$ such that Pu is an extension of u from Ω to $\mathbb{R}^d$), then (6.17) follows from the embedding of $H^1(\mathbb{R}^d)$ into $L^{2^*}(\mathbb{R}^d)$. In that case, every closed subspace V of $H^1(\Omega)$ satisfies (6.17).[3] Note that (6.18) for arbitrary open subset Ω follows from the same inequality when $\Omega = \mathbb{R}^d$ (one uses the classical fact that for every $u \in H_0^1(\Omega)$, the function $\tilde{u}$ which extends u by 0 on $\mathbb{R}^d \setminus \Omega$ is in $H^1(\mathbb{R}^d)$). The inequality (6.18) with $\Omega = \mathbb{R}^d$ follows from Theorem 6.4 applied to the Gaussian semigroup, that is, the semigroup generated by the Laplacian on $L^2(\mathbb{R}^d, dx)$.

Note that (6.17) means that

$$\int_\Omega |\nabla u|^2 + \int_\Omega |u|^2 \geq c\|u\|_{2^*}^2 \text{ for all } u \in V, \qquad (6.19)$$

where $c > 0$ is a constant.

---

[3]There are several geometrical conditions on Ω that imply (6.17); see Maz'ya [Maz85], Section 4.9

If $d \leq 2$, it is more convenient to work with Nash or Gagliardo-Nirenberg inequalities. Recall that the Gagliardo-Nirenberg inequality is the inequality

$$c\|u\|_q \leq \|u\|_2^{1-d\frac{q-2}{2q}} (\|\nabla u\|_2 + \|u\|_2)^{d\frac{q-2}{2q}} \text{ for all } u \in V \qquad (6.20)$$

for all $q \in (2, \infty]$ such that $d\frac{q-2}{2q} < 1$. If $V = H^1(\Omega)$, (6.20) holds if and only if (6.19) holds (see Theorems 6.2 and 6.4). As in (6.18), if $V = H_0^1(\Omega)$, the following stronger inequality holds:

$$c\|u\|_q \leq \|u\|_2^{1-d\frac{q-2}{2q}} \|\nabla u\|_2^{d\frac{q-2}{2q}} \text{ for all } u \in H_0^1(\Omega). \qquad (6.21)$$

We will also need the assumption (4.12) on V. That is,

$$u \in V \Rightarrow (1 \wedge |u|)\text{sign } u \in V. \qquad (6.22)$$

In the sequel, w_p denotes the same constant as in Theorem 4.28.

LEMMA 6.7 *Suppose that V satisfies (6.19) and (6.22). Assume that (U.Ell) holds and that the coefficients a_{kj} are real-valued for $1 \leq k, j \leq d$.*
1) If $d \geq 3$, then for every $t > 0$, e^{-tA_V} is bounded from $L^2(\Omega)$ into $L^{2^*}(\Omega)$. Moreover, for every $\varepsilon > 0$

$$\|e^{-tA_V}\|_{\mathcal{L}(L^2, L^{2^*})} \leq C_\varepsilon e^{w_{2^*}t} e^{\varepsilon t} t^{-1/2} \text{ for all } t > 0,$$

where C_ε is a positive constant depending only on η, d, ε and the constant c in (6.19).
2) If $d \leq 2$, then for every $q \in (2, \infty)$ and every $\varepsilon > 0$

$$\|e^{-tA_V}\|_{\mathcal{L}(L^2, L^q)} \leq C_\varepsilon e^{w_q t} e^{\varepsilon t} t^{-d\frac{q-2}{4q}} \text{ for all } t > 0,$$

where C_ε is a positive constant depending only on η, d, ε and the constant c in (6.20).
If $V = H_0^1(\Omega)$, the estimates in both assertions hold with $\varepsilon = 0$.

Proof. Assume that $d \geq 3$. Note that (6.19) implies that for every $\varepsilon > 0$, there exists a constant c_ε such that

$$\int_\Omega |\nabla u|^2 + \varepsilon \int_\Omega |u|^2 \geq c_\varepsilon \|u\|_{2^*}^2 \text{ for all } u \in V. \qquad (6.23)$$

The assumption (U.Ell) implies:

$$\eta\|\nabla u\|_2^2 \leq \Re \sum_{k,j} \int_\Omega a_{kj} D_k u \overline{D_j u} \, dx$$

$$= \Re\left[a_V(u,u) - \sum_k \int_\Omega b_k \bar{u} D_k u + c_k u \overline{D_k u} \, dx - \int_\Omega a_0 |u|^2 dx \right]$$

$$\leq \Re a_V(u,u) + \sum_k \int_\Omega (|\Re(b_k + c_k)| + |\Im(c_k - b_k)|)|D_k u||u| dx$$

$$+ \int_\Omega (\Re a_0)^- |u|^2 dx$$

$$\leq \Re a_V(u,u) + \frac{\eta}{2} \sum_k \|D_k u\|_2^2 + w\|u\|_2^2,$$

where $w = \frac{1}{2\eta} \sum_{k=1}^d \||\Re(b_k + c_k)| + |\Im(c_k - b_k)|\|_\infty^2 + \|(\Re a_0)^-\|_\infty$. Using this, it follows from (6.23) and the expression of w_{2^*} in Theorem 4.28 that for every $\varepsilon > 0$

$$\Re a_V(u,u) + (w_{2^*} + \varepsilon)\int_\Omega |u|^2 dx \geq c_\varepsilon' \|u\|_{2^*}^2 \text{ for all } u \in V, \qquad (6.24)$$

where c_ε' is a constant depending only on η, d, ε, and the constant c in (6.19). Note that (6.24) holds with $\varepsilon = 0$ if $V = H_0^1(\Omega)$ (apply (6.18) instead of (6.19) in the proof).

We now define the semigroup $T(t) := e^{-tA_V} e^{-w_{2^*}t} e^{-\varepsilon t}$. By Theorem 4.28, $e^{-tA_V} e^{-w_{2^*}t}$ is a contraction operator on $L^{2^*}(\Omega)$ (and so is $T(t)$). This and (6.24) imply that for every $f \in L^2(\Omega) \cap L^{2^*}(\Omega)$ and $t > 0$

$$c_\varepsilon' t \|T(t)f\|_{2^*}^2 \leq c_\varepsilon' \int_0^t \|T(s)f\|_{2^*}^2 ds$$

$$\leq \int_0^t \Re\left[a_V(T(s)f, T(s)f) + (w_{2^*} + \varepsilon)(T(s)f; T(s)f) \right] ds$$

$$= \int_0^t -\frac{d}{ds}\|T(s)f\|_2^2 ds$$

$$= \|f\|_2^2 - \|T(t)f\|_2^2$$

$$\leq \|f\|_2^2.$$

Hence, we have proved

$$\|e^{-tA_V} f\|_{2^*} \leq \frac{1}{c_\varepsilon'} t^{-1/2} e^{w_{2^*}t} e^{\varepsilon t} \|f\|_2$$

and if $V = H_0^1(\Omega)$, this estimate holds without the extra term $e^{\varepsilon t}$.

The proof of assertion 2) is similar; one uses (6.20) instead of (6.19) (or (6.21) instead of (6.18) if $V = H_0^1(\Omega)$). The analog of (6.24) is

$$\left[\Re a_V(u, u) + (w_q + \varepsilon)\|u\|_2^2\right]^{d\frac{q-2}{4q}} \|u\|_2^{1-d\frac{q-2}{2q}} \geq c_\varepsilon'\|u\|_q \text{ for all } u \in V.$$
(6.25)

Thus, as previously, if $T(t) := e^{-tA_V}e^{-w_q t}e^{-\varepsilon t}$ then

$$c_\varepsilon' t \|T(t)f\|_q^{\frac{4q}{d(q-2)}}$$

$$\leq c_\varepsilon' \int_0^t \|T(s)f\|_q^{\frac{4q}{d(q-2)}} ds$$

$$\leq \int_0^t \|T(s)f\|_2^{\frac{4q}{d(q-2)}-2} \Re\left[a_V(T(s)f, T(s)f) + (w_q + \varepsilon)\|T(s)f\|_2^2\right] ds$$

$$\leq \int_0^t \|f\|_2^{\frac{4q}{d(q-2)}-2} \Re\left[a_V(T(s)f, T(s)f) + (w_q + \varepsilon)\|T(s)f\|_2^2\right] ds$$

$$= \|f\|_2^{\frac{4q}{d(q-2)}-2} \int_0^t -\frac{d}{ds}\|T(s)f\|_2^2 ds$$

$$= \|f\|_2^{\frac{4q}{d(q-2)}-2}\left[\|f\|_2^2 - \|T(t)f\|_2^2\right]$$

$$\leq \|f\|_2^{\frac{4q}{d(q-2)}}.$$

This proves the lemma. $\square$

Define

$$s(A_V) := \inf\{\Re a_V(u, u), u \in V \text{ and } \|u\|_2 = 1\}.$$
(6.26)

Clearly, the form $a_V - s(A_V)$ is accretive. Hence $\|e^{-t(A_V-s(A_V))}\|_{\mathcal{L}(L^2)} \leq 1$ for all $t \geq 0$. That is,

$$\|e^{-tA_V}\|_{\mathcal{L}(L^2)} \leq e^{-s(A_V)t} \ t \geq 0.$$
(6.27)

If $b_V := \frac{1}{2}(a_V + a_V^*)$ denotes the symmetric part of a_V, then

$$s(A_V) = \inf\{b_V(u, u), u \in V \text{ and } \|u\|_2 = 1\},$$

which is by the min-max principle the spectral bound of the self-adjoint operator B_V associated with b_V. That is, $s(A_V) = \inf \sigma(B_V)$.

THEOREM 6.8 *Suppose that the assumptions of Lemma 6.7 are satisfied. Then, for every $t > 0$, e^{-tA_V} is bounded from $L^2(\Omega)$ into $L^\infty(\Omega)$ and*

$$\|e^{-tA_V}\|_{\mathcal{L}(L^2, L^\infty)} \leq C_\varepsilon t^{-d/4} e^{-s(A_V)t} \left[1 + \alpha_1 t + c'\alpha_2 t + \varepsilon t + s(A_V)t \right]^{d/4}$$

for all $t > 0$, $\varepsilon > 0$, where C_ε is a positive constant depending only on ε, d, the ellipticity constant η, and the constant c in (6.19), $c' \geq 1$ is a constant depending only on d. The constants α_1 and α_2 are given by

$$\alpha_1 := \|(\Re a_0)^-\|_\infty + \frac{1}{\eta} \sum_{k=1}^d \|b_k - c_k\|_\infty^2, \quad \alpha_2 := \frac{1}{\eta} \sum_{k=1}^d \|\Re c_k\|_\infty^2.$$

If $V = H_0^1(\Omega)$, the above estimate holds with $\varepsilon = 0$.

Proof. Using (4.49) and (4.50) one may choose a constant $c' \geq 1$ such that $\alpha_1 + c'\alpha_2 + s(A_V) \geq 0$. Therefore, by Lemma 6.5 and (6.27), it is enough to prove the estimate

$$\|e^{-tA_V}\|_{\mathcal{L}(L^2, L^\infty)} \leq C_\varepsilon' t^{-d/4} e^{\alpha_1 t} e^{\alpha_2 c' t} e^{\varepsilon t}. \tag{6.28}$$

Assume first that $d \geq 3$. For every $r \geq 2$, we have by Theorem 4.28

$$\|e^{-tA_V}\|_{\mathcal{L}(L^r)} \leq e^{w_r t} \text{ for all } t \geq 0.$$

Using this and Lemma 6.7, we obtain by the Riesz-Thorin interpolation theorem

$$\|e^{-tA_V}\|_{\mathcal{L}(L^{p_\theta}, L^{q_\theta})} \leq C_\varepsilon^\theta e^{w_r(1-\theta)t} e^{w_{2^*}\theta t} e^{\varepsilon\theta t} t^{-\theta/2} \text{ for all } t > 0, \tag{6.29}$$

where $\frac{1}{p_\theta} = \frac{\theta}{2} + \frac{1-\theta}{r}$, $\frac{1}{q_\theta} = \frac{\theta}{2^*} + \frac{1-\theta}{r}$ for $\theta \in [0, 1]$.

Fix $p \in (2, \infty)$ and choose $\theta = \frac{1}{p}$ and $r = 2(p-1)$. Thus, $p_\theta = p$ and $q_\theta = p\frac{d}{d-1}$. In addition,

$$(1-\theta)w_r + \theta w_{2^*} \leq \alpha_1 + \alpha_2 \frac{2(p-1)^2 + 2d/d - 2}{p} := \alpha_1 + \alpha_2 \gamma_p.$$

Inserting this in (6.29) gives that

$$\|e^{-tA_V}\|_{\mathcal{L}(L^p, L^{pd/d-1})} \leq C_\varepsilon^{1/p} e^{\alpha_1 t} e^{\alpha_2 \gamma_p t} e^{t\varepsilon/p} t^{-1/(2p)} \text{ for all } t > 0. \tag{6.30}$$

This estimate holds with $\varepsilon = 0$ if $V = H_0^1(\Omega)$.

Set $R := \frac{d}{d-1}$, $t_k := \frac{d+1}{2d}(2R)^{-k}$ and $p_k = 2R^k$ for all integer $k \geq 0$. We have

$$\sum_{k \geq 0} t_k = 1, \quad \sum_{k \geq 0} \frac{1}{p_k} = \frac{d}{2}, \quad c' := \max\left(1, \sum_{k \geq 0} t_k \gamma_{p_k}\right),$$

where c' is a positive constant depending only on d. Applying now (6.30) (with p_k in place of p) yields for all $t > 0$

$$\|e^{-tA_V}\|_{\mathcal{L}(L^2, L^\infty)} \leq \prod_{k \geq 0} \|e^{-tt_k A_V}\|_{\mathcal{L}(L^{p_k}, L^{p_{k+1}})}$$

$$\leq \prod_{k \geq 0} C_\varepsilon^{1/p_k} e^{\alpha_1 t t_k} e^{\alpha_2 \gamma_{p_k} t t_k} e^{t\varepsilon/p_k} t^{-1/(2p_k)} t_k^{-1/(2p_k)}$$

$$= C_\varepsilon' t^{-d/4} e^{\varepsilon t d/2} e^{\alpha_1 t} e^{c' \alpha_2 t},$$

which gives the estimate (6.28).

Assume now that $d \leq 2$. We use assertion 2) of the previous lemma with $q = \frac{2n}{n-2}$ where n is any constant in $(2, \infty)$. We have

$$\|e^{-tA_V}\|_{\mathcal{L}(L^2, L^q)} \leq C_\varepsilon t^{-1/2} t^{1/2 - d\frac{q-2}{4q}} e^{w_q t} e^{t\varepsilon} \text{ for all } t > 0.$$

We are in a position to apply the same proof as in the previous case with n in place of d. We obtain

$$\|e^{-tA_V}\|_{\mathcal{L}(L^2, L^\infty)} \leq C_\varepsilon t^{-n/4} t^{(\frac{1}{2} - d\frac{q-2}{4q})n/2} e^{\varepsilon t n/2} e^{\alpha_1 t} e^{c' \alpha_2 t}$$

$$= C_\varepsilon t^{-d/4} e^{\varepsilon t n/2} e^{\alpha_1 t} e^{c' \alpha_2 t}.$$

This proves (6.28). □

For complex-valued coefficients a_{kj}, one has by Theorems 4.25 and 4.27 the following corollary.

COROLLARY 6.9 *Assume that $\sum_{j=1}^d D_j \Im a_{kj} \in L^\infty(\Omega)$, $\Im(a_{kj} + a_{jk}) = 0$ for all $1 \leq k, j \leq d$, and (4.47) holds if $V \neq H_0^1(\Omega)$. Assume that V satisfies (6.22), (6.17) and that $(\Re u)^+ \in V$ for all $u \in V$. Then the conclusions of the above theorem hold with $\|(\Re a_0)^-\|_\infty$ replaced by $\|(\Re a_0 - m)^-\|_\infty$ where m is the function given by (4.38).*

The previous results applied to the adjoint semigroup $(e^{-tA_V^*})_{t \geq 0}$ give an estimate for the $L^1 - L^2$ norm of $(e^{-tA_V})_{t \geq 0}$. Thus, under the assumptions of the last corollary or those of Theorem 6.8 (if the a_{kj} are real), one obtains

$$\|e^{-tA_V}\|_{\mathcal{L}(L^1, L^\infty)} \leq C_\varepsilon t^{-d/2} e^{-s(A_V)t} \left[1 + \varepsilon t + s(A_V)\frac{t}{2} + c_0 \alpha t \right]^{d/2},$$

(6.31)

for all $t > 0, \varepsilon > 0$, where C_ε is a positive constant depending only on η, ε, d, and the constant c in (6.19), and c_0 is a constant depending only on d and η (the precise expression of c_0 in terms of d and η can be obtained easily from the proof of the previous theorem). The constant α is given by

$$\alpha = \|(\Re a_0 - m)^-\|_\infty + \sum_{k=1}^d \|b_k - c_k\|_\infty^2 + \sum_{k=1}^d (\|\Re b_k\|_\infty^2 + \|\Re c_k\|_\infty^2). \quad (6.32)$$

In particular, $\alpha = \|m\|_\infty$ if $\Re b_k = \Re c_k = \Im(b_k - c_k) = (\Re a_0)^- = 0$, $1 \leq k \leq d$. Again, (6.31) holds with $\varepsilon = 0$ if $V = H_0^1(\Omega)$.

The estimate (6.31) implies that e^{-tA_V} is given by a kernel $p_V(t, x, y)$[4] such that

$$|p_V(t, x, y)| \leq C_\varepsilon t^{-d/2} e^{-s(A_V)t} \left[1 + \varepsilon t + s(A_V)\frac{t}{2} + c_0\alpha t \right]^{d/2} \quad (6.33)$$

for all $t > 0$ and $\varepsilon > 0$, where the constants are the same as in (6.31).

Using a perturbation technique due to E.B. Davies, (6.33) can be converted into a Gaussian upper bound. This is possible because of the good control of the constants involved in (6.33).

We shall first apply the following weaker inequality than (6.33):

$$|p_V(t, x, y)| \leq C_\varepsilon t^{-d/2} e^{\varepsilon t + c_0\alpha t} \text{ for all } t > 0, \varepsilon > 0. \quad (6.34)$$

We make the following additional assumption on V :

$$u \in V \Rightarrow e^\psi u \in V \quad (6.35)$$

for every real-valued $C^\infty(\mathbb{R}^d)$-function ψ such that ψ and $|\nabla\psi|$ are bounded on $\mathbb{R}^d$.

Fix $\lambda \in \mathbb{R}$ and ϕ a real-valued bounded C^∞-function on $\mathbb{R}^d$ such that $|\nabla\phi| \leq 1$ on $\mathbb{R}^d$. Under assumption (6.35), one can define the form

$$\mathfrak{b}_V(u, v) := \mathfrak{a}_V(e^{\lambda\phi}u, e^{-\lambda\phi}v) \text{ for } u, v \in V.$$

The form $\mathfrak{b}_V$ has a similar expression as that of $\mathfrak{a}_V$, but with the terms $b_k - \lambda \sum_{j=1}^d a_{kj} D_j\phi$ in place of b_k, $c_k + \lambda \sum_{j=1}^d a_{jk} D_j\phi$ in place of c_k and $a_0 - \lambda^2 \sum_{k,j=1}^d a_{kj} D_k\phi . D_j\phi + \lambda \sum_{k=1}^d (b_k - c_k) D_k\phi$ in place of a_0. Hence, if the assumptions of Theorem 6.8 or those of Corollary 6.9 are satisfied, one can apply (6.34) to the kernel of the semigroup $T(t) := e^{-\lambda\phi} e^{-tA_V} e^{\lambda\phi}$ (generated by (minus) the operator associated with $\mathfrak{b}_V$). This and the assumption $|\nabla\phi| \leq 1$ give

$$\|e^{-\lambda\phi} e^{-tA_V} e^{\lambda\phi}\|_{\mathcal{L}(L^1, L^\infty)} \leq C_\varepsilon t^{-d/2} e^{\varepsilon t} e^{\delta(\alpha + \lambda^2)t}, \quad (6.36)$$

---

[4]This is the heat kernel of A_V.

for all $t > 0$, $\varepsilon > 0$, where C_ε is as in (6.31), δ is a constant depending only on d, η and $\|a_{kj}\|_\infty$; and α is as in (6.32). This implies that

$$|p_V(t, x, y)| \le C_\varepsilon t^{-d/2} e^{\varepsilon t} e^{\delta \alpha t} e^{\lambda(\phi(x) - \phi(y)) + \delta \lambda^2 t} \text{ for all } t > 0, \varepsilon > 0.$$
(6.37)

Taking $\lambda = \frac{\phi(y) - \phi(x)}{2\delta t}$ and optimizing over ϕ yields the Gaussian upper bound

$$|p_V(t, x, y)| \le C_\varepsilon t^{-d/2} e^{\varepsilon t + \delta \alpha t} \exp[-\frac{|x - y|^2}{4\delta t}] \text{ for all } t > 0, \text{ a.e. } x, y \in \Omega.$$
(6.38)

We have proved the following theorem.

THEOREM 6.10 *Assume that (U.Ell), (6.17), (6.22), and (6.35) hold. Assume in addition that one of the following conditions is satisfied:*
i) The coefficients a_{kj} are real-valued for $1 \le k, j \le d$.
ii) $\sum_{j=1}^d D_j \Im a_{kj} \in L^\infty(\Omega), \Im(a_{kj} + a_{jk}) = 0$ for $1 \le j, k \le d$, (4.47)
holds and that $(\Re u)^+ \in V$ for all $u \in V$.
Then, the semigroup e^{-tA_V} is given by a kernel $p_V(t, x, y)$ which satisfies the Gaussian upper bound:

$$|p_V(t, x, y)| \le C_\varepsilon t^{-d/2} e^{\varepsilon t + \delta \alpha t} \exp\left[-\frac{|x - y|^2}{4\delta t}\right]$$

for all $\varepsilon > 0, t > 0$, and a.e. $(x, y) \in \Omega \times \Omega$. Here α is as in (6.32), C_ε is a positive constant depending only on η, ε, d and the constant c in (6.19), δ is a constant depending only on η, d and $\|a_{kj}\|_\infty$.
If $V = H_0^1(\Omega)$, this estimate holds with $\varepsilon = 0$.

Remark. We point out that when we apply the estimate (6.31) to the semigroup $(e^{-\lambda\phi} e^{-tA_V} e^{\lambda\phi})_{t \ge 0}$, we can actually obtain a better estimate in (6.36). Simple calculations show that one can replace the term $e^{\delta \alpha t}$ by $e^{(1+\varepsilon')\alpha t}$ for every $\varepsilon' > 0$ (δ will depend then on ε'). Using this, the Gaussian upper bound in the previous theorem can be replaced by

$$|p_V(t, x, y)| \le C_\varepsilon t^{-d/2} e^{\varepsilon t} e^{(\varepsilon'+1)\alpha t} \exp\left[-\frac{|x - y|^2}{4\delta t}\right]$$

for all $\varepsilon > 0, \varepsilon' > 0$, and $t > 0$. The constants $C_\varepsilon, \alpha, \delta$ are as in the theorem, with the additional condition that δ depends on ε'.

We already know from Theorem 4.28 that the semigroup $(e^{-tA_V})_{t \ge 0}$ acts on $L^p(\Omega)$ for all $p \in (1, \infty)$. We deduce from the previous result that the same holds for $p = 1$. More precisely, using the above remark, we have for every $\varepsilon, \varepsilon' > 0$

$$\|e^{-tA_V}\|_{\mathcal{L}(L^1(\Omega))} \leq C_{\varepsilon,\varepsilon'} e^{\varepsilon t} e^{(1+\varepsilon')\alpha t} \text{ for all } t \geq 0, \qquad (6.39)$$

where $C_{\varepsilon,\varepsilon'}$ is a positive constant and α is as above. This estimate holds with $\varepsilon = 0$ if $V = H_0^1(\Omega)$. In addition, the semigroup $(e^{-tA_V})_{t\geq 0}$ is strongly continuous on $L^1(\Omega)$ (see Theorem 6.16 below).

As mentioned previously, the assumptions on V in Theorem 6.10 are satisfied for $V = H_0^1(\Omega)$, with Ω an arbitrary open set of $\mathbb{R}^d$. They are satisfied for $V = H^1(\Omega)$ or V as in (4.5) if, for example, Ω satisfies the extension property. (The fact that (6.35) holds for V as in (4.5) follows from the corresponding result for $H^1(\Omega)$ and the fact that V is an ideal of $H^1(\Omega)$; see the proof of Theorem 4.21.)

The following example shows, however, that the previous theorem cannot hold for mixed boundary conditions when Ω is an arbitrary domain (in particular, it cannot hold for the Neumann nor the good Neumann boundary conditions on arbitrary domains).

Example 6.3.1   *Consider the bounded open set*

$$\Omega = \cup_{n\geq 0}]2^{-2n-1}, 2^{-2n}[.$$

Let $\Gamma \subseteq \partial\Omega$ be a finite set. For n large enough, the function $u_n :=$ $\chi_{]2^{-2n-1},2^{-2n}[}$ belongs to $V := \overline{\{u_{|\Omega} \in C_c^\infty(\mathbb{R} \setminus \Gamma)\}}^{H^1(\Omega)}$. Indeed, let $v_n \in C_c^\infty(\mathbb{R})$, be such that $v_n = 1$ on $]2^{-2n-1}, 2^{-2n}[$ with support contained in $]2^{-2n-1} - \varepsilon_n, 2^{-2n} + \varepsilon_n[$. Clearly, if $\varepsilon_n > 0$ is small enough, then $v_{n|\Omega} = u_n$. When n is large enough, $v_n \in C_c^\infty(\mathbb{R} \setminus \Gamma)$ since Γ is finite. Define the form

$$\mathfrak{a}_V(u,v) = \int_\Omega u'\overline{v'}dx, \quad D(\mathfrak{a}_V) = V = \overline{\{u_{|\Omega} \in C_c^\infty(\mathbb{R} \setminus \Gamma)\}}^{H^1(\Omega)}.$$

Each u_n (for n large enough) is an eigenfunction of the operator A_V, with eigenvalue 0. Thus, 0 is an eigenvalue of infinite multiplicity. Hence, A_V does not have compact resolvent. As a consequence, e^{-tA_V} cannot be bounded from $L^2(\Omega)$ into $L^\infty(\Omega)$ (otherwise, it will be a Hilbert-Schmidt operator and hence compact). This shows in particular that the Gaussian upper bound cannot hold for $(e^{-tA_V})_{t\geq 0}$.
We can also arrange to take Γ to be infinite in this example.

We turn now to another problem. We have assumed in Theorem 6.10 that the space V satisfies (6.35). This plays an important rôle in the proof of the

Gaussian bound. Unfortunately, this assumption is not satisfied by several spaces V; and this reduces the range of boundary conditions to which the previous theorem applies. If V does not satisfy (6.35), one can still prove a Gaussian upper bound by using another metric that takes into account the boundary conditions.

Let us say that a real-valued function $\phi \in W^{1,\infty}(\mathbb{R}^d)$ is V-admissible if

$$u \in V \Rightarrow e^{\lambda\phi}u \in V \text{ for all } \lambda \in \mathbb{R}.$$

We denote by W the set of all V-admissible functions ϕ such that

$$\phi \in W^{1,\infty}(\mathbb{R}^d) \text{ and } |\nabla\phi| \le 1.$$

Now we define a metric ϱ_V on Ω by

$$\varrho_V(x, y) := \sup\{\phi(x) - \phi(y), \phi \in W\}.$$

We can repeat the same proof as previously with now $\phi \in W$. We obtain (6.37) with $\phi \in W$. We optimize over ϕ and obtain (6.38) with $\varrho_V(x, y)^2$ instead of $|x - y|^2$. Hence, we have

THEOREM 6.11 *Assume that (U.Ell), (6.22), and (6.17) hold. Assume in addition that one of the following conditions satisfied:*
i) The coefficients a_{kj} are real-valued for $1 \le k, j \le d$.
ii) $\sum_{j=1}^{d} D_j \Im a_{kj} \in L^\infty(\Omega), \Im(a_{kj} + a_{jk}) = 0$ for $1 \le j, k \le d$, (4.47) holds, and $(\Re u)^+ \in V$ for all $u \in V$.
Then the semigroup e^{-tA_V} is given by a kernel $p_V(t, x, y)$ that satisfies the Gaussian upper bound

$$|p_V(t, x, y)| \le C_\varepsilon t^{-d/2} e^{\varepsilon t + \delta \alpha t} \exp\left[-\frac{\varrho_V(x, y)^2}{4\delta t}\right]$$

for all $t > 0, \varepsilon > 0$ and a.e. $(x, y) \in \Omega \times \Omega$. Here α is as in (6.32), C_ε is a positive constant depending only on η, ε, d, and the constant c in (6.19), and δ is a constant depending only on η, d, and $\|a_{kj}\|_\infty$.

For every V as in the above theorem, ϱ_V defines a metric on Ω. This follows easily from the fact that $C_c^\infty(\Omega) \subseteq W$. In order to prove this inclusion, let $\phi \in C_c^\infty(\Omega)$ and $u \in V \cap L^\infty(\Omega)$. By Theorem 2.25, we have $(e^\phi - 1) \in H_0^1(\Omega)$ and thus $(e^\phi - 1)^+ \in H_0^1(\Omega)$. The fact that $H_0^1(\Omega)$ is an ideal of $H^1(\Omega)$ implies that $(e^\phi - 1)^+ u \in H_0^1(\Omega) \subseteq V$. Similarly, $(e^\phi - 1)^- \in H_0^1(\Omega)$. Thus, $e^\phi u \in V$ for $u \in V \cap L^\infty(\Omega)$. Approximating u by $(k \wedge |u|)\text{sign } u^5$ as $k \to \infty$, we obtain $e^\phi u \in V$ for all $u \in V$ and $\phi \in C_c^\infty(\Omega)$.

---

[5]By (6.22), $(k \wedge |u|)\text{sign } u \in V$ for every $k \in \mathbb{N}$.

6.4 SHARPER GAUSSIAN UPPER BOUNDS

We assume in this section that the leading coefficients a_{kj} are real-valued and symmetric. In this case we prove sharper Gaussian upper bounds on the heat kernel $p_V(t, x, y)$. These estimates will be obtained by using (6.31) (or (6.33)) rather than the weaker estimate (6.34) used in the proof of Theorem 6.10.

Recall the definition of $s(A_V)$ in (6.26):

$$s(A_V) := \inf\{\Re a_V(u, u),\ u \in V \text{ and } \|u\|_2 = 1\}.$$

Define on Ω the metric

$$\rho(x, y) := \sup\left\{\phi(x) - \phi(y), \phi \in C_c^\infty(\mathbb{R}^d),\ \sum_{k,j=1}^d a_{kj}(x)D_k\phi D_j\phi \leq 1\right\}.$$
(6.40)

We have

THEOREM 6.12 *Assume that (U.Ell), (6.17), (6.22), and (6.35) hold. Assume in addition that $a_{kj} = a_{jk}$ and that a_{kj} are real-valued for $1 \leq k, j \leq d$. Then for every $\varepsilon > 0, t > 0$ and a.e. $x, y \in \Omega$:*

$$|p_V(t, x, y)| \leq M_\varepsilon t^{-d/2} \exp\left\{-\frac{\rho(x, y)^2}{4t} + \frac{\rho(x, y)}{2\sqrt{\eta}}\sum_{k=1}^d \|\Re b_k - \Re c_k\|_\infty\right\}$$

$$\times e^{-s(A_V)t}\left[1 + \varepsilon t + \frac{s(A_V)}{2}t + \alpha t + \frac{\rho(x, y)}{2}\beta + \frac{\rho(x, y)^2}{4t}\gamma\right]^{d/2},$$

where

$$\alpha = \frac{1}{2}\|(\Re a_0)^-\|_\infty + \frac{1}{2}\eta^{-1}\sum_{k=1}^d (\|b_k - c_k\|_\infty^2 + \max(\|\Re c_k\|_\infty^2, \|\Re b_k\|_\infty^2)),$$

$$\beta = 2\eta^{-3/2}\sum_{k,j=1}^d \|a_{kj}\|_\infty\|b_k - c_k\|_\infty$$

$$+\eta^{-3/2}\sum_{k,j=1}^d \|a_{kj}\|_\infty \max(\|\Re c_k\|_\infty, \|\Re b_k\|_\infty),$$

$$\gamma = \frac{5}{2}\eta^{-2}\sum_{k=1}^d\left(\sum_{j=1}^d \|a_{kj}\|_\infty\right)^2.$$

The constant M_ε depends only on the constant c in (6.19), η, ε and d.
The above Gaussian upper bound holds with $\varepsilon = 0$ if $V = H_0^1(\Omega)$.

Proof. Fix $\lambda \in \mathbb{R}$ and $\phi \in C_c^\infty(\mathbb{R}^d, \mathbb{R})$ such that

$$\sum_{k,j=1}^d a_{kj} D_k \phi D_j \phi \le 1 \text{ a.e. on } \Omega.$$

Define the form

$$\mathfrak{b}_V(u, v) := \mathfrak{a}_V(e^{\lambda\phi} u, e^{-\lambda\phi} v).$$

As already mentioned in the proof of Theorem 6.10,

$$\mathfrak{b}_V(u, v) = \int_\Omega \left[\sum_{k,j=1}^d a_{kj} D_k u \overline{D_j v} + \sum_{k=1}^d (b_{k,\lambda,\phi} D_k u . \overline{v} + c_{k,\lambda,\phi} u \overline{D_k v}) \right. $$
$$\left. + a_{0,\lambda,\phi} u \overline{v} \right] dx,$$

where

$$b_{k,\lambda,\phi} = b_k - \lambda \sum_{j=1}^d a_{kj} D_j \phi, \quad c_{k,\lambda,\phi} = c_k + \lambda \sum_{j=1}^d a_{jk} D_j \phi,$$

$$a_{0,\lambda,\phi} = a_0 - \lambda^2 \sum_{k,j=1}^d a_{kj} D_k \phi D_j \phi + \lambda \sum_{j=1}^d (b_j - c_j) D_j \phi.$$

Using the fact that

$$\eta \sum_k |D_k \phi|^2 \le \sum_{k,j} a_{kj} D_k \phi D_j \phi \le 1$$

we see that

$$\Re \mathfrak{b}_V(u, u) \ge w_\lambda \int_\Omega |u|^2 dx,$$

where

$$w_\lambda := s(A_V) - \lambda^2 - |\lambda| \eta^{-1/2} \sum_{k=1}^d \|\Re b_k - \Re c_k\|_\infty.$$

Theorem 6.8 applied to the semigroup $(e^{-\lambda\phi} e^{-tA_V} e^{\lambda\phi})_{t\ge0}$ (associated with the form $\mathfrak{b}_V$) gives for all $t > 0$

$$\|e^{-\lambda\phi} e^{-tA_V} e^{\lambda\phi}\|_{\mathcal{L}(L^2, L^\infty)} \le C_\varepsilon t^{-d/4} [1 + \max(w_\lambda + \alpha_\lambda + \varepsilon, 0)t]^{d/4} e^{-w_\lambda t},$$
$$(6.41)$$

where

$$\alpha_\lambda = \|(\Re a_{0,\lambda,\phi})^-\|_\infty + \eta^{-1}\sum_{k=1}^d \|b_{k,\lambda,\phi} - c_{k,\lambda,\phi}\|_\infty^2 + c'\eta^{-1}\sum_{k=1}^d \|\Re c_{k,\lambda,\phi}\|_\infty^2.$$

Here $c' \geq 1$ is a constant depending only on d.[6] Note that we may take $\varepsilon = 0$ in (6.41) if $V = H_0^1(\Omega)$.

A similar estimate holds for the adjoint semigroup $(e^{\lambda\phi}e^{-tA_V^*}e^{-\lambda\phi})_{t\geq 0}$ (now with $\overline{b_{k,\lambda,\phi}}$ in place of $c_{k,\lambda,\phi}$, $\overline{c_{k,\lambda,\phi}}$ in place of $b_{k,\lambda,\phi}$, and $\overline{a_{0,\lambda,\phi}}$ in place of $a_{0,\lambda,\phi}$). This allows to estimate the $L^1 - L^2$ norm of $e^{-\lambda\phi}e^{-tA_V}e^{\lambda\phi}$ as follows:

$$\|e^{-\lambda\phi}e^{-tA_V}e^{\lambda\phi}\|_{\mathcal{L}(L^1,L^2)} \leq C_\varepsilon t^{-d/4}[1 + \max(w_\lambda + \alpha'_\lambda + \varepsilon, 0)t]^{d/4}e^{-w_\lambda t},$$
$$(6.42)$$

with

$$\alpha'_\lambda = \|(\Re a_{0,\lambda,\phi})^-\|_\infty + \eta^{-1}\sum_{k=1}^d \|b_{k,\lambda,\phi} - c_{k,\lambda,\phi}\|_\infty^2 + c'\eta^{-1}\sum_{k=1}^d \|\Re b_{k,\lambda,\phi}\|_\infty^2.$$

Therefore,

$$\|e^{-\lambda\phi}e^{-tA_V}e^{\lambda\phi}\|_{\mathcal{L}(L^1,L^\infty)}$$
$$\leq \|e^{-\lambda\phi}e^{-\frac{t}{2}A_V}e^{\lambda\phi}\|_{\mathcal{L}(L^1,L^2)}\|e^{-\lambda\phi}e^{-\frac{t}{2}A_V}e^{\lambda\phi}\|_{\mathcal{L}(L^2,L^\infty)}$$
$$\leq C_\varepsilon^2 2^{d/2}t^{-d/2}[1 + \max(w_\lambda + \alpha_\lambda + \varepsilon, 0)t/2]^{d/4}$$
$$\times [1 + \max(w_\lambda + \alpha'_\lambda + \varepsilon, 0)t/2]^{d/4}e^{-w_\lambda t}.$$

Using again the fact that $|D_k\phi|^2 \leq \eta^{-1}$, we estimate α_λ as follows:

$$\alpha_\lambda \leq \|(\Re a_0)^-\|_\infty + \eta^{-1}\sum_{k=1}^d (\|b_k - c_k\|_\infty^2 + c'\|\Re c_k\|_\infty^2)$$

$$+ |\lambda|\left[\eta^{-1/2}\sum_{k=1}^d \|\Re(b_k - c_k)\|_\infty + 4\eta^{-3/2}\sum_{k,j=1}^d \|a_{kj}\|_\infty\|b_k - c_k\|_\infty\right.$$

$$\left. + 2c'\eta^{-3/2}\sum_{k,j=1}^d \|a_{kj}\|_\infty\|\Re c_k\|_\infty\right]$$

$$+ \lambda^2\left[1 + 4\eta^{-2}\sum_{k=1}^d\left(\sum_{j=1}^d \|a_{kj}\|_\infty\right)^2 + c'\eta^{-2}\sum_{k=1}^d\left(\sum_{j=1}^d \|a_{kj}\|_\infty\right)^2\right].$$

---

[6]The value of c' in terms of d can be calculated; see the proof of Theorem 6.8.

A similar estimate holds for α'_λ.

It follows from this and the previous $L^1 - L^\infty$ estimate for $e^{-\lambda\phi}e^{-tA_V}e^{\lambda\phi}$ that

$$\|e^{-\lambda\phi}e^{-tA_V}e^{\lambda\phi}\|_{\mathcal{L}(L^1, L^\infty)}$$

$$\leq C'_\varepsilon c'^{d/2}t^{-d/2}e^{-w_\lambda t}\left[1 + \varepsilon t + \frac{s(A_V)}{2}t + \alpha t + |\lambda|\beta t + \lambda^2 \gamma t\right]^{d/2},$$

$$(6.43)$$

where $C'_\varepsilon = C^2_\varepsilon 2^{d/2}$ and α, β, and γ are the constants in Theorem 6.12. Thus, if we let $M_\varepsilon := C'_\varepsilon c'^{d/2}$ and use the expression of w_λ, the heat kernel $p_V(t, x, y)$ of A_V satisfies for a.e. $x, y \in \Omega$ and every $t > 0$

$$|p_V(t, x, y)|$$

$$\leq M_\varepsilon t^{-d/2}e^{-w_\lambda t}e^{\lambda(\phi(x) - \phi(y))}\left[1 + \varepsilon t + \frac{s(A_V)}{2}t + \alpha t + |\lambda|\beta t + \lambda^2 \gamma t\right]^{d/2}$$

$$= M_\varepsilon t^{-d/2}e^{-s(A_V)t}e^{\lambda(\phi(x) - \phi(y))}\exp\left\{\lambda^2 t + \frac{|\lambda|t}{\sqrt{\eta}}\sum_{k=1}^d \|\Re b_k - \Re c_k\|_\infty\right\}.$$

$$\times\left[1 + \varepsilon t + \frac{s(A_V)}{2}t + \alpha t + |\lambda|\beta t + \lambda^2 \gamma t\right]^{d/2}.$$

Choosing $\lambda = \frac{\phi(y) - \phi(x)}{2t}$, we obtain

$$|p_V(t, x, y)|$$

$$\leq M_\varepsilon t^{-d/2}e^{-s(A_V)t}$$

$$\times \exp\left\{-\frac{|\phi(x) - \phi(y)|^2}{4t} + \frac{|\phi(x) - \phi(y)|}{2\sqrt{\eta}}\sum_{k=1}^d \|\Re b_k - \Re c_k\|_\infty\right\}$$

$$\times\left[1 + \varepsilon t + \frac{s(A_V)}{2}t + \alpha t + \frac{|\phi(x) - \phi(y)|}{2}\beta + \frac{|\phi(x) - \phi(y)|^2}{4t}\gamma\right]^{d/2}$$

$$\leq M_\varepsilon t^{-d/2}e^{-s(A_V)t}$$

$$\times \exp\left\{-\frac{|\phi(x) - \phi(y)|^2}{4t} + \frac{\rho(x, y)}{2\sqrt{\eta}}\sum_{k=1}^d \|\Re b_k - \Re c_k\|_\infty\right\}$$

$$\times\left[1 + \varepsilon t + \frac{s(A_V)}{2}t + \alpha t + \frac{\rho(x, y)}{2}\beta + \frac{\rho(x, y)^2}{4t}\gamma\right]^{d/2}.$$

Optimizing over ϕ yields

$$|p_V(t,x,y)| \leq M_\varepsilon t^{-d/2} e^{-s(A_V)t}$$

$$\times \exp\left\{ -\frac{\rho(x,y)^2}{4t} + \frac{\rho(x,y)}{2\sqrt{\eta}} \sum_{k=1}^{d} \|\Re b_k - \Re c_k\|_\infty \right\}$$

$$\times \left[1 + \varepsilon t + \frac{s(A_V)}{2}t + \alpha t + \frac{\rho(x,y)}{2}\beta + \frac{\rho(x,y)^2}{4t}\gamma \right]^{d/2}.$$

This estimate holds with $\varepsilon = 0$ if $V = H_0^1(\Omega)$. $\qquad\qquad\square$

We make now some comments on the previous theorem, which we summarize in the following corollaries.

COROLLARY 6.13 *If $\Re b_k = \Re c_k$ for $1 \leq k \leq d$, then*

$$|p_V(t,x,y)| \leq M_\varepsilon t^{-d/2} e^{-s(A_V)t} \exp\left\{ -\frac{\rho(x,y)^2}{4t} \right\}$$

$$\times \left[1 + \varepsilon t + \frac{s(A_V)t}{2} + \alpha t + \frac{\rho(x,y)\beta}{2} + \frac{\rho(x,y)^2\gamma}{4t} \right]^{d/2}.$$

Theorem 6.12 and the trivial inequality

$$[1+x]^{d/2} \leq C_\delta e^{\delta x} \text{ for all } x > 0, \delta > 0$$

imply the following corollary:

COROLLARY 6.14 *For every $\varepsilon > 0$ and every $\delta > 0$, there exist constants $C_{\varepsilon,\delta}$ and w_δ such that*

$$|p_V(t,x,y)| \leq C_{\varepsilon,\delta} t^{-d/2} e^{-(1-\delta)s(A_V)t} e^{\varepsilon t} e^{w_\delta t} \exp\left\{ -(1-\delta)\frac{\rho(x,y)^2}{4t} \right\}$$

for every $t > 0$ and a.e. $x, y \in \Omega$. The constant w_δ equals 0 if $\Re b_k = \Re c_k = (\Re a_0)^- = 0$ and $\Im b_k = \Im c_k$ for $1 \leq k \leq d$.
The above estimate holds with $\varepsilon = 0$ if $V = H_0^1(\Omega)$.

If we let μ be a positive constant such that

$$\sum_{k,j=1}^{d} a_{kj}(x)\xi_k\xi_j \leq \mu|\xi|^2 \text{ for all } \xi \in \mathbb{R}^d, \text{ a.e. } x \in \Omega,$$

then we have

COROLLARY 6.15

$$p_V(t, x, y)| \leq M_\varepsilon t^{-d/2} e^{-s(A_V)t}$$

$$\times \left[1 + \varepsilon t + \frac{s(A_V)t}{2} + \alpha t + \frac{|x-y|\beta}{2\sqrt{\eta}} + \frac{|x-y|^2\gamma}{4\eta t} \right]^{d/2}$$

$$\times \exp \left\{ -\frac{|x-y|^2}{4\mu t} + \frac{|x-y|}{2\eta} \sum_{k=1}^{d} \|\Re b_k - \Re c_k\|_\infty \right\}.$$

Again, the bound holds with $\varepsilon = 0$ if $V = H_0^1(\Omega)$.

Proof. Observe that

$$\rho(x, y) = \sup \left\{ \phi(x) - \phi(y), \sum_{k,j=1}^{d} a_{kj} D_k \phi D_j \phi \leq 1 \right\}$$

$$\geq \sup \left\{ \phi(x) - \phi(y), \sum_{k=1}^{d} |D_k \phi|^2 \leq \frac{1}{\mu} \right\}$$

$$= \mu^{-1/2} |x - y|.$$

Similarly, the ellipticity assumption implies

$$\rho(x, y) \leq \eta^{-1/2} |x - y|.$$

The corollary follows from this and Theorem 6.12. □

Remark. 1) The following example shows that one cannot get rid of the term $\exp\{\frac{\rho(x,y)}{2\sqrt{\eta}} \sum_{k=1}^{d} \|\Re b_k - \Re c_k\|_\infty\}$ in the upper bound given by Theorem 6.12.

Let $Au := -u'' - u'$ acting on $L^2(\mathbb{R})$. The heat kernel of this operator is given by

$$p(t, x, y) = (4\pi t)^{-1/2} e^{-t/4} \exp \left\{ -\frac{|x-y|^2}{4t} - \frac{(x-y)}{2} \right\}.$$

2) In the case of real-valued but nonsymmetric coefficients a_{kj}, one obtains from the proof of Theorem 6.12 an estimate where the leading term $\exp\{-\frac{\rho(x,y)^2}{4t}\}$ is replaced by $\exp\{-\frac{\rho(x,y)^2}{4wt}\}$ with $w = 1 + \sum_k (\sum_j \|a_{kj} - a_{jk}\|_\infty)^2$; which takes into account the nonsymmetric part of the matrix (a_{kj}).

3) For complex-valued a_{kj}, the following elementary example shows that one cannot obtain a sharper upper bound by using the metric

$$\rho(x, y) = \sup \left\{ \phi(x) - \phi(y), \phi \in C_c^\infty(\mathbb{R}^d, \mathbb{R}), \Re \sum_{k,j=1}^d a_{kj} D_k \phi D_j \phi \le 1 \right\}.$$
$$(6.44)$$

Indeed, let $a = 1 + is$ where $s \in \mathbb{R}$, $s \ne 0$ and consider $A = -a\Delta = -\nabla a \nabla$ on $L^2(\mathbb{R}^d)$. In this case, the metric ρ defined by (6.44) is simply the Euclidean one. The heat kernel is of course

$$p(t, x, y) = (4\pi(1 + is)t)^{-d/2} \exp \left\{ -\frac{|x - y|^2}{4(1 + is)t} \right\}.$$

Therefore,

$$|p(t, x, y)| = (4\pi|1 + is|t)^{-d/2} \exp \left\{ -\frac{|x - y|^2}{4(1 + s^2)t} \right\},$$

from which we see that the bound in Theorem 6.12 cannot hold with ρ defined by (6.44).

6.5 GAUSSIAN BOUNDS FOR COMPLEX TIME AND L^p-ANALYTICITY

The next result shows that Gaussian upper bounds for $t > 0$ extend to complex t. As a consequence, the semigroup is holomorphic on $L^p(\Omega)$ for all $1 \le p < \infty$.

Let Ω be an open set of $\mathbb{R}^d$ and let $(e^{-tA})_{t \ge 0}$ be a bounded holomorphic semigroup on the sector $\Sigma(\psi) := \{z \in \mathbb{C}, z \ne 0, |\arg z| < \psi\}$ on $L^2(\Omega)$, where $\psi \in (0, \frac{\pi}{2}]$. Suppose that e^{-tA} is given by a kernel $p(t, x, y)$. We have

THEOREM 6.16 *Assume that*

$$|p(t, x, y)| \le Ct^{-d/2} \exp \left[-\frac{c|x - y|^2}{t} \right]$$

for all $t > 0$ and a.e. *$(x, y) \in \Omega \times \Omega$, where C and c are positive constants. Then for every $\nu \in [0, \psi)$, we have*

$$|p(z, x, y)| \le C_\nu (\Re z)^{-d/2} \exp \left[-\frac{c_\nu |x - y|^2}{|z|} \right]$$

for all $z \in \Sigma(\nu)$ and a.e. *$(x, y) \in \Omega \times \Omega$, where C_ν and c_ν are positive constants (depending on ν). In addition, $(e^{-tA})_{t \ge 0}$ extends to a bounded holomorphic semigroup on the sector $\Sigma(\psi)$ on $L^p(\Omega)$ for all $p \in [1, \infty)$.*

In the case of self-adjoint operators, the proof below gives more precise constants in the Gaussian estimate for complex time. This gives, in particular, precise estimates for $\|e^{-zA}\|_{\mathcal{L}(L^p)}$ in terms of $\arg z$. Such precise estimates have several interesting consequences. For example, they allow us to establish a functional calculus and also to prove L^p properties of the Schrödinger group $(e^{itA})_{t\in\mathbb{R}}$. We will turn to these questions in the next chapter.

Proof. Let E and F be compact disjoint subsets of $\mathbb{R}^d$. Fix $f, g \in L^1(\Omega) \cap L^2(\Omega)$ and define the function

$$H(z) := (\chi_E e^{-zA}\chi_F f; g) := \int_\Omega \chi_E(x)(e^{-zA}\chi_F f)(x)\overline{g(x)}dx, \quad (6.45)$$

where χ_E denotes the characteristic function of E.

It follows from the holomorphy of the semigroup on $L^2(\Omega)$ that the function H is holomorphic on $\Sigma(\nu)$. On the other hand, it follows from the Gaussian upper bound that

$$|H(t)| \leq Ct^{-d/2}e^{-c\varrho(E,F)^2/t}\|f\|_1\|g\|_1 \text{ for all } t > 0$$

where $\varrho(E,F)$ denotes the Euclidean distance from E to F. In addition

$$|H(z)| \leq C'(\Re z)^{-d/2}\|f\|_1\|g\|_1 \text{ for all } z \in \Sigma(\nu). \quad (6.46)$$

Indeed, fix $\nu' \in (\nu, \psi)$ and let $\varepsilon > 0$ such that $(1 - \varepsilon)t + is \in \Sigma(\nu')$ for all $t + is \in \Sigma(\nu)$. For every $z = t + is \in \Sigma(\nu)$, write $e^{-zA} = e^{-\varepsilon\frac{t}{2}A}e^{-((1-\varepsilon)t+is)A}e^{-\varepsilon\frac{t}{2}A}$ and use the fact that $\|e^{-zA}\|_{\mathcal{L}(L^2)}$ is bounded (for $z \in \Sigma(\nu')$) to obtain:

$$\begin{aligned}
&\|e^{-zA}\|_{\mathcal{L}(L^1,L^\infty)} \\
&\leq \|e^{-\varepsilon t/2A}\|_{\mathcal{L}(L^2,L^\infty)}\|e^{-((1-\varepsilon)t+is)A}\|_{\mathcal{L}(L^2)}\|e^{-\varepsilon t/2A}\|_{\mathcal{L}(L^1,L^2)} \\
&\leq C_{\nu'}t^{-d/2}.
\end{aligned}$$

This proves (6.46).

An application of Lemma 6.18 below gives for all $z = |z|e^{i\theta} \in \Sigma(\nu)$:

$$|H(z)| \leq C''(\Re z)^{-d/2}\|f\|_1\|g\|_1 \exp\left[-c'\varrho(E,F)^2\frac{\sin(\nu' - |\theta|)}{|z|}\right]. \quad (6.47)$$

Since this is true for all f and g in $L^1(\Omega) \cap L^2(\Omega)$, we conclude that $\chi_E e^{-zA}\chi_F$ is a bounded operator from $L^1(\Omega)$ into $L^\infty(\Omega)$ and

$$\|\chi_E e^{-zA}\chi_F\|_{\mathcal{L}(L^1,L^\infty)} \leq C''(\Re z)^{-d/2} \exp\left[-c'\varrho(E,F)^2\frac{\sin(\nu' - |\theta|)}{|z|}\right]$$

for all $z \in \Sigma(\nu)$.

In other words, for a.e. $x, y \in \Omega$ and all $z \in \Sigma(\nu')$:

$$|\chi_E(x)p(z,x,y)\chi_F(y)| \leq C''(\Re z)^{-d/2} \exp\left[-c'\varrho(E,F)^2 \frac{\sin(\nu' - |\theta|)}{|z|}\right].$$

Choosing appropriate E and F we obtain the Gaussian upper estimate for $p(z, x, y)$.

We prove the second assertion of the theorem. It is enough to prove this for $p = 1$; the result for $p \in (1, 2)$ follows then by interpolation, and for $p \in (2, \infty)$ by duality. Let $f \in L^1(\Omega)$, and $f_n \in L^1(\Omega) \cap L^2(\Omega)$ be a sequence that converges to f in $L^1(\Omega)$. Let $(\Omega_n)_n$ be a sequence of subsets of Ω such that Ω_n has finite Lebesgue measure for each n, $\Omega_n \subseteq \Omega_{n+1}$ and $\Omega = \cup_{n \geq 0} \Omega_n$. Using the Gaussian upper estimate for $z \in \Sigma(\nu)$, we find a constant M_ν such that

$$\|\chi_{\Omega_n} e^{-zA} f_n\|_1 \leq M_\nu \text{ for all } n, z \in \Sigma(\nu).$$

Since the semigroup is holomorphic on $L^2(\Omega)$, it follows that for each n, the function $z \to \chi_{\Omega_n} e^{-zA} f_n$ is holomorphic on $\Sigma(\nu)$ with values in $L^1(\Omega)$. By Vitali's theorem,[7] the limit $z \to e^{-zA} f$ is holomorphic on $\Sigma(\nu)$ (with values in $L^1(\Omega)$).

It remains to prove that $\|e^{-zA} f - f\|_1 \to 0$ as $z \to 0$ (with $z \in \Sigma(\nu)$). Since $(e^{-zA})_{z \in \Sigma(\nu)}$ is uniformly bounded on $L^1(\Omega)$, it is engough to consider f in a dense subset.

Let $f \in L^1(\Omega) \cap L^2(\Omega)$ with compact support and let B be a bounded subset which contains $\mathrm{supp}(f)$ and such that $\mathrm{dist}(\mathrm{supp} f, \Omega \setminus B) =: \delta > 0$. We write

$$\int_\Omega |e^{-zA} f - f| dx = \int_{\Omega \setminus B} |e^{-zA} f| dx + \int_{B \cap \Omega} |e^{-zA} f - f| dx.$$

Using the above estimate for $p(z, x, y)$ and the Cauchy-Schwarz inequality,

---

[7]See Hille and Phillips [HiPh57], Theorem 3.14.1 or Arendt et al. [ABHN01], Theorem A.5.

we obtain

$$\int_{\Omega} |e^{-zA}f - f| dx$$

$$\leq C_\nu (\Re z)^{-d/2} \int_{\Omega \setminus B} \int_{\mathrm{supp}(f)} \exp\left[-c_\nu \frac{|x-y|^2}{|z|} \right] |f(y)| dy dx$$

$$+ |B|^{1/2} \|e^{zA}f - f\|_2$$

$$\leq C_\nu (\Re z)^{-d/2} \exp\left[-\frac{c_\nu \delta^2}{2|z|} \right] \int_B \int_\Omega \exp\left[-\frac{c_\nu}{2} \frac{|x-y|^2}{|z|} \right] dx |f(y)| dy$$

$$+ |B|^{1/2} \|e^{zA}f - f\|_2$$

$$\leq C'_\nu \exp\left[-\frac{c_\nu \delta^2}{2|z|} \right] \|f\|_1 + |B|^{1/2} \|e^{zA}f - f\|_2,$$

where C'_ν is a positive constant independent of z. From this and the fact that e^{-zA} is strongly continuous on $L^2(\Omega)$, we obtain $\|e^{-zA}f - f\|_1 \to 0$ as $z \to 0$. $\qquad \square$

As an application of the previous theorem, one obtains under the assumptions of Theorem 6.10 that $(e^{-tA_V})_{t \geq 0}$ is a holomorphic semigroup on $L^p(\Omega), 1 \leq p < \infty$ on some sector that contains at least $\Sigma(\frac{\pi}{2} - \arctan M)$, where M is the smallest possible constant for which the inequality

$$|\Im \mathfrak{a}_V(u,u)| \leq M \Re (\mathfrak{a}_V(u,u) + w(u;u)) \text{ for all } u \in V$$

holds for some constant w (see Theorem 1.54). In particular, if A_V is self-adjoint, then $(e^{-tA_V})_{t \geq 0}$ is holomorphic on $\Sigma(\frac{\pi}{2})$ on $L^p(\Omega)$ for all $p \in [1, +\infty)$.

The following result shows that Gaussian estimates for the time derivatives of the heat kernel follow from the Gaussian estimate. This follows from (6.47) and the classical Cauchy formula (applied to the function $H(z)$ given by (6.45)), arguing as in the previous proof. We have

THEOREM 6.17 *Under the assumptions of the previous theorem we have for every $k \in \mathbb{N}$*

$$\left| \frac{\partial^k}{\partial t^k} p(t,x,y) \right| \leq C_k t^{-d/2-k} \exp\left[-\frac{c|x-y|^2}{t} \right]$$

for all $t > 0$ and a.e. $(x,y) \in \Omega \times \Omega$, where C_k and c are positive constants.

Note that the last two results hold in a more general setting. We do not use the fact that Ω is an open subset of $\mathbb{R}^d$. They are stated in the present form only for simplicity and because the present chapter is mainly devoted to elliptic operators A_V.

LEMMA 6.18 *Let $\psi \in (0, \pi/2]$ and assume that $F : \Sigma(\psi) = \{z \in \mathbb{C}, z \neq 0, |\arg z| < \psi\} \to \mathbb{C}$ is a holomorphic function such that*

$$|F(re^{i\theta})| \leq a(r \cos \theta)^{-\beta} \; for \; all \; re^{i\theta} \in \Sigma(\psi),$$

$$|F(r)| \leq ar^{-\beta}e^{-br^{-\alpha}} \; for \; all \; r > 0,$$

where a, b are positive constants, $\beta \geq 0$, and $0 \leq \alpha \leq 1$. Then for every $r > 0$ and $\theta \in (-\psi, \psi)$

$$|F(re^{i\theta})| \leq a2^{\beta}(r \cos \theta)^{-\beta} \exp\left[-\frac{b\alpha}{2}r^{-\alpha}\sin(\psi - |\theta|)\right].$$

Proof. Fix $\gamma \in (0, \psi)$ and define the function

$$G(z) := z^{-\beta}F(1/z)\exp[be^{i(\frac{\pi}{2}-\gamma\alpha)}z^{\alpha}/\sin(\gamma\alpha)].$$

The function G is holomorphic on $\Sigma(\psi)$. Using the assumptions on F, we obtain

$$|G(r)| \leq a \text{ and } |G(re^{i\gamma})| \leq a(\cos\gamma)^{-\beta} \text{ for all } r > 0.$$

In addition, G is exponentially bounded on $\Sigma(\gamma)$. The Phragmen-Lindelöf theorem implies that

$$|G(re^{i\theta})| \leq a(\cos\gamma)^{-\beta} \text{ for all } \theta \in [0, \gamma].$$

Similar arguments show that this estimate holds also for $\theta \in [-\gamma, 0]$. Hence,

$$|F(re^{i\theta})| \leq a(r\cos\gamma)^{-\beta}\exp[-br^{-\alpha}\sin(\gamma\alpha - |\theta|\alpha)/\sin(\gamma\alpha)],$$

for all $\theta \in [-\gamma, \gamma]$. We choose now $\gamma := (\psi + |\theta|)/2$. We have

$$\cos\gamma \geq \cos\left(\frac{\pi}{4} + \frac{|\theta|}{2}\right)$$

$$= \sin\left(\frac{\pi}{4} - \frac{|\theta|}{2}\right)$$

$$\geq \frac{1}{2}\sin\left(\frac{\pi}{2} - |\theta|\right)$$

$$= \frac{\cos\theta}{2}.$$

We also have the bound

$$\frac{\sin[(\gamma - |\theta|)\alpha]}{\sin(\gamma\alpha)} \geq \sin[(\gamma - |\theta|)\alpha]$$

$$= \sin\left[\left(\frac{\psi}{2} - \frac{|\theta|}{2}\right)\alpha\right]$$

$$\geq \frac{\alpha}{2}\sin(\psi - |\theta|).$$

This proves the lemma. □

6.6 WEIGHTED GRADIENT ESTIMATES

Assume that the assumptions of Theorem 6.10 hold. Observe that for t and $\varepsilon > 0$

$$e^{-(t+\varepsilon)A_V^*}f(y) = \int_\Omega \overline{p_V(t + \varepsilon, x, y)}f(x)dx = (f; p_V(t + \varepsilon, ., y))$$

and

$$\begin{aligned} e^{-(t+\varepsilon)A_V^*}f(y) &= e^{-tA_V^*}e^{-\varepsilon A_V^*}f(y) \\ &= (e^{-tA_V^*}f; p_V(\varepsilon, ., y)) \\ &= (f; e^{-tA_V}(p_V(\varepsilon, ., y))). \end{aligned}$$

Hence,

$$p_V(t + \varepsilon, ., y) = e^{-tA_V}p_V(\varepsilon, ., y) \text{ for all } t > 0, \varepsilon > 0 \text{ and a.e. } y \in \Omega. \tag{6.48}$$

This implies in particular that for a.e. y, $p_V(t, ., y) \in V \subseteq H^1(\Omega)$ for all $t > 0$. In particular, $\nabla_x p_V(t, x, y)$ (the gradient with respect to the x variable) is in $L^2(\Omega)$. In Theorem 6.17 we obtained Gaussian upper bounds for the time derivatives of $p_V(t, x, y)$. In this section, we prove weighted L^p-estimates for its space derivatives.

THEOREM 6.19 *Assume that V is either $H_0^1(\Omega)$ (with Ω an arbitrary open set of $\mathbb{R}^d$) or $H^1(\Omega)$ or as in (4.5) and that the assumptions of Theorem 6.10 are satisfied.*
1) There exists a constant $\beta > 0$ such that the following estimate holds for every $p \in [1, 2]$:

$$\left[\int_\Omega |\nabla_x p_V(t, x, y)|^p e^{\beta p|x-y|^2/t}dx\right]^{1/p} \leq C_\varepsilon t^{-\frac{d}{2}[1-1/p]-\frac{1}{2}}e^{\varepsilon t + \delta\alpha t}$$

for all $t > 0, \varepsilon > 0$ a.e. $y \in \Omega$. Here δ and α are as in Theorem 6.10.
2) If $V = H_0^1(\Omega)$ and $b_k = c_k = a_0 = 0$ for all $k, 1 \le k \le d$, then

$$\left[\int_\Omega |\nabla_x p_{H_0^1}(t,x,y)|^p e^{\beta p|x-y|^2/t} dx\right]^{1/p} \le Ct^{-\frac{d}{2}[1-1/p]-\frac{1}{2}}$$

for all $t > 0$ and a.e. $y \in \Omega$.

Proof. Let $\varepsilon' > 0$. By Hölder's inequality we have for $p \in [1,2)$

$$\int_\Omega |\nabla_x p_V(t,x,y)|^p e^{\beta p|x-y|^2/t} dx$$

$$= \int_\Omega |\nabla_x p_V(t,x,y)|^p e^{(\beta+\varepsilon')p|x-y|^2/t}.e^{-\varepsilon' p|x-y|^2/t} dx$$

$$\le \left[\int_\Omega |\nabla_x p_V(t,x,y)|^2 e^{2(\beta+\varepsilon')|x-y|^2/t} dx\right]^{p/2}$$

$$\times \left[\int_\Omega e^{-\varepsilon' p\frac{2}{2-p}|x-y|^2/t} dx\right]^{(2-p)/p},$$

from which we see that it is enough to prove the theorem for $p = 2$.
By Theorem 6.10, we have

$$|p_V(t,x,y)| \le Ct^{-d/2} e^{-c|x-y|^2/2} \text{ for } 0 < t \le 1, \qquad (6.49)$$

where C and c are positive constants. If the assumptions of assertion 2) hold, then (6.49) holds for all $t > 0$.

Let $\beta \in (0, c)$ be any fixed constant. We want to estimate

$$I_t := \int_\Omega |\nabla_x p_V(t,x,y)|^2 e^{\beta|x-y|^2/t} dx.$$

Let $\psi \in C_c^\infty(\mathbb{R}^d)$ such that $0 \le \psi \le 1$ and set

$$I_t(\psi) := \int_\Omega |\nabla_x p_V(t,x,y)|^2 e^{\beta|x-y|^2/t}\psi(x) dx.$$

Using (6.48), one obtains that $p_V(t, ., y)e^{\beta|.-y|^2/t}\psi(.) \in V$. Thus,

$$\eta I_t(\psi) \leq \Re \sum_{k,j=1}^{d} \int_\Omega a_{kj} D_k p_V(t, x, y)\overline{D_j p_V(t, x, y)}e^{\beta|x-y|^2/t}\psi(x)dx$$

$$= \Re a_V(p_V(t, ., y), p_V(t, ., y)e^{\beta|.-y|^2/t}\psi)$$

$$-\Re \sum_{k,j=1}^{d} \int_\Omega a_{kj} D_k p_V(t, x, y)\overline{p_V(t, x, y)}D_j\left[e^{\beta|x-y|^2/t}\psi(x)\right]dx$$

$$-\Re \sum_{k=1}^{d} \int_\Omega b_k D_k p_V(t, x, y)\overline{p_V(t, x, y)}e^{\beta|x-y|^2/t}\psi(x)dx$$

$$-\Re \sum_{k=1}^{d} \int_\Omega c_k p_V(t, x, y)D_k\left[\overline{p_V(t, x, y)}e^{\beta|x-y|^2/t}\psi(x)\right]dx$$

$$-\Re \int_\Omega a_0 p_V(t, x, y)\overline{p_V(t, x, y)}e^{\beta|x-y|^2/t}\psi(x)dx$$

$$:= J_1(\psi) + J_2(\psi) + J_3(\psi) + J_4(\psi) + J_5(\psi).$$

Using (6.48), the first term $J_1(\psi) = \Re a_V(p_V(t, ., y), p_V(t, ., y)e^{\beta|.-y|^2/t}\psi)$ satisfies

$$J_1(\psi) = \Re(A_V p_V(t, ., y); p_V(t, ., y)e^{\beta|.-y|^2/t}\psi)$$

$$= \Re(-\frac{\partial}{\partial t}p_V(t, ., y); p_V(t, ., y)e^{\beta|.-y|^2/t}\psi).$$

From (6.49) and Theorem 6.17, we see that for some constant C_1

$$|J_1(\psi)| \leq C_1 t^{-\frac{d}{2}-1} \text{ for all } t \in (0, 1]. \tag{6.50}$$

We want to estimate

$$J_2(\psi) = -\Re \sum_{k,j=1}^{d} \int_\Omega a_{kj} D_k p_V(t, x, y)\overline{p_V(t, x, y)}D_j\left[e^{\beta|x-y|^2/t}\psi(x)\right]dx.$$

We write $J_2(\psi) = J_2'(\psi) + J_2''(\psi)$, where

$$J_2'(\psi) = -\Re \sum_{k,j=1}^{d} \int_\Omega a_{kj} D_k p_V(t, x, y)\overline{p_V(t, x, y)}2\beta(x_j - y_j)t^{-1}$$

$$\times e^{\beta|x-y|^2/t}\psi(x)dx$$

and

$$J_2''(\psi) = -\Re \sum_{k,j=1}^{d} \int_{\Omega} a_{kj} D_j \psi(x) D_k p_V(t,x,y) \overline{p_V(t,x,y)} e^{\beta|x-y|^2/t} dx.$$

The term $J_2'(\psi)$ is estimated as follows:

$$|J_2'(\psi)| \le \sum_{k,j=1}^{d} \|a_{kj}\|_\infty t^{-1/2} \int_{\Omega} \left[|D_k p_V(t,x,y)| |p_V(t,x,y)| \psi(x) \right.$$

$$\left. \times 2 \frac{\beta|x_j - y_j|}{\sqrt{t}} e^{\beta|x-y|^2/t} \right] dx$$

$$\le M t^{-1/2} \sum_{k=1}^{d} \int_{\Omega} |D_k p_V(t,x,y)| \psi(x) |p_V(t,x,y)| e^{\frac{3}{2}\beta|x-y|^2/t} dx,$$

where M is a constant. By (6.49) and the Cauchy-Schwarz inequality

$$|J_2'(\psi)|$$

$$\le MC t^{-1/2} \sum_{k=1}^{d} \int_{\Omega} |D_k p_V(t,x,y)| |p_V(t,x,y)| e^{\frac{1}{2}\beta|x-y|^2/t} \psi(x)$$

$$\times e^{\beta|x-y|^2/t} dx$$

$$\le MC t^{-1/2} I_t(\psi)^{1/2} \left[\int_{\Omega} t^{-d} e^{2(\beta-c)|x-y|^2/t} dx \right]^{1/2}.$$

Thus

$$|J_2'(\psi)| \le C_2 t^{-d/4-1/2} I_t(\psi)^{1/2} \text{ for all } t \in (0,1]$$

and hence

$$|J_2(\psi)| \le C_2 t^{-d/4-1/2} I_t(\psi)^{1/2} + J_2''(\psi) \text{ for all } t \in (0,1]. \qquad (6.51)$$

The term

$$J_3(\psi) = -\Re \sum_{k=1}^{d} \int_{\Omega} b_k D_k p_V(t,x,y) \overline{p_V(t,x,y)} e^{\beta|x-y|^2/t} \psi(x) dx$$

can be estimated as above by using (6.49) and the Cauchy-Schwarz inequality. One obtains

$$|J_3(\psi)| \le C_3 t^{-d/4} I_t(\psi)^{1/2} \text{ for all } t \in (0,1]. \qquad (6.52)$$

We estimate

$$J_4(\psi) = -\Re \sum_{k=1}^{d} \int_{\Omega} c_k p_V(t,x,y) D_k \left[\overline{p_V(t,x,y)} e^{\beta|x-y|^2/t} \psi(x) \right] dx.$$

We write

$$J_4(\psi) := -\Re \sum_{k=1}^{d} \int_{\Omega} c_k p_V(t,x,y) D_k [\overline{p_V(t,x,y)} e^{\beta|x-y|^2/t} \psi(x)] dx$$

$$= -\Re \sum_{k=1}^{d} \int_{\Omega} c_k p_V(t,x,y) \overline{D_k p_V(t,x,y)} e^{\beta|x-y|^2/t} \psi(x) dx$$

$$-\Re \sum_{k=1}^{d} \int_{\Omega} c_k |p_V(t,x,y)|^2 2\beta(x_k - y_k) t^{-1} e^{\beta|x-y|^2/t} \psi(x) dx$$

$$-\Re \sum_{k=1}^{d} \int_{\Omega} c_k D_k \psi(x) |p_V(t,x,y)|^2 e^{\beta|x-y|^2/t} dx.$$

As above, using (6.49) and the Cauchy-Schwarz inequality, we can estimate the first term by $t^{-d/4} I_t(\psi)^{1/2}$ (up to a constant). The second one is estimated by

$$\sum_{k=1}^{d} \int_{\Omega} |c_k| t^{-1/2} |p_V(t,x,y)|^2 e^{2\beta|x-y|^2/t} \psi(x) dx$$

and hence by $t^{-d/2-1/2}$ (up to a constant). We obtain

$$|J_4(\psi)| \leq C_4 [t^{-d/4-1/2} I_t(\psi)^{1/2} + t^{-d/2-1/2}] + J_4''(\psi) \text{ for all } t \in (0,1], \tag{6.53}$$

where C_4 is a constant and

$$J_4''(\psi) = \sum_{k=1}^{d} \|c_k\|_{\infty} \int_{\Omega} |D_k \psi(x)| |p_V(t,x,y)|^2 e^{\beta|x-y|^2/t} dx.$$

Using (6.49), it is clear that

$$J_5(\psi) = -\Re \int_{\Omega} a_0 p_V(t,x,y) \overline{p_V(t,x,y)} e^{\beta|x-y|^2/t} \psi(x) dx$$

is bounded by

$$|J_5(\psi)| \leq C_5 t^{-d/2} \text{ for all } t \in (0,1]. \tag{6.54}$$

We apply the above estimates with ψ_j in place of ψ, where $\psi_j(x) := \psi(x/j)$ and $\psi \in C_c^\infty(\mathbb{R}^d)$, such that $0 \le \psi \le 1$, $\psi(x) = 1$ for $|x| \le 1$. Simple computations show that $J_2''(\psi_j)$ and $J_4''(\psi_j)$ converge to 0 as $j \to \infty$. This together with the estimates (6.50)−(6.54) and Fatou's lemma gives

$$\eta I_t \le C_1 t^{-d/2-1} + C_2 t^{-d/4-1/2} I_t^{1/2} + C_3 t^{-d/4} I_t^{1/2}$$
$$+ C_4 [t^{-d/4-1/2} I_t^{1/2} + t^{-d/2-1/2}] + C_5 t^{-d/2}.$$

This implies that

$$I_t \le C t^{-d/2-1} \text{ for all } t \in (0,1]$$

and proves assertion 1).

If $V = H_0^1(\Omega)$ and $b_k = c_k = a_0 = 0$, then we only have $J_1(\psi)$ and $J_2(\psi)$ in the proof. In addition, (6.49) holds for all $t > 0$. The above proof gives in this case

$$\eta I_t \le C_1 t^{-\frac{d}{2}-1} + C_2 t^{-d/4-1/2} I_t^{1/2} \text{ for all } t > 0,$$

from which we deduce the assertion 2). $\square$

Notes
Section 6.1. This section follows the books of Davies [Dav89], Robinson [Rob91], and Varopoulos, Saloff-Coste, and Coulhon [VSC92]. As mentioned previously, Lemma 6.1 holds without the L^∞-contractivity assumption; see Coulhon [Cou92] and Varopoulos et al. [VSC92]. Theorem 6.2 and other extensions can be found in Coulhon [Cou92]. Theorem 6.3 is due to Fabes and Stroock [FaSt86] (see also Carlen, Kusuoka, and Stroock [CKS87], Davies [Dav89]). The implication 1) $\Rightarrow$ 2) in Theorem 6.4 was proved by Bénilan [Bén78] and the converse by Yoshikawa [Yosh71]. The equivalence of the two parts was rediscovered by Varopoulos [Var85]. Theorem 6.3 is not limited to polynomial decay. Characterizations of the estimates $\|e^{-tA}\|_{\mathcal{L}(L^1,L^\infty)} \le \theta(t)$ (for more general functions θ) in terms of Nash type inequalities are given in Coulhon [Cou96]. Such estimates can also be characterized by logarithmic Sobolev inequalities; see Davies [Dav89] and the references there. Lemma 6.5 is extracted from Coulhon [Cou93]. See also Ouhabaz [Ouh03a].

Section 6.2. Theorem 6.6 is shown in Ouhabaz [Ouh98]. In a recent work, Coulhon [Cou03] has proved that Hölder continuity of the heat kernel is equivalent to certain Sobolev inequalities and also to some embedding of Besov type spaces. Hölder continuity of the heat kernel is intimately related to Gaussian upper and lower bounds; see Saloff-Coste [SaC95], Coulhon [Cou03].

Section 6.3. Gaussian upper bounds were studied by several authors. For uniformly elliptic operators $A = -\sum_{k,j} D_j(a_{kj} D_k)$ (on $L^2(\mathbb{R}^d)$) with real-valued coefficients, Gaussian upper and lower bounds were proved by Aronson [Aro68], see also Porper and Eidel'man [PrEi84]. The subject of Gaussian bounds was revived by Davies [Dav87] who introduced the perturbation technique used here and proved such bounds for symmetric operators on domains under Dirichlet or Neumann boundary conditions; see [Dav89] and the references there. The Gaussian upper bounds proved in [Dav87] and [Dav89] for $A = -\sum_{k,j} D_j(a_{kj} D_k)$ are the same as those in Corollary 6.14 (but without the term $e^{-s(A_V)t}$).

Gaussian bounds for second-order operators with terms of order 1, acting on $L^2(\mathbb{R}^d)$, have been proved by Stroock [Str88] and Norris and Stroock [NoSt91]. Some smoothness of the coefficients is required there.

Gaussian bounds for Laplacians on Riemannian manifolds or sub-Laplacians on Lie groups have been studied by several authors. We refer reader to the monographs of Davies [Dav89], Varopoulos et al. [VSC92], Robinson [Rob91] and Saloff-Coste [SaC02] for a systematic account and references.

There is a big difference between second order uniformly elliptic operators with complex-valued coefficients and their relatives with real-valued ones. We have seen in Chapter 4 that the L^∞-contractivity property fails to hold if the coefficients are complex-valued. Also the Gaussian upper bounds do not hold in general in this case.

For operators with complex-valued coefficients, Gaussian upper bounds were obtained by Auscher, McIntosh, and Tchamitchian [AMcT98] in the case where $\Omega = \mathbb{R}^d$ and the coefficients are smooth (see also Auscher and Tchamitchian [AuTc98] and [AuTc01]). Note that if $d \le 2$, the smoothness of the coefficients is not needed. If $d \ge 5$, a counterexample was given by Auscher, Coulhon, and Tchamitchian [ACT96]. See also Davies [Dav97a]. The example used in [ACT96] is taken from Maz'ya, Nazarov, and Plamenevskii [MNP85], who proved that De Giorgi type result do not hold for operators with complex-valued coefficients.

The situation for uniformly elliptic operators with complex-valued coefficients on arbitrary domains is more complicated. Proposition 6 in [AuTc01] shows that the regularity of the coefficients does not ensure the validity of the Gaussian bound if the operator is considered on a domain and is subject to Dirichlet boundary conditions. The subject of complex-valued coefficients on arbitrary domains is not very well understood yet.

For operators with terms of order one, $A = -\sum_{k,j} D_j(a_{kj} D_k + c_j) + \sum_k b_k D_k + a_0$ and subject to boundary conditions, Gaussian bounds are studied in Ouhabaz [Ouh02] where Theorems 6.10 and 6.11 are proved. The proof follows a similar strategy as in Coulhon [Cou93] and Robinson [Rob91]. Previous results for such operators but with real-valued coefficients (under Dirichlet, Neumann, or mixed boundary conditions) were given by Arendt and ter Elst [ArEl97] (assuming some smoothness on the first order terms), Daners [Dan00], and Karrmann [Kar01].

Note that the results in this section extend to operators of type bA_V, where $b : \Omega \to \mathbb{C}$ is a bounded measurable function with real-part bounded from below by a positive constant. More generally, it is shown in Duong and Ouhabaz [DuOu99] that Gaussian upper bounds carry over from operators A to mutiplicative perturbations bA.

Section 6.4. Sharp bounds as in Theorem 6.12 where proved by Davies and Pang [DaPa89] for symmetric operators. A weaker result was proved previously by Varopoulos [Var88] in the setting of Lie groups; see also Coulhon [Cou93] and Grigor'yan [Gri97]. Sharper estimates have been proved by Sikora [Sik96] for a class of sub-Laplacians on Lie groups. Theorem 6.12 as stated here for complex-valued coefficients is proved in Ouhabaz [Ouh03a]. Related results in Euclidean space can be found in Norris and Stroock [NoSt91] and Stroock [Str88].

Section 6.5. The extension of the Gaussian upper bound from $t > 0$ to complex t was first proved by Davies [Dav89]. Lemma 6.18 and its proof are taken from Davies [Dav95c].

The L^p-analyticity result (Theorem 6.16) was first proved in Ouhabaz [Ouh92a] and [Ouh95] in the symmetric case. It was then extended to nonsymmetric operators by Hieber [Hie96] and Arendt and ter Elst [ArEl97]. Similar results are also proved by Davies [Dav95b] and Duong and Robinson [DuRo96]. A previous result concerning analyticity of semigroups generated by some elliptic operators on $L^1(\Omega)$ was given by Amann [Ama83], who assumed that the coefficients as well as the domain are smooth enough. According to Theorems 6.10 and 6.16, no smoothness assumption is needed if the coefficients are real-valued and the operator is subject to Dirichlet boundary conditions. For Neumann boundary conditions, one needs some smoothness of the domain (e.g., Ω has the extension property, which holds for domains with Lipschitz boundary). An example of a domain Ω for which the semigroup generated by the Laplacian with Neumann boundary conditions is not holomorphic on $L^1(\Omega)$ is given by Kunstmann [Kun02]. Results on the holomorphy of semigroups generated by elliptic operators on $C_0(\Omega)$ can be found in Ouhabaz [Ouh95].

Section 6.6. Weighted gradient estimates as in Theorem 6.19 were proved in the Riemannian manifold setting by Grigor'yan [Gri95]. For uniformly elliptic operators with nonsmooth coefficients, results of the same type were proved by Auscher and Tchamitchian [AuTc98], Duong and McIntosh [DuMc99b], and Coulhon and Duong [CoDu99].

Chapter Seven

GAUSSIAN UPPER BOUNDS AND

L^p-SPECTRAL THEORY

The present chapter is devoted to applications of Gaussian upper bounds to L^p-spectral theory. Although this monograph is mainly concerned with second-order differential operators on domains of the Euclidean space, this chapter is written in a general setting of operators on metric spaces. The framework includes Laplacians on Riemannian manifolds or arbitrary domains of manifolds as well as some higher-order differential operators on domains of Euclidean space.

Let (X, ρ, μ) be a metric σ-finite measured space. Denote by $B(x, r)$ the open ball in X for the distance ρ, of center $x \in X$ and radius $r > 0$, that is,

$$B(x, r) := \{y \in X, \rho(x, y) < r\},$$

and by $V(x, r)$ its volume, that is,

$$V(x, r) := \mu\left(B(x, r)\right).$$

We shall always assume $V(x, r) < \infty$ for every $x \in X$ and $r \in (0, \infty)$.

Throughout this chapter, we shall assume that X has regular volume growth (or it satisfies the doubling property), that is, there exists a positive constant C such that

$$V(x, 2r) \leq CV(x, r) \text{ for all } x \in X, \ r > 0. \tag{7.1}$$

Note that (7.1) implies easily that there exists $d > 0$ such that

$$V(x, s) \leq C\left(\frac{s}{r}\right)^d V(x, r) \text{ for all } x \in X, \ s \geq r > 0. \tag{7.2}$$

Let Ω be an open subset of X and let A be a non-negative self-adjoint operator on $L^2(\Omega, \mu)$. Assume that the semigroup $(e^{-tA})_{t \geq 0}$ has a kernel $p(t, x, y)$[1], that is, a measurable function: $\Omega \times \Omega \to \mathbb{C}$, such that

$$e^{-tA} f(x) = \int_\Omega p(t, x, y) f(y) \, d\mu(y) \text{ for every } f \in L^2(\Omega) \text{ and a.e. } x \in \Omega.$$

---

[1]We again call $p(t, x, y)$ the heat kernel of A.

The main assumption on $p(t, x, y)$ is the Gaussian upper bound:

$$|p(t, x, y)| \leq \frac{C}{\sqrt{V(x, t^{1/m})V(y, t^{1/m})}} \exp\left\{-c\left(\frac{\rho^m(x, y)}{t}\right)^{\frac{1}{m-1}}\right\} \quad (7.3)$$

for all $t > 0$ and a.e. $x, y \in \Omega$, where $C, c > 0$, and $m > 1$ are constants.
In (7.3), $V(x, t^{1/m})$ is the volume of the ball in X and not in Ω. The open
subset Ω is not supposed to satisfy the doubling volume property.

Since $B(x, r) \subseteq B(y, r + \rho(x, y))$ for all $x, y \in X$ and $r > 0$, one has,
thanks to (7.2),

$$V(x, r) \leq V\left(y, r(1 + \frac{\rho(x, y)}{r})\right) \leq C\left(1 + \frac{\rho(x, y)}{r}\right)^d V(y, r). \quad (7.4)$$

Using this, it follows that one can replace in (7.3) $\sqrt{V(x, t^{1/m})V(y, t^{1/m})}$
by $V(x, t^{1/m})$ provided one changes the constants c, C. In other words,
(7.3) is equivalent to

$$|p(t, x, y)| \leq \frac{C}{V(x, t^{1/m})} \exp\left\{-c\left(\frac{\rho^m(x, y)}{t}\right)^{\frac{1}{m-1}}\right\} \quad (7.5)$$

for some positive constants c, C. Similarly, $V(x, t^{1/m})$ can be replaced by
$V(y, t^{1/m})$ in (7.5).

We describe now some situations where both properties (7.1) and (7.3)
are satisfied.

1) Second-order operators on domains
Let $X = \mathbb{R}^d$ endowed with the flat metric, Ω is an arbitrary open subset of
$\mathbb{R}^d$ and $A = -\sum_{k,j=1}^d D_j(a_{kj} D_k)$ be a self-adjoint uniformly elliptic operator with real-valued coefficients and subject to Dirichlet boundary conditions. By Theorem 6.10, the heat kernel of A satisfies a Gaussian upper
bound. The bound given there coincides with (7.3) for $m = 2$. Note that in
the present case $V(x, t^{1/2}) = c_d t^{d/2}$.

For other boundary conditions, the Gaussian upper bound holds with an
additional term $e^{\varepsilon t}$ for every $\varepsilon > 0$, see Theorem 6.10.

2) Schrödinger operators
The Gaussian upper bound (7.3) holds also for Schrödinger operators $-\Delta + V$.

If $V \in L^1_{\text{loc}}(\mathbb{R}^d, dx)$ is non-negative, the operator $-\Delta + V$ is defined as
the operator associated with the form

$$\mathfrak{a}(u, v) := \int_{\mathbb{R}^d} \nabla u \nabla v \, dx + \int_{\mathbb{R}^d} V u v \, dx,$$

$$D(\mathfrak{a}) := \left\{ u \in H^1(\mathbb{R}^d), \int_{\mathbb{R}^d} V|u|^2 dx < \infty \right\}.$$

The semigroup $(e^{-t(-\Delta+V)})_{t\geq 0}$ is dominated by the Gaussian semigroup $(e^{t\Delta})_{t\geq 0}$ (this follows immediately from Theorem 2.24). Hence, $-\Delta + V$ has a heat kernel $p(t, x, y)$ which satisfies

$$0 \leq p(t, x, y) \leq \frac{1}{(4\pi t)^{d/2}} e^{-\frac{|x-y|^2}{4t}} \text{ for all } t > 0, x, y \in \mathbb{R}^d. \qquad (7.6)$$

Now if V changes sign, we write $V = V^+ - V^-$ and define $-\Delta + V :=$ $(-\Delta+V^+) - V^-$ as the perturbation of $-\Delta+V^+$ by $-V^-$. If the negative part V^- is in the Kato class (see Kato [Kat73] or Simon [Sim82]), the heat kernel of $-\Delta + V$ satisfies

$$0 \leq p(t, x, y) \leq e^{\alpha t} \frac{C}{t^{d/2}} e^{-c\frac{|x-y|^2}{t}} \text{ for all } t > 0, x, y \in \mathbb{R}^d, \qquad (7.7)$$

for some constant α. This and other information on Schrödinger semigroups can be found in Simon [Sim82].

3) Higher-order operators
For higher-order elliptic operators

$$Hf(x) = \sum_{|\alpha|=|\beta|=n} (-1)^{|\alpha|} D^{\alpha}(a_{\alpha,\beta}(x) D^{\beta} f(x))$$

acting on $L^2(\mathbb{R}^d)$, the Gaussian bound, when it holds, takes the form

$$|p(t, x, y)| \leq Ct^{-d/2n} \exp\left\{ -c \left(\frac{\rho^{2n}(x, y)}{t} \right)^{\frac{1}{2n-1}} \right\}.$$

This is again (7.3) with $m = 2n$ and $V(x, t^{1/2n}) = c_d t^{d/2n}$. This estimate holds if the coefficients $a_{\alpha\beta}$ are smooth enough or if $d \leq 2n$, see the survey of Davies [Dav97b].

4) Laplacians on manifolds
Let $X = \Omega = M$ be a Riemannian manifold and $A = -\Delta$, where Δ is the Laplace-Beltrami operator. Let ρ be the geodesic distance and μ the Riemannian measure.

If the manifold has non-negative Ricci curvature, then (7.3) with $m = 2$ was proved by Li and Yau [LiYa86]. See also Davies [Dav89]. The same estimate holds if Δ is considered on an arbitrary domain of M with Dirichlet boundary conditions. The reason is that the heat kernel on a domain is pointwise dominated by the heat kernel on M. This holds as in Proposition 4.23, stated for Euclidean domains.

Note that if the manifold M satisfies the doubling condition (7.1) and the heat kernel $p(t, x, y)$ of Δ satisfies the diagonal upper bound

$$p(t, x, x) \leq \frac{C}{V(x, \sqrt{t})} \text{ for } x \in M, \ t > 0, \tag{7.8}$$

then, according to a result of Grigor'yan [Gri97], the heat kernel satisfies (7.3) with $m = 2$.

Note that by the symmetry of the semigroup, one has

$$p(t, x, y) = p(t, y, x)$$

and thus

$$p(t, x, x) = \int_M p\left(\frac{t}{2}, x, y\right) p\left(\frac{t}{2}, y, x\right) d\mu(y)$$

$$= \int_M \left| p(\frac{t}{2}, x, y) \right|^2 d\mu(y).$$

Thus, the diagonal estimate (7.8) is precisely the L^2-norm estimate:

$$\left\| p(\frac{t}{2}, x, \cdot) \right\|_2^2 \leq \frac{C}{V(x, \sqrt{t})}, \ x \in M, \ t > 0.$$

There are many other interesting situations where (7.3) holds, for example, sub-Laplacians on Lie groups (see Robinson [Rob91] and Varopoulos, Saloff-Coste and Coulhon [VSC92]) or Laplacians on some fractals. In the latter case, (7.3) may hold with some $m > 2$ although the operator is of second order, see Barlow [Bar98].

7.1 L^p-BOUNDS AND HOLOMORPHY

Let (X, ρ, μ) be as above, Ω is an open subset of X and A is a non-negative self-adjoint operator on $L^2(\Omega, \mu) := L^2(\Omega, \mu, \mathbb{C})$. We assume that X satisfies the doubling condition (7.1) and that the heat kernel $p(t, x, y)$ of A satisfies the Gaussian upper bound (7.3). Under these assumptions, the semigroup $(e^{-tA})_{t \geq 0}$ extends to $L^p(\Omega, \mu)$ for all $p \in [1, \infty)$. More precisely,

PROPOSITION 7.1 *If (7.1) and (7.3) are satisfied, then for each $p \in [1, \infty]$, e^{-tA} extends from $L^2(\Omega, \mu) \cap L^p(\Omega, \mu)$ to $L^p(\Omega, \mu)$ for all $t \geq 0$ and $\sup_{t \geq 0} \|e^{-tA}\|_{\mathcal{L}(L^p)} < \infty$.*

Proof. By the Riesz-Thorin interpolation theorem, we only have to prove the result for $p = 1$. Since (7.3) is equivalent to (7.5), it is enough to show that there exists a constant K such that

$$\sup_{y \in \Omega, t > 0} \int_\Omega \frac{\exp\{-c(\frac{\rho^m(x,y)}{t})^{\frac{1}{m-1}}\}}{V(y, t^{1/m})} d\mu(x) \le K. \tag{7.9}$$

Using (7.2), we have

$$\int_\Omega \frac{\exp\left\{-c(\frac{\rho^m(x,y)}{t})^{\frac{1}{m-1}}\right\}}{V(y, t^{1/m})} d\mu(x)$$

$$\le \sum_{k \ge 0} \int_{\{kt^{1/m} \le \rho(x,y) \le (k+1)t^{1/m}\}} \frac{\exp\left\{-c(\frac{\rho^m(x,y)}{t})^{\frac{1}{m-1}}\right\}}{V(y, t^{1/m})} d\mu(x)$$

$$\le \sum_{k \ge 0} e^{-ck^{\frac{m}{m-1}}} \frac{V(y, (k+1)t^{1/m})}{V(y, t^{1/m})}$$

$$\le C \sum_{k \ge 0} e^{-ck^{\frac{m}{m-1}}} (k+1)^d < +\infty.$$

This shows (7.9). □

By a density argument, the family $(e^{-tA})_{t \ge 0}$ obtained in $L^p(\Omega, \mu)$ satisfies the semigroup property. It is even a strongly continuous semigroup on $L^p(\Omega, \mu)$ for $1 \le p < \infty$. The strong continuity is proved in Corollary 7.5 below.

Our aim now is to extend the Gaussian upper bound for $t > 0$ to complex t. Let $\mathbb{C}^+ := \{z \in \mathbb{C}; \Re z > 0\}$ be the open right half-plane. Note first that by the semigroup property,

$$p(z, x, y) = \int_\Omega p\left(\frac{z}{2}, x, u\right) p\left(\frac{z}{2}, u, y\right) d\mu(u)$$

and hence if $t = \Re z$,

$$|p(z, x, y)| \le \left(\int_\Omega |p(z/2, x, u)|^2 d\mu(u)\right)^{1/2} \left(\int_\Omega |p(z/2, u, y)|^2 d\mu(u)\right)^{1/2}$$

$$= \left(\int_\Omega p(z/2, x, u) p(\overline{z}/2, u, x) d\mu(u)\right)^{1/2}$$

$$\times \left(\int_\Omega p(z/2, u, y) p(\overline{z}/2, y, u) d\mu(u)\right)^{1/2}$$

$$= \sqrt{p(t, x, x).p(t, y, y)}.$$

Here the symmetry of the semigroup is used in order to write

$$\overline{p(z/2, x, u)} = p(\overline{z}/2, u, x).$$

We have proved that

$$|p(z, x, y)| \le \sqrt{p(t, x, x) \cdot p(t, y, y)} \le \frac{C}{\sqrt{V(x, t^{1/m}) V(y, t^{1/m})}}. \quad (7.10)$$

The Gaussian bound extends from $t > 0$ to complex t. The following result is more precise than the bound in Theorem 6.16 in the sense that it holds uniformly on $\mathbb{C}^+$.

THEOREM 7.2 *Assume that the doubling volume condition (7.2) holds. Assume that A is a non-negative self-adjoint operator on $L^2(\Omega, \mu)$ having a heat kernel $p(t, x, y)$ which satisfies the Gaussian upper bound (7.3). Then there exist positive constants C, c such that*

$$|p(z, x, y)| \le \frac{C(\cos \theta)^{-d} \exp \left\{ -c \left(\frac{\rho^m(x,y)}{|z|} \right)^{\frac{1}{m-1}} \cos \theta \right\}}{\sqrt{V(x, (\frac{|z|}{(\cos \theta)^{m-1}})^{1/m}) V(y, (\frac{|z|}{(\cos \theta)^{m-1}})^{1/m})}},$$

for all $z \in \mathbb{C}^+$ and a.e. $x, y \in \Omega$, where $\theta = \arg z$.

Proof. It is enough to prove that there exist positive constants C, c such that

$$|p(z, x, y)| \le \frac{C e^{\frac{\Re z}{\lambda}} (\frac{\lambda}{\Re z})^{d/m}}{\sqrt{V(x, \lambda^{1/m}) V(y, \lambda^{1/m})}} \exp \left\{ -c \left(\frac{\rho^m(x, y)}{|z|} \right)^{\frac{1}{m-1}} \cos \theta \right\}, \quad (7.11)$$

for all $\lambda > 0$, all $z \in \mathbb{C}^+$ and a.e. $x, y \in \Omega$. Indeed, if (7.11) holds, then choosing $\lambda = \frac{|z|}{(\cos \theta)^{m-1}}$ yields the theorem.

Let us now prove (7.11). Fix $\lambda \in]0, +\infty[$. If z is such that $\Re z \in]0, \lambda]$, apply (7.10) and write

$$V(x, \lambda^{1/m}) \le C \left(\frac{\lambda}{\Re z} \right)^{d/m} V(x, (\Re z)^{1/m}),$$

which gives

$$|p(z, x, y)| \le \frac{C}{\sqrt{V(x, (\Re z)^{1/m}) V(y, (\Re z)^{1/m})}}$$

$$\le \frac{C'}{\sqrt{V(x, \lambda^{1/m}) V(y, \lambda^{1/m})}} \left(\frac{\lambda}{\Re z} \right)^{d/m}.$$

If moreover $z = t$ belongs to $]0, \lambda[$, use (7.3) and write

$$|p(t,x,y)| \leq \frac{C'(\frac{\lambda}{t})^{d/m}}{\sqrt{V(x,\lambda^{1/m})V(y,\lambda^{1/m})}} \exp\left\{-c(\frac{\rho^m(x,y)}{t})^{\frac{1}{m-1}}\right\}.$$

If on the contrary $\Re z > \lambda$, write

$$|p(z,x,y)| \leq \frac{C}{\sqrt{V(x,(\Re z)^{1/m})V(y,(\Re z)^{1/m})}}$$

$$\leq \frac{C'}{\sqrt{V(x,\lambda^{1/m})V(y,\lambda^{1/m})}} \left(\frac{\lambda}{\Re z}\right)^{d/m} e^{\frac{\Re z}{\lambda}},$$

and if moreover $z = t$ belongs to $]\lambda, +\infty[$,

$$|p(t,x,y)|$$

$$\leq \frac{C}{\sqrt{V(x,t^{1/m})V(y,t^{1/m})}} \exp\left\{-c(\frac{\rho^m(x,y)}{t})^{\frac{1}{m-1}}\right\}$$

$$\leq \frac{C'e^{t/\lambda}}{\sqrt{V(x,\lambda^{1/m})V(y,\lambda^{1/m})}} \left(\frac{\lambda}{t}\right)^{d/m} \exp\left\{-c(\frac{\rho^m(x,y)}{t})^{\frac{1}{m-1}}\right\}.$$

Finally, for all z with $\Re z > 0$ and all $\lambda > 0$, we have

$$|p(z,x,y)| \leq \frac{C}{\sqrt{V(x,\lambda^{1/m})V(y,\lambda^{1/m})}} \left(\frac{\lambda}{\Re z}\right)^{d/m} e^{\frac{\Re z}{\lambda}}, \qquad (7.12)$$

and if moreover $z = t$ belongs to $]0, +\infty[$,

$$|p(t,x,y)| \leq \frac{Ce^{t/\lambda}}{\sqrt{V(x,\lambda^{1/m})V(y,\lambda^{1/m})}} \left(\frac{\lambda}{t}\right)^{d/m} \exp\left\{-c(\frac{\rho^m(x,y)}{t})^{\frac{1}{m-1}}\right\}$$

$$(7.13)$$

for some constants $C, c > 0$.

If the heat kernel $p(t,x,y)$ is continuous with respect to x and y, then $z \mapsto p(z,x,y)$ is a holomorphic function on $\mathbb{C}^+$ for every $x, y \in \Omega$ (this follows from the holomorphy of the semigroup and (6.48) applied to $p(t,x,y)$ and A). Now (7.11) follows from (7.12) and (7.13) by applying Lemma 6.18 to $F(z) := p(z,x,y)e^{-z/\lambda}$ for fixed $x, y \in \Omega$.

In order to prove (7.11) in the general case, we fix $x_0, y_0 \in \Omega$ and let $E_0 := B(x_0, \lambda^{1/m})$ and $F_0 := B(y_0, \lambda^{1/m})$. Since $E_0 \subseteq B(x, 2\lambda^{1/m})$ for $x \in E_0$, it follows from the doubling volume property (7.1) that

$$V(x_0, \lambda^{1/m}) \leq CV(x, \lambda^{1/m}) \text{ for all } x \in E_0. \qquad (7.14)$$

Using this, one obtains from (7.12) and (7.13)

$$|\chi_{E_0}(x)p(z,x,y)\chi_{F_0}(y)| \leq \frac{C}{\sqrt{V(x_0,\lambda^{1/m})V(y_0,\lambda^{1/m})}} \left(\frac{\lambda}{\Re z}\right)^{d/m} e^{\frac{\Re z}{\lambda}},$$

$$(7.15)$$

and

$$|\chi_{E_0}(x)p(t,x,y)\chi_{F_0}(y)| \leq \frac{C \exp\left\{-c\left(\frac{\rho^m(x,y)}{t}\right)^{\frac{1}{m-1}}\right\}}{\sqrt{V(x_0,\lambda^{1/m})V(y_0,\lambda^{1/m})}} \left(\frac{\lambda}{t}\right)^{d/m} e^{t/\lambda}$$

$$(7.16)$$

for all $z \in \mathbb{C}^+$ and all $t > 0$, where C, c are positive constants.

Using the last two estimates, we can now repeat the proof of Theorem 6.16, with $\chi_{E_0}(x)p(z,x,y)\chi_{F_0}(y)e^{-z/\lambda}$ in place of $p(z,x,y)$ and apply Lemma 6.18 to obtain

$$|\chi_{E_0}(x)p(z,x,y)\chi_{F_0}(y)|$$

$$\leq \frac{C'e^{\Re z/\lambda} \exp\left\{-c'\left(\frac{\rho^m(x,y)}{|z|}\right)^{\frac{1}{m-1}}\cos\theta\right\}}{\sqrt{V(x_0,\lambda^{1/m})V(y_0,\lambda^{1/m})}} \left(\frac{\lambda}{\Re z}\right)^{d/m} \quad (7.17)$$

for all $z \in \mathbb{C}^+$ and (a.e.) $x, y \in \Omega$, where the positive constants C', c' are independent of λ. We switch x_0 and x in (7.14) to obtain from (7.17)

$$|\chi_{E_0}(x)p(z,x,y)\chi_{F_0}(y)|$$

$$\leq \frac{C''e^{\Re z/\lambda} \exp\left\{-c'\left(\frac{\rho^m(x,y)}{|z|}\right)^{\frac{1}{m-1}}\cos\theta\right\}}{\sqrt{V(x,\lambda^{1/m})V(y,\lambda^{1/m})}} \left(\frac{\lambda}{\Re z}\right)^{d/m} \quad (7.18)$$

for all $z \in \mathbb{C}^+$ and a.e. $x, y \in \Omega$. Again C'' is a positive constant independent of λ. Since this inequality holds for all x_0 and y_0 and a.e. $(x, y) \in \Omega \times \Omega$ it implies (7.11). $\square$

The above proof shows a family of upper bounds for $|p(z,x,y)|$. Different choices of λ in (7.11) yield different bounds. Taking for example $\lambda = \Re z$, one obtains

THEOREM 7.3 *Assume that the doubling condition (7.2) holds. Assume that A is a non-negative self-adjoint operator on $L^2(\Omega,\mu)$ having a heat kernel $p(t,x,y)$ which satisfies (7.3). Then there exist positive constants*

C, c such that

$$|p(z,x,y)| \leq \frac{C \exp\left\{ -c\left(\frac{\rho^m(x,y)}{|z|} \right)^{\frac{1}{m-1}} \cos\theta \right\}}{\sqrt{V(x, (\Re z)^{1/m})V(y, (\Re z)^{1/m})}}$$

for all $z \in \mathbb{C}^+$ and a.e. $x, y \in \Omega$, where $\theta = \arg z$.

The pointwise bounds in Theorems 7.2 and 7.3 have the following inte-grated form.

THEOREM 7.4 *Assume that the assumptions of Theorem 7.2 hold. The fol-lowing bounds hold for all $\varepsilon > 0$:*

$$\|e^{-zA}\|_{\mathcal{L}(L^p(\Omega))} \leq C_\varepsilon \left(\frac{|z|}{\Re z} \right)^{d|\frac{1}{2} - \frac{1}{p}| + \varepsilon} \quad for\ all\ p \in [1, +\infty], z \in \mathbb{C}^+, \tag{7.19}$$

where C_ε is a positive constant which depends on ε.

Proof. By the semigroup property one has for all $r > 0$ and $\theta \in (-\frac{\pi}{2}, \frac{\pi}{2})$,

$$\int_\Omega |p(re^{i\theta}, x, y)|^2 \, d\mu(x) = \int_\Omega p(re^{i\theta}, x, y)p(re^{-i\theta}, y, x) \, d\mu(x)$$
$$= p(2r\cos\theta, y, y).$$

We then obtain from assumption (7.3) that

$$\int_\Omega |p(re^{i\theta}, x, y)|^2 \, d\mu(x) \leq \frac{C}{V(y, (r\cos\theta)^{1/m})}. \tag{7.20}$$

Fix $\varepsilon \in]0, 1[$ and let $p_\varepsilon \in]0, 1[$ be such that $1 = \frac{1-\varepsilon}{2} + \frac{\varepsilon}{p_\varepsilon}$. Write

$$\int_\Omega |p(re^{i\theta}, x, y)| \, d\mu(x) \leq \left(\int_\Omega |p(re^{i\theta}, x, y)|^2 \, d\mu(x) \right)^{\frac{1-\varepsilon}{2}}$$
$$\times \left(\int_\Omega |p(re^{i\theta}, x, y)|^{p_\varepsilon} \, d\mu(x) \right)^{\frac{1+\varepsilon}{2}} \tag{7.21}$$

Using (7.4), it follows from Theorem 7.2 that

$$\int_\Omega |p(re^{i\theta}, x, y)|^{p_\varepsilon} \, d\mu(x) \leq \frac{C}{V(y, (\frac{r}{(\cos\theta)^{m-1}})^{1/m})^{p_\varepsilon}(\cos\theta)^{p_\varepsilon d}}$$
$$\times \int_\Omega \exp\left\{ -c'(\frac{\rho^m(x,y)}{r})^{\frac{1}{m-1}} \cos\theta \right\} d\mu(x).$$

From (7.9) with $t = \frac{r}{(\cos\theta)^{m-1}}$ we deduce the estimate

$$\int_\Omega |p(re^{i\theta}, x, y)|^{p_\varepsilon} \, d\mu(x) \le \frac{C'}{(\cos\theta)^{p_\varepsilon d}} V(y, (\frac{r}{(\cos\theta)^{m-1}})^{1/m})^{1-p_\varepsilon}.$$

The doubling property (7.2) implies that

$$V\left(y, \left(\frac{r}{(\cos\theta)^{m-1}}\right)^{1/m}\right) = V\left(y, (r\cos\theta)^{1/m}\frac{1}{\cos\theta}\right)$$
$$\le C(\cos\theta)^{-d} V(y, (r\cos\theta)^{1/m}),$$

therefore

$$\int_\Omega |p(re^{i\theta}, x, y)|^{p_\varepsilon} \, d\mu(x) \le \frac{C}{(\cos\theta)^d} V(y, (r\cos\theta)^{1/m})^{1-p_\varepsilon}. \qquad (7.22)$$

Inserting (7.20) and (7.22) in (7.21) we obtain

$$\int_\Omega |p(re^{i\theta}, x, y)| \, d\mu(x) \le \frac{C''}{(\cos\theta)^{\frac{d(1+\varepsilon)}{2}}},$$

that is,

$$\|e^{-zA}\|_{\mathcal{L}(L^1(\Omega))} = \sup_{y\in\Omega} \int_\Omega |p(z, x, y)| \, d\mu(x) \le C'' \left(\frac{|z|}{\Re z}\right)^{\frac{d(1+\varepsilon)}{2}}.$$

The estimate of $\|e^{-zA}\|_{p\to p}$ follows now for $p \in [1, 2]$ by interpolation with the fact that, since A is self-adjoint,

$$\|e^{-zA}\|_{\mathcal{L}(L^2(\Omega))} \le 1$$

and for $p \in [2, +\infty]$ by duality. $\square$

COROLLARY 7.5 *Under the assumptions of the previous theorem, the semi-group $(e^{-tA})_{t\ge 0}$ extends to a bounded holomorphic semigroup on $\mathbb{C}^+$ on $L^p(\Omega, \mu)$ for all $p \in [1, \infty)$.*

Proof. The proof is the same as for Theorem 6.16. For each fixed $f \in L^1(\Omega, \mu)$ one can approximate $e^{-zA} f$ by $\chi_{\Omega_n} e^{-zA} f_n$, where the sequence $f_n \in L^1(\Omega, \mu) \cap L^2(\Omega, \mu)$ converges to f in $L^1(\Omega, \mu)$, Ω_n is a nondecreasing sequence of subsets of Ω such that $\mu(\Omega_n) < \infty$ and $\Omega = \cup_n \Omega_n$. By Theorem 7.4, $\|\chi_{\Omega_n} e^{-zA} f_n\|_1 \le M_\psi$ for all n and all $z \in \Sigma(\psi)$, where $\psi \in (0, \frac{\pi}{2})$ is fixed and M_ψ is a constant depending only on ψ. Using the fact that the semigroup is holomorphic on $L^2(\Omega, \mu)$, one concludes by Vitali's theorem that $z \mapsto e^{-zA} f$ is holomorphic on $\Sigma(\psi)$ with values in $L^1(\Omega, \mu)$.

In order to prove the strong continuity on $L^1(\Omega, \mu)$, we let $f \in L^1(\Omega, \mu) \cap L^2(\Omega, \mu)$ with support contained in a ball $B_0 = B(x_0, r)$. Denote by B the ball $B(x_0, r + \delta)$, for some $\delta > 0$. Using Theorem 7.2 we obtain

$$\int_{\Omega \setminus B} |e^{-zA} f(x)| d\mu(x)$$

$$\leq \int_{\Omega \setminus B} \int_{B_0} \frac{C(\cos \theta)^{-d} \exp\left\{ -c\left(\frac{\rho^m(x,y)}{|z|}\right)^{\frac{1}{m-1}} \cos \theta \right\}}{\sqrt{V(x, (\frac{|z|}{(\cos \theta)^{m-1}})^{1/m}) V(y, (\frac{|z|}{(\cos \theta)^{m-1}})^{1/m})}}$$
$$\times |f(y)| d\mu(y) d\mu(x)$$

$$\leq C(\cos \theta)^{-d} \exp\left\{ -\frac{c}{2}\left(\frac{\delta^m}{|z|}\right)^{\frac{1}{m-1}} \cos \theta \right\}$$

$$\times \int_{B_0} \int_{\Omega} \frac{\exp\left\{ -\frac{c}{2}\left(\frac{\rho^m(x,y)}{|z|}\right)^{\frac{1}{m-1}} \cos \theta \right\}}{\sqrt{V(x, (\frac{|z|}{(\cos \theta)^{m-1}})^{1/m}) V(y, (\frac{|z|}{(\cos \theta)^{m-1}})^{1/m})}} d\mu(x)$$
$$\times |f(y)| d\mu(y).$$

As we have seen in the proof of Proposition 7.1, the term

$$\sup_{y \in \Omega} \int_{\Omega} \frac{\exp\left\{ -\frac{c}{2}\left(\frac{\rho^m(x,y)}{|z|}\right)^{\frac{1}{m-1}} \cos \theta \right\}}{\sqrt{V(x, (\frac{|z|}{(\cos \theta)^{m-1}})^{1/m}) V(y, (\frac{|z|}{(\cos \theta)^{m-1}})^{1/m})}} d\mu(x)$$

is bounded by a positive constant independent of of z. Therefore, there exists a positive constant C' such that

$$\|e^{zA} f - f\|_1 = \int_{\Omega \setminus B} |e^{-zA} f(x)| d\mu(x) + \int_{B \cap \Omega} |e^{-zA} f(x) - f(x)| d\mu(x)$$

$$\leq C'(\cos \theta)^{-d} \exp\left\{ -\frac{c}{2}\left(\frac{\delta^m}{|z|}\right)^{\frac{1}{m-1}} \cos \theta \right\} \|f\|_1$$

$$+ \sqrt{\mu(B)} \|e^{-zA} f - f\|_2.$$

From this and the strong continuity of e^{-zA} on $L^2(\Omega, \mu)$, we obtain that

$$\|e^{zA} f - f\|_1 \to 0 \text{ as } z \to 0.$$

All of this holds for $z \in \Sigma(\psi)$ for every $\psi \in (0, \frac{\pi}{2})$. $\square$

7.2 L^p-SPECTRAL INDEPENDENCE

7.2.1 Functional calculus

In this section, we give a short introduction to a functional calculus which will allow us to prove that for a self-adjoint operator whose heat kernel satisfies a Gaussian upper bound, its spectrum in L^p is the same as in L^2.

Let $\beta \in \mathbb{R}$ and $n \in \mathbb{N}$. A functional $f : \mathbb{R} \to \mathbb{C}$ belongs to the class S_n^β if f is of class C^n on $\mathbb{R}$ and for every $k \in \{1, ..., n\}$, the derivative of order k satisfies:

$$f^{(k)}(x) = O(<x>^{-\beta-k}) \text{ as } x \to \infty.$$

Here and in what follows we denote $<z> := \sqrt{1 + |z|^2}$ for all $z \in \mathbb{C}$.

Let $f \in S_n^\beta$ for some $n \geq 0$ and $\beta > 0$. Define $\tilde{f} : \mathbb{C} \to \mathbb{C}$ by

$$\tilde{f}(x, y) := \left(\sum_{k=0}^{n} \frac{1}{k!} f^{(k)}(x)(iy)^k \right) \sigma(x, y), \qquad (7.23)$$

where $\sigma(x, y) := \tau(\frac{y}{<x>})$ with $\tau \in C_c^\infty(\mathbb{R})$ such that $0 \leq \tau \leq 1$, $\tau(t) = 1$ for $t \in [-1, 1]$ and $= 0$ for $|t| \geq 2$.

DEFINITION 7.6 *The function $\tilde{f}$ defined by (7.23) is called an almost analytic extension of f.*

Let E be a Banach space and A be an operator on E. We assume that the following two conditions are satisfied:

$$\sigma(A) \subseteq \mathbb{R}, \qquad (7.24)$$

$$\|(zI - A)^{-1}\|_{\mathcal{L}(E)} \leq C \frac{<z>^\alpha}{|\Im z|^{\alpha+1}} \text{ for all } z \notin \mathbb{R} \qquad (7.25)$$

for some $\alpha \geq 0$ and some positive constant C. Under these two assumptions, one can define the functional calculus $f(A)$ for an appropriate function f.

DEFINITION 7.7 *Assume that A is an operator on a Banach space E for which (7.24) and (7.25) are satisfied for some $\alpha \geq 0$. If $f \in S_{n+1}^\beta$ for some $\beta > 0$ and some $n > \alpha$, one defines*

$$f(A) := -\frac{1}{\pi} \int_{\mathbb{C}} \frac{\partial \tilde{f}}{\partial \bar{z}}(z)(zI - A)^{-1} dx dy \qquad (7.26)$$

for $z = x + iy$, where $\frac{\partial \tilde{f}}{\partial \bar{z}}(z) := \frac{1}{2}[\frac{\partial \tilde{f}}{\partial x} + \frac{\partial \tilde{f}}{\partial y}](z)$.

The formula (7.26) defines a functional calculus. We refer the reader to Davies [Dav95a], [Dav95d]. The following result summarizes the most important properties of this functional calculus (see [Dav95a]).

THEOREM 7.8 *Under the assumptions of Definition 7.7 one has:*
(i) $f(A) \in \mathcal{L}(E)$ and there exists a positive constant c such that

$$\|f(A)\|_{\mathcal{L}(E)} \leq c \sum_{k=0}^{n+1} \int_{\mathbb{R}} |f^{(k)}(x)| < x >^{k-1} dx.$$

(ii) $(fg)(A) = f(A)g(A)$ for all $f, g \in S_{n+1}^{\beta}$ $(n > \alpha)$.
(iii) For $w \notin \mathbb{R}$ and $f_w(x) := (w - x)^{-1}$ for all $x \in \mathbb{R}$, we have $f_w(A) = (wI - A)^{-1}$.
(iv) Let $f \in C_c^{n+1}(\mathbb{R})$, $n > \alpha$, have support disjoint from $\sigma(A)$. Then $f(A) = 0$.

Note also that for self-adjoint operators on Hilbert spaces, the two assumptions (7.24) and (7.25) are satisfied (with $\alpha = 0$). Moreover, the functional calculus defined above coincides with the usual one defined in terms of spectral measures. See Helffer-Sjöstrand [HeSj89] or Davies [Dav95a], [Dav95d].

Let us mention that in Davies [Dav95a], [Dav95d] functions f are assumed to be in S_n^{β} for every n and some $\beta > 0$. The condition $f \in S_{n+1}^{\beta}$ for some $n > \alpha$ is enough to define the functional calculus and obtain Theorem 7.8.

7.2.2 Interpolation of the spectrum

Assume that (X, ρ, μ) is a σ-finite measured space, endowed with a metric ρ and satisfying the doubling condition (7.1). Let Ω be an open subset of X. Assume that A is a non-negative self-adjoint operator on $L^2(\Omega, \mu)$ with a heat kernel $p(t, x, y)$ which satisfies the Gaussian upper bound (7.3). It is proved in Corollary 7.5 that the semigroup $(e^{-tA})_{t \geq 0}$, initially defined on $L^2(\Omega, \mu)$, extends to a bounded holomorphic semigroup on $L^p(\Omega, \mu)$ for all $p \in [1, \infty)$. Denote now by $-A_p$ the corresponding generator on $L^p(\Omega, \mu)$ ($A_2 := A$). In the sequel, $\sigma(A_p)$ denotes the spectrum of the operator A_p on $L^p(\Omega, \mu)$. Again, Corollary 7.5 says that $-A_p$ generates a bounded holomorphic semigroup on the sector $\Sigma(\frac{\pi}{2})$ on $L^p(\Omega, \mu)$, and therefore $\sigma(A_p) \subseteq [0, \infty)$ by Theorem 1.45. In addition, the resolvent satisfies the following estimate.

PROPOSITION 7.9 *If $p \in [1, +\infty[$, then for any $\beta > d|\frac{1}{2} - \frac{1}{p}|$ we have*

$$\|(A_p - zI)^{-1}\|_{\mathcal{L}(L^p)} \leq C \frac{|z|^\beta}{|\operatorname{Im} z|^{\beta+1}} \ \text{for all} \ z \in \mathbb{C} \setminus \mathbb{R}.$$

Here d is the constant appearing in (7.2).

Proof. Set $z = re^{-i\theta}$ and assume that $\theta \in]0, \pi]$. Since the resolvent is the Laplace transform of the semigroup, we can write

$$(A_p - zI)^{-1} = e^{-i\frac{\pi-\theta}{2}} (A_p e^{-i\frac{\pi-\theta}{2}} + re^{i\frac{\pi-\theta}{2}} I)^{-1}$$

$$= e^{-i\frac{\pi-\theta}{2}} \int_0^{+\infty} e^{-rte^{i\frac{\pi-\theta}{2}}} e^{-te^{-i\frac{\pi-\theta}{2}} A_p} dt.$$

Now we apply Theorem 7.4 to obtain

$$\|(A_p - zI)^{-1}\|_{p \to p} \leq \frac{C}{(\sin \frac{\theta}{2})^\beta} \int_0^{+\infty} e^{-rt \sin \frac{\theta}{2}} dt,$$

which gives the desired estimate. If $\theta \in [-\pi, 0[$, we change π to $-\pi$ in the above argument and argue similarly. $\qquad\square$

THEOREM 7.10 *Assume that X satisfies the doubling property (7.1) and let Ω be an arbitrary open subset of X and A a non-negative self-adjoint operator on $L^2(\Omega, \mu)$. Assume that A has a heat kernel $p(t, x, y)$ which satisfies the Gaussian upper bound (7.3). Then $\sigma(A_p) = \sigma(A_2)$ for all $p \in [1, \infty)$. In other words, the spectrum of A_p is p-independent.*

Proof. We have seen previously that $\sigma(A_p) \subseteq [0, \infty)$ for all $p \in [1, \infty)$. By Proposition 7.9, we can define the functional calculus $f(A_p)$ for functions $f \in S_{n+1}^\beta$ for $n > d|\frac{1}{2} - \frac{1}{p}|$ and $\beta > 0$.

Fix $p, q \in [1, \infty)$ and assume that $\lambda \in [0, \infty)$ is such that $\lambda \notin \sigma(A_p)$. There exists $\varepsilon > 0$ such that $(\lambda - \varepsilon, \lambda + \varepsilon) \cap \sigma(A_p) = \emptyset$. Take $f \in C_c^\infty(\lambda - \varepsilon, \lambda + \varepsilon)$ with $f(x) = 1$ for $x \in [\lambda - \frac{\varepsilon}{2}, \lambda + \frac{\varepsilon}{2}]$. By assertion (iv) of Theorem 7.8, $f(A_p) = 0$. Since $e^{-tA_p}\phi = e^{-tA_q}\phi$ for all $\phi \in L^p(\Omega, \mu) \cap L^q(\Omega, \mu)$, the same equality holds for the resolvents $(A_p - zI)^{-1}$ and $(A_q - zI)^{-1}$ for $z \notin [0, \infty)$. This implies $f(A_q)\phi = f(A_p)\phi$ for all $\phi \in L^p(\Omega, \mu) \cap L^q(\Omega, \mu)$. It follows by density that $f(A_q) = 0$. Now we prove that this implies $\lambda \notin \sigma(A_q)$. For each $\delta \in \mathbb{R}, \delta \neq 0$, we define $g_\delta(x) := \frac{1-f(x)}{\lambda + i\delta - x}$ for $x \in \mathbb{R}$. Since $f(A_q) = 0$, we have

$$g_\delta(A_q) = ((\lambda + i\delta)I - A_q)^{-1}. \tag{7.27}$$

Using assertion (i) of Theorem 7.8, we obtain

$$\|((\lambda + i\delta)I - A_q)^{-1}\|_{\mathcal{L}(L^q)} = \|g_\delta(A_q)\|_{\mathcal{L}(L^q)}$$

$$\leq c \sum_{k=0}^{n+1} \int_{\mathbb{R}} \left| \frac{d^k g_\delta(x)}{dx^k} \right| < x >^{k-1} dx$$

$$= c \sum_{k=0}^{n+1} \int_{I_\varepsilon} \left| \frac{d^k g_\delta(x)}{dx^k} \right| < x >^{k-1} dx,$$

where $I_\varepsilon := \mathbb{R} \setminus [\lambda - \frac{\varepsilon}{2}, \lambda + \frac{\varepsilon}{2}]$, since $g_\delta(x) = 0$ for all $x \in [\lambda - \frac{\varepsilon}{2}, \lambda + \frac{\varepsilon}{2}]$. A simple calculation shows that the term

$$\sum_{k=0}^{n+1} \int_{I_\varepsilon} \left| \frac{d^k g_\delta(x)}{dx^k} \right| < x >^{k-1} dx$$

is bounded independently of δ. Thus, $\|((\lambda + i\delta)I - A_q)^{-1}\|_{\mathcal{L}(L^q)}$ is bounded uniformly with respect to δ for $\delta \neq 0$. This implies that $\lambda \notin \sigma(A_q)$. $\qquad \square$

If no assumption is made on the operator, the spectrum in L^p may depend on p. This is shown in the following example.

Example 7.2.1   *Consider on $L^p((0, \infty), dx), 1 \leq p < \infty$, the group $(T_p(t))_{t \in \mathbb{R}}$ defined by*

$$T_p(t)f(x) := f(e^{-t}x).$$

Denote by A_p the generator of $(T_p(t))_{t \in \mathbb{R}}$. Then

$$\sigma(A_p) := \left\{ \lambda \in \mathbb{C}, \Re\lambda = \frac{1}{p} \right\} \quad \text{for all } p, 1 \leq p < \infty.$$

Indeed, for every $f \in L^p((0, \infty), dx)$ and $t \in \mathbb{R}$, $\|T_p(t)f\|_p = e^{t/p}\|f\|_p$. Thus, $A_p - \frac{1}{p}I$ is the generator of a group of isometries on $L^p((0, \infty), dx)$ and this implies that $\sigma(A_p - \frac{1}{p}) \subseteq i\mathbb{R}$, i.e., $\sigma(A_p) \subseteq \frac{1}{p} + i\mathbb{R}$. Fix $p \in [1, 2)$ and assume that $\lambda \in \rho(A_p)$ with $\Re\lambda = \frac{1}{p}$. Since $T_p(t)f = T_2(t)f$ for all $f \in L^p((0, \infty), dx) \cap L^2((0, \infty), dx)$, it follows that $(zI - A_p)^{-1}f = (zI - A_2)^{-1}f$ for $z > \frac{1}{p}$. By analytic continuation and the fact that $\lambda \in \rho(A_p)$, this equality holds for $z \in (\frac{1}{2}, \frac{1}{p})$. For $z \in (\frac{1}{2}, \frac{1}{p})$ and every non-negative $f \in L^2 \cap L^q$, we have $(zI - A_2)^{-1}f = \int_0^\infty e^{-zt}T_2(t)f\,dt \geq 0$ and

$$(zI - A_p)^{-1}f = -(-zI + A_p)^{-1}f = -\int_0^\infty e^{-zt}T_p(-t)f\,dt \leq 0,$$

which shows that the equality $(zI - A_p)^{-1}f = (zI - A_2)^{-1}f$ *cannot hold on* $L^2 \cap L^p$. *Thus,* $\sigma(A_p) = \{\lambda \in \mathbb{C}, \Re\lambda = \frac{1}{p}\}$. *By similar arguments, the same equality holds for* $p \geq 2$.

The generator of the group $(T_p(t))_{t \in \mathbb{R}}$ considered in the previous example is a first-order differential operator. Taking the square $B_p := A_p^2$ one obtains a second-order operator whose spectrum, as an operator on $L^p((0, \infty), dx)$, depends on p.

7.3 RIESZ MEANS AND REGULARIZATION OF THE SCHRÖDINGER GROUP

It is a classical result that for every self-adjoint operator A on a Hilbert space, iA generates a strongly continuous group $(e^{itA})_{t \in \mathbb{R}}$. If $A = \Delta$ is the Laplacian, then the Schrödinger group $(e^{it\Delta})_{t \in \mathbb{R}}$ is not defined on $L^p(\mathbb{R}^d, dx)$ for every $p \neq 2$. See Hörmander [Hör60]. The next result shows that appropriate integration of the group yields a bounded family on L^p. More precisely,

THEOREM 7.11 *Assume that* (X, ρ, μ) *satisfies the doubling property (7.2). Let* Ω *be an open subset of* X *and* A *a self-adjoint operator on* $L^2(\Omega, \mu)$. *Assume that* A *has a heat kernel satisfying the Gaussian upper bound (7.3). For all* $p \in [1, +\infty]$ *and every* $\alpha > d|\frac{1}{2} - \frac{1}{p}|$, *the Riesz mean operator defined by*

$$I_\alpha(t) := t^{-\alpha} \int_0^t (t - s)^{\alpha-1} e^{-isA} \, ds$$

for $t > 0$, *and* $I_\alpha(t) = \overline{I_\alpha(-t)}$ *for* $t < 0$, *acts continuously on* $L^p(\Omega, \mu)$ *and one has*

$$\|I_\alpha(t)\|_{\mathcal{L}(L^p)} \leq C_\alpha \text{ for all } t \in \mathbb{R}^*. \tag{7.28}$$

Proof. It relies on the estimate (7.19). For $z \in \mathbb{C}^+$, consider the operator

$$J_\alpha(z) = \int_{[0,z]} (z - \zeta)^{\alpha-1} e^{-\zeta A} \, d\zeta.$$

We are going to show that, if $\alpha > d|\frac{1}{2} - \frac{1}{p}|$,

$$\|J_\alpha(z)\|_{\mathcal{L}(L^p)} \leq C_\alpha |z|^\alpha. \tag{7.29}$$

This will imply by strong continuity on $L^2(\Omega, \mu)$ that, for $z = it$, $t \in \mathbb{R}$, $i^\alpha t^\alpha I_\alpha(t) = J_\alpha(it)$, which is well defined on $L^2(\Omega, \mu) \cap L^p(\Omega, \mu)$, extends continuously to a bounded operator on $L^p(\Omega, \mu)$. Finally, (7.29) yields (7.28).

Because $\zeta \mapsto e^{-\zeta A}$ is holomorphic on $\mathbb{C}^+$, $J_\alpha(z)$ doesn't depend on the path chosen to integrate from 0 to z. We choose the following path: first, follow the straight line $[0, |z|]$ and then follow the circle with radius $|z|$ and center 0. Let $J^1_\alpha(z)$ denote the first integral, $J^2_\alpha(z)$ the second one. One has

$$J^1_\alpha(z) = \int_0^{|z|} (z - s)^{\alpha-1} e^{-sA} \, ds.$$

By Proposition 7.1, the assumptions (7.1) and (7.3) imply

$$\|e^{-tA}\|_{\mathcal{L}(L^p)} \leq M \text{ for all } t > 0,$$

for some constant $M > 0$. Therefore

$$\|J^1_\alpha(z)\|_{\mathcal{L}(L^p)} \leq M \int_0^{|z|} |z - s|^{\alpha-1} \, ds,$$

and if $z = |z|e^{i\theta}$,

$$\|J^1_\alpha(z)\|_{\mathcal{L}(L^p)} \leq M \int_0^{|z|} \left| |z|e^{i\theta} - s \right|^{\alpha-1} ds = M|z|^\alpha \int_0^1 |e^{i\theta} - u|^{\alpha-1} \, du,$$

Finally

$$\|J^1_\alpha(z)\|_{\mathcal{L}(L^p)} \leq C_\alpha |z|^\alpha,$$

where

$$C_\alpha = M \sup_{\theta \in [-\frac{\pi}{2}, \frac{\pi}{2}]} \int_0^1 |e^{i\theta} - u|^{\alpha-1} \, du.$$

Now

$$J^2_\alpha(z) = \int_0^{\arg z} (z - |z|e^{i\varphi})^{\alpha-1} e^{-|z|e^{i\varphi}A} i|z|e^{i\varphi} \, d\varphi.$$

To fix ideas, assume that $\arg z \geq 0$. Using the estimate (7.19), and setting $\beta = d|\frac{1}{2} - \frac{1}{p}| + \varepsilon$, one obtains

$$\|J^2_\alpha(z)\|_{\mathcal{L}(L^p)} \leq C'_\alpha \int_0^{\arg z} |z - |z|e^{i\varphi}|^{\alpha-1} \frac{|z|}{(\cos \varphi)^\beta} \, d\varphi$$

$$\leq C'_\alpha |z|^\alpha \int_0^{\arg z} \frac{[\sin(\frac{\arg z - \varphi}{2})]^{\alpha-1}}{(\cos \varphi)^\beta} \, d\varphi,$$

but the function sin is increasing on $[0, \pi/2]$ so that

$$\sin\left(\frac{\arg z - \varphi}{2}\right) \le \sin\left(\frac{\pi}{2} - \varphi\right) = \cos\varphi.$$

Hence,

$$\int_0^{\arg z} \frac{[\sin(\frac{\arg z - \varphi}{2})]^{\alpha-1}}{(\cos\varphi)^\beta}\, d\varphi \le \int_0^{\arg z} \left[\sin(\frac{\arg z - \varphi}{2})\right]^{\alpha-\beta-1} d\varphi$$

$$\le \int_0^{\frac{\pi}{4}} (\sin\theta)^{\alpha-\beta-1}\, d\theta.$$

The last term is finite if ε is chosen such that $\beta < \alpha$, which is possible if $\alpha > d|\frac{1}{2} - \frac{1}{p}|$. This proves that $\|J_\alpha(z)\|_{p\to p} \le C_\alpha''|z|^\alpha$ and finishes the proof of the theorem. $\qquad\square$

Theorem 7.11 shows that the Riesz mean operator $I_\alpha(t) = t^{-\alpha}\int_0^t (t - s)^{\alpha-1}e^{-isA}\, ds$ is bounded on $L^p(\Omega, \mu)$ for $\alpha > d|\frac{1}{2} - \frac{1}{p}|$ and all $p \in [1, \infty]$. We look now at another family of operators associated with the Schrödinger group.

THEOREM 7.12 *Assume that the assumptions of Theorem 7.11 hold. For all $p \in [1, \infty]$ and $\alpha > d|\frac{1}{2} - \frac{1}{p}|$, $(I + A)^{-\alpha}e^{itA}$ extends from $L^2(\Omega, \mu) \cap L^p(\Omega, \mu)$ to a bounded operator on $L^p(\Omega, \mu)$, and*

$$\|(I + A)^{-\alpha}e^{itA}\|_{\mathcal{L}(L^p)} \le C_\alpha(1 + |t|)^\alpha \text{ for all } t \in \mathbb{R}. \qquad (7.30)$$

In addition, the mapping $t \mapsto (I + A)^{-\alpha}e^{itA}$ is strongly continuous on $L^p(\Omega, \mu)$ for $p \in [1, \infty)$ $(t \in \mathbb{R})$.

Proof. Let us start by proving the estimate (7.30). If $f \in L^2 \cap L^p$ and $\alpha > 0$, one has

$$(I + A)^{-\alpha}e^{itA}f = \frac{1}{\Gamma(\alpha)}\int_0^{+\infty} e^{-s}s^{\alpha-1}e^{(-s+it)A}f\, ds,$$

where Γ is the Euler gamma function. Using (7.19), for $\alpha > d|\frac{1}{2} - \frac{1}{p}|$, one can bound the L^p-norm of the right-hand side by

$$C_{\alpha,\varepsilon}\|f\|_p \int_0^{+\infty} e^{-s}s^{\alpha-1}\left(\sqrt{\frac{s^2 + t^2}{s^2}}\right)^{d|\frac{1}{2} - \frac{1}{p}|+\varepsilon} ds,$$

for every $\varepsilon > 0$. Then, cutting the integral at $s = |t|$, we can estimate it, for ε small enough, by

$$|t|^\alpha \int_0^1 u^{\alpha-1-d|\frac{1}{2} - \frac{1}{p}|-\varepsilon}\, du + 2^{\frac{d}{2}|\frac{1}{2} - \frac{1}{p}|+\frac{\varepsilon}{2}}\int_{|t|}^{+\infty} e^{-s}s^{\alpha-1}\, ds \le C'|t|^\alpha + C'',$$

which yields (7.30).

Set $W(t) := (I + A)^{-\alpha} e^{itA}$. We have just seen that for every $t \in \mathbb{R}$, the operator $W(t)$ is bounded on $L^p(\Omega, \mu)$, $1 \leq p \leq +\infty$. Let us now prove that the mapping $t \to W(t)$ is strongly continuous on $L^p(\Omega, \mu)$ for $p \in [1, +\infty)$. One can write, for $t > 0$, $s \in \mathbb{R}$, and $f \in L^p$,

$$\|e^{-(t+is)A}(I+A)^{-\alpha}f - e^{-isA}(I+A)^{-\alpha}f\|_p \leq \|W(s)\|_{\mathcal{L}(L^p)}\|f - e^{-tA}f\|_p.$$

According to (7.30), the first factor on the right-hand side is bounded by $C(1 + |s|)^{\alpha}$, and the second goes to zero with t, because of strong continuity of the semigroup on $L^p(\Omega, \mu)$, for $1 \leq p < +\infty$ (see Corollary 7.5). Therefore, $e^{-(t+is)A}(I + A)^{-\alpha}f$ tends to $e^{-isA}(I + A)^{-\alpha}f = W(s)f$ in L^p, uniformly in $s \in [-T, T]$, as $t \to 0^+$. Again by Corollary 7.5, since $z \mapsto e^{-zA}$ is holomorphic on $L^p(\Omega, \mu)$ on the right half-plane, $s \mapsto e^{-(t+is)A}(I + A)^{-\alpha}f$ is continuous from $\mathbb{R}$ to $L^p(\Omega, \mu)$, for every $t > 0$ and $f \in L^p(\Omega, \mu)$. This implies that $W(s)$ is strongly continuous on $L^p(\Omega, \mu)$. $\square$

Theorem 7.12 allows one to measure smoothness properties of the solution to the Schrödinger equation

$$\begin{cases} \frac{\partial u}{\partial t} = iAu, \\ u(0) = f. \end{cases}$$

If the initial data $f \in D(A_p^{\alpha})$ for $\alpha > d|\frac{1}{2} - \frac{1}{p}|$, then $e^{itA}f \in L^p(\Omega, \mu)$ and

$$\begin{aligned} \|e^{itA}f\|_p &= \|(I + A_p)^{-\alpha} e^{itA}(I + A_p)^{\alpha}f\|_p \\ &\leq C_\alpha \|(I + A_p)^{\alpha}f\|_p. \end{aligned}$$

If $f \in D(A_p^{\beta})$ for some $\beta > \alpha > d|\frac{1}{2} - \frac{1}{p}|$, then

$$e^{itA}f = (I + A_p)^{-(\beta-\alpha)}(I + A_p)^{-\alpha} e^{itA}(I + A_p)^{\beta}f$$

and hence $e^{itA}f \in D(A_p^{\beta-\alpha})$ for all $t \in \mathbb{R}$. In some situations, $D(A_p^{\beta-\alpha}) \subseteq W^{2(\beta-\alpha),p}(\Omega)$. In that case, one obtains from the above estimates $e^{itA}f \in W^{2(\beta-\alpha),p}(\Omega)$.

The above theorem can also be used to prove existence of solutions (in L^p-spaces) to Schrödinger equations with initial data f in the domain of some power of A_p. It can also be reformulated in terms of generation of C-regularized groups. We shall not develop this here, the interested reader is referred to de Laubenfels [deL93].

Likewise, Theorem 7.11 can be reformulated in terms of generation of an α-integrated group by iA. Integrated (semi-)groups have been introduced in order to study Cauchy problems associated with operators that are not generators of semigroups. A typical example is $i\Delta$ on $L^p(\mathbb{R}^d, dx)$ for $p \neq 2$. We refer the reader to Arendt et al. [ABHN01] for a detailed study of integrated semigroups.

A special feature of the last two theorems is that they allow one to derive solvability of Schrödinger type equations from that of heat equations. These results can be applied to a wide class of operators, which includes all the examples discussed in the beginning of this chapter. In particular, they apply to uniformly elliptic operator with real-valued coefficients on arbitrary domains (and subject to Dirichlet boundary conditions).

The previous theorem can be extended as follows.

THEOREM 7.13 *Assume that the assumptions of Theorem 7.11 hold. Let $p \in [1, +\infty[$. If $f \in S_{n+1}^\alpha$ for some $\alpha > d|\frac{1}{2} - \frac{1}{p}|$ and $n > d|\frac{1}{2} - \frac{1}{p}|$, then $f(A)e^{itA}$ extends from $L^2(\Omega, \mu) \cap L^p(\Omega, \mu)$ to a bounded operator on $L^p(\Omega, \mu)$ and*

$$\|f(A)e^{itA}\|_{\mathcal{L}(L^p)} \leq C_\varepsilon (1 + |t|)^{d|\frac{1}{2} - \frac{1}{p}| + \varepsilon} \ for \ all \ t \in \mathbb{R}$$

for every $\varepsilon > 0$.

Proof. As seen previously, by Proposition 7.9, we can define the functional calculus $f(A_p)$ for functions $f \in S_{n+1}^\varepsilon$ with $n > d|\frac{1}{2} - \frac{1}{p}|$ and $\varepsilon > 0$. Now if f is as in the theorem, then $(1 + \lambda)^\beta f \in S_{n+1}^{\alpha - \beta}$. Thus, if $\alpha > \beta > d|\frac{1}{2} - \frac{1}{p}|$, $f(A)(I + A)^\beta$ extends to a bounded operator on $L^p(\Omega, \mu)$. Now we write

$$f(A)e^{itA} = f(A)(I + A)^\beta (I + A)^{-\beta} e^{itA},$$

and apply Theorem 7.12 to finish the proof. $\qquad\square$

Remark. The results in this section are based on the L^p-estimate of e^{-zA} which we proved in Theorem 7.4. It is interesting to notice that if one has a better estimate

$$\|e^{-zA}\|_{\mathcal{L}(L^p)} \leq C \left(\frac{|z|}{\Re z} \right)^{\gamma_p} \ for \ all \ z \in \mathbb{C}^+,$$

for some constant $\gamma_p < d|\frac{1}{2} - \frac{1}{p}|$, then the results in this section hold (with the same proofs) for $\alpha > \gamma_p$.

The final observation in this section is that the order $d|\frac{1}{2} - \frac{1}{p}|$ in Theorems 7.11 and 7.12 can be improved if the open subset Ω has finite measure and a Sobolev inequality holds in Ω. More precisely,

THEOREM 7.14 *Let A be a self-adjoint operator on $L^2(\Omega, \mu)$ such that the semigroup e^{-tA} is bounded from $L^2(\Omega, \mu)$ into $L^\infty(\Omega, \mu)$ with*

$$\|e^{-tA}\|_{\mathcal{L}(L^2, L^\infty)} \le Ct^{-d/4} \; for \; all \; t > 0, \qquad (7.31)$$

where C and d are positive constants. Assume in addition that e^{-tA} extends from $L^2(\Omega, \mu) \cap L^p(\Omega, \mu)$ to a bounded operator on $L^p(\Omega, \mu)$ with $\sup_{t \ge 0} \|e^{-tA}\|_{\mathcal{L}(L^p)} < \infty$ for all $p \in [1, \infty]$. Assume finally that $\mu(\Omega) < \infty$. Then:

1) For every $p \in [1, \infty)$, the Riesz mean operator $I_\alpha(t)$ acts as a bounded operator on $L^p(\Omega, \mu)$ for all $\alpha > \frac{d}{2}|\frac{1}{2} - \frac{1}{p}|$ and

$$\|I_\alpha(t)\|_{\mathcal{L}(L^p)} \le C_\alpha(1 + |t|^{-\frac{d}{2}|\frac{1}{2} - \frac{1}{p}|}) \; for \; all \; t \in \mathbb{R}^*. \qquad (7.32)$$

2) For all $p \in [1, \infty]$ and every $\alpha > \frac{d}{2}|\frac{1}{2} - \frac{1}{p}|$, $(I + A)^{-\alpha}e^{itA}$ extends from $L^2(\Omega, \mu) \cap L^p(\Omega, \mu)$ to a bounded operator on $L^p(\Omega, \mu)$, and

$$\|(I + A)^{-\alpha}e^{itA}\|_{\mathcal{L}(L^p)} \le C_\alpha \; for \; all \; t \in \mathbb{R}. \qquad (7.33)$$

Proof. By self-adjointness on $L^2(\Omega, \mu)$, one has

$$\|e^{-zA}\|_{\mathcal{L}(L^1, L^2)} = \|e^{-\Re zA}e^{-i\Im zA}\|_{\mathcal{L}(L^1, L^2)}$$
$$\le \|e^{-\Re zA}\|_{\mathcal{L}(L^1, L^2)}\|e^{-i\Im zA}\|_{\mathcal{L}(L^2)}$$
$$\le C(\Re z)^{-d/4}.$$

By the Riesz-Thorin interpolation theorem, it follows that for all $z \in \mathbb{C}^+$ and $p \in [1, 2]$

$$\|e^{-zA}\|_{\mathcal{L}(L^p, L^2)} \le C'(\Re z)^{-\frac{d}{2}|\frac{1}{2} - \frac{1}{p}|}.$$

Using this and the fact that $\mu(\Omega) < \infty$ it follows by Hölder's inequality that

$$\|e^{-zA}\|_{\mathcal{L}(L^p)} \le C_\Omega(\Re z)^{-\frac{d}{2}|\frac{1}{2} - \frac{1}{p}|}. \qquad (7.34)$$

Here C_Ω is a constant depending on $\mu(\Omega)$ and C'. Note that the same estimate holds for $p > 2$ by duality.

The proof of assertion 1) is similar to that of Theorem 7.11. We use (7.34) instead of (7.19). The family $J_\alpha^1(z)$ introduced in that proof satisfies

$$\|J_\alpha^1(z)\|_{\mathcal{L}(L^p)} \le C_\alpha|z|^\alpha.$$

The proof of this estimate uses only the fact that $(e^{-tA})_{t \ge 0}$ is uniformly bounded on $L^p(\Omega, \mu)$, which is the case here by assumptions. The term

$J_\alpha^2(z)$ can be estimated as follows:

$$\|J_\alpha^2(z)\|_{\mathcal{L}(L^p)} \le \int_0^{\arg z} |z - |z|e^{i\varphi}|^{\alpha-1} \frac{|z|}{(|z|\cos\varphi)^{\frac{d}{2}|\frac{1}{2}-\frac{1}{p}|}}\, d\varphi$$

$$\le C_\alpha' |z|^{\alpha-\frac{d}{2}|\frac{1}{2}-\frac{1}{p}|} \int_0^{\arg z} \frac{[\sin(\frac{\arg z - \varphi}{2})]^{\alpha-1}}{(\cos\varphi)^{\frac{d}{2}|\frac{1}{2}-\frac{1}{p}|}}\, d\varphi$$

$$\le C_\alpha'' |z|^{\alpha-\frac{d}{2}|\frac{1}{2}-\frac{1}{p}|}.$$

In order to prove assertion 2), we write as in the proof of Theorem 7.12

$$(I + A)^{-\alpha} e^{itA} f = \frac{1}{\Gamma(\alpha)} \int_0^{+\infty} e^{-s} s^{\alpha-1} e^{(-s+it)A} f\, ds,$$

for $f \in L^2(\Omega, \mu) \cap L^p(\Omega, \mu)$. Using (7.34), we estimate the L^p norm of the right-hand side by

$$C_\alpha \|f\|_p \int_0^{+\infty} e^{-s} s^{\alpha-1} s^{\frac{d}{2}|\frac{1}{2}-\frac{1}{p}|}\, ds,$$

which is finite for all $\alpha > \frac{d}{2}|\frac{1}{2} - \frac{1}{p}|$. $\square$

The previous theorem can be applied to uniformly elliptic operators with real-valued coefficients on bounded domains (see Theorem 6.10). It can also be applied to the Laplace-Beltrami operators on compact manifolds. For such manifolds, the Sobolev inequality holds (see, e.g., Chavel [Cha84] Theorem 7, p. 101) and this implies the estimate (7.31) by Theorem 6.4.

7.4 L^p-ESTIMATES FOR WAVE EQUATIONS

7.4.1 The general case

Let (X, ρ, μ) satisfy the doubling volume property (7.1) and denote again by $V(x, r) := \mu(B(x, r))$ the volume of the ball $B(x, r)$. Throughout this section d denotes a constant for which (7.2) is satisfied.

Let Ω be an open subset of X and A a non-negative self-adjoint operator on $L^2(\Omega, \mu)$. We assume that A has a heat kernel $p(t, x, y)$ which satisfies the Gaussian upper bound (7.3).

Let $f : [0, \infty) \to \mathbb{C}$ be a bounded measurable function. By spectral theory,

$$f(A) := \int_0^\infty f(\lambda) dE_\lambda(A) \tag{7.35}$$

is well defined and acts as a bounded operator on $L^2(\Omega, \mu)$. Here $E_\lambda(A)$ denotes the associated spectral measure with the self-adjoint operator A. Recall also that

$$\|f(A)\|_{\mathcal{L}(L^2)} \leq \|f\|_\infty.$$

In this section, we examine the question of extending the operator $f(A)$ from $L^2(\Omega, \mu) \cap L^1(\Omega, \mu)$ to a bounded operator on $L^1(\Omega, \mu)$. In order to do this, one needs to impose conditions on A and f.

We shall be working with an auxiliary nontrivial function φ of compact support. The choice of φ that will appear in the statements is not unique.

Let φ be a $C_c^\infty(0, \infty)$ non-negative function such that

$$\operatorname{supp}(\varphi) \subseteq [\tfrac{1}{4}, 1] \text{ and } \sum_{n \in \mathbb{Z}} \varphi(2^{-n}\lambda) = 1 \text{ for all } \lambda > 0. \tag{7.36}$$

In order to see that such a function φ exists, let $\psi \in C_c^\infty(0, \infty)$ be a non-negative function with support $\operatorname{supp}(\psi) \subseteq [\tfrac{1}{4}, 1]$ and such that $\psi(\lambda) > 0$ for all $\lambda \in [\tfrac{1}{3}, \tfrac{2}{3}]$. We obtain the function φ with the desired properties by putting

$$\varphi(\lambda) := \psi(\lambda) \left(\sum_{k=-\infty}^{+\infty} \psi(2^{-k}\lambda) \right)^{-1}.$$

The following theorem gives a condition on f under which $f(A)$ extends from $L^1(\Omega, \mu) \cap L^2(\Omega, \mu)$ to a bounded linear operator on $L^1(\Omega, \mu)$. In this result, C^s denotes the classical Lipschitz space of functions on $\mathbb{R}$.[2]

THEOREM 7.15 *Assume that X satisfies the doubling property (7.2) and let Ω be an open subset of X. Let A be a non-negative self-adjoint operator on $L^2(\Omega, \mu)$ whose heat kernel satisfies the Gaussian upper bound (7.3). Let φ be a non-negative C_c^∞ function satisfying (7.36). If the bounded measurable function $f : [0, \infty) \to \mathbb{C}$ satisfies*

$$\alpha_\varphi := \sum_{n \in \mathbb{Z}} \|\varphi(.)[f(2^n.) - f(0)]\|_{C^s} < \infty \tag{7.37}$$

for some $s > d/2$, then $f(A^{\frac{1}{m}})$ extends to a bounded operator on $L^1(\Omega, \mu)$ with norm bounded by $C(\alpha_\varphi + |f(0)|)$, where C is a positive constant independent of f.

---

[2]More precisely, C^s is the space of functions f of class $C^{[s]}$ with all bounded derivatives $f^{(k)}$ for $0 \leq k \leq [s]$ and such that the $[s]$-derivative $f^{([s])}$ satisfies
$$\sup_{x,y \in \mathbb{R}, x \neq y} \frac{f^{([s])}(x) - f^{([s])}(y)}{|x - y|^{s-[s]}} < \infty. \text{ Here } [s] \text{ denotes the integer part of } s.$$

Before we start the proof, we first make some remarks.

First, if for all $u \in L^2(\Omega, \mu)$, we have $\|e^{-tA}u\|_2 \to 0$ as $t \to +\infty$, then 0 can be neglected in the spectral resolution of A. Therefore, we can consider that f is defined on $(0, \infty)$. We can then ignore the term $f(0)$ in the theorem.

The second observation is that the sum over $n \in \mathbb{Z}$ in (7.37) can be reduced to the sum over $n \geq 0$. The reason is that one can decompose f as $f = f_1 + f_2$, with f_1 having compact support and f_2 having support in $[1, \infty)$. The first term $f_1(\sqrt[m]{A})$ can be handled directly by Lemma 7.18 below. The second term $f_2(\sqrt[m]{A})$ is treated by Theorem 7.15, but (7.37) for f_2 is precisely

$$\sum_{n \geq 0} \|\varphi(.)f_2(2^n.)\|_{C^s} < \infty.$$

This strategy will be used in the proof of the Theorem 7.19, where we consider wave operators on $L^p(\Omega, \mu)$.

The proof of Theorem 7.15 will be achieved in several steps. We start by some simple properties which will be needed latter.

Let f be a bounded and compactly supported function and define $g(\lambda) := f(\sqrt[m]{\lambda})e^{\lambda}$. Since

$$f(\sqrt[m]{A}) = g(A)e^{-A},$$

it follows that $f(\sqrt[m]{A})$ is given by a kernel $f(\sqrt[m]{A})(x, y)$, that is,

$$f(\sqrt[m]{A})u(x) = \int_{\Omega} f(\sqrt[m]{A})(x, y)u(y)d\mu(y)$$

for all $u \in L^2(\Omega, \mu)$ and a.e. $x \in \Omega$. The kernel $f(\sqrt[m]{A})(x, y)$ is given by

$$f(\sqrt[m]{A})(x, y) = (g(A)e^{-A})(x, y)$$
$$= (g(A)p(1, x, .))(y) = (g(A)p(1, ., y))(x). \quad (7.38)$$

This makes sense because $g(A) \in \mathcal{L}(L^2(\Omega, \mu))$ and $p(1, x, .) \in L^2(\Omega)$. The latter property can be shown as follows:

$$\int_{\Omega} |p(1, x, y)|^2 d\mu(y) = \int_{\Omega} p(1, x, y)p(1, y, x)d\mu(y)$$
$$= p(2, x, x)$$
$$\leq \frac{C}{V(x, \sqrt[m]{2})}.$$

LEMMA 7.16 *Assume that the assumptions of Theorem 7.15 are satisfied. Let $s > 0$. There exists a constant $\delta \geq 1$ such that the following estimate holds for a.e. $x \in \Omega$:*

$$\int_\Omega |f(\sqrt[m]{A})(x,y)|^2 (1 + \rho(x,y))^s d\mu(y) \leq \frac{C_2}{V(x,1)} \|f\|^2_{H^{s/2+\delta}}$$

for all $f \in H^{s/2+\delta}$ with $\mathrm{supp}(f) \subset [0,1]$. C_2 is a positive constant.

Proof. As previously, we define $g(\lambda) := f(\sqrt[m]{\lambda})e^\lambda$. If $\hat{g}$ denotes the Fourier transform of g, then

$$f(\sqrt[m]{A}) = g(A)e^{-A} = \frac{1}{2\pi}\int_\mathbb{R} \hat{g}(\zeta)e^{i\zeta A}d\zeta e^{-A}. \tag{7.39}$$

Hence, the kernel $f(\sqrt[m]{A})(x,y)$ satisfies

$$f(\sqrt[m]{A})(x,y) := \frac{1}{2\pi}\int_\mathbb{R} \hat{g}(\zeta)p(1 - i\zeta, x, y)d\zeta. \tag{7.40}$$

This follows from (7.39) and Fubini's theorem. Indeed, for every $u \in L^2(\Omega, \mu)$

$$\int_\Omega |p(1 - i\zeta, x, y)u(y)|d\mu(y)$$

$$\leq \left[\int_\Omega |p(1 - i\zeta, x, y)|^2 d\mu(y)\right]^{1/2} \|u\|_2$$

$$= \|u\|_2 \left[\int_\Omega p(1 - i\zeta, x, y)p(1 + i\zeta, y, x)d\mu(y)\right]^{1/2}$$

$$= \|u\|_2\sqrt{p(2, x, x)}.$$

Therefore,

$$\int_\mathbb{R} |\hat{g}(\zeta)| \int_\Omega |p(1 - i\zeta, x, y)u(y)|d\mu(y)d\zeta$$

can be estimated by

$$\|u\|_2\sqrt{p(2, x, x)} \left[\int_\mathbb{R} |\hat{g}(\zeta)|^2(1+\zeta^2)^{\varepsilon+1/2}d\zeta\right]^{1/2} \left[\int_\mathbb{R} (1+\zeta^2)^{-\varepsilon-1/2}d\zeta\right]^{1/2}.$$

The term $\int_\mathbb{R} |\hat{g}(\zeta)|^2(1 + \zeta^2)^{\varepsilon+1/2}d\zeta$ is finite since $f \in H^{1/2+\varepsilon}$ for $\varepsilon > 0$ (which we can choose to be small enough). Thus, $\int_\mathbb{R} |\hat{g}(\zeta)| \int_\Omega |p(1 - i\zeta, x, y)u(y)|d\mu(y)d\zeta$ is finite, too, and we can then apply Fubini's theorem to obtain (7.40).

Using (7.40) one has

$$
\begin{aligned}
I &:= \left(\int_\Omega |f(\sqrt[m]{A})(x,y)|^2 (1+\rho(x,y))^s d\mu(y) \right)^{1/2} \\
&= \|f(\sqrt[m]{A})(x,.)\|_{L^2(\Omega,(1+\rho(x,.))^s d\mu)} \\
&\le \frac{1}{2\pi} \int_{\mathbb{R}} |\hat{g}(\zeta)| \|p(1-i\zeta,x,.)\|_{L^2(\Omega,(1+\rho(x,.))^s d\mu)} d\zeta.
\end{aligned}
$$

We apply now Theorem 7.3 to obtain

$$
\begin{aligned}
J &:= \int_\Omega |p(1-i\zeta,x,y)|^2 (1+\rho(x,y))^s d\mu(y) \\
&\le \int_X \frac{C^2}{V(x,1)V(y,1)} \exp\{-2c\rho(x,y)^{m/(m-1)} (\sqrt{1+\zeta^2})^{-m/(m-1)}\} \\
&\qquad\qquad\qquad\qquad\qquad\qquad\qquad \times (1+\rho(x,y))^s d\mu(y).
\end{aligned}
$$

By the doubling property (7.2) one has

$$
\frac{1}{V(y,1)} \le \frac{C(1+\rho(x,y))^d}{V(x,1)}.
$$

Set $\alpha := (\sqrt{1+\zeta^2})^{-m/(m-1)}$ and

$$
U_k := \{y \in \Omega, k^{(m-1)/m} \le \rho(x,y) \le (k+1)^{(m-1)/m}\}.
$$

We have

$$
\begin{aligned}
J &\le \frac{C'}{V(x,1)^2} \int_X \exp\{-2c\rho(x,y)^{m/(m-1)}\alpha\}(1+\rho(x,y))^{s+d} d\mu(y) \\
&= \frac{C'}{V(x,1)^2} \sum_{k\ge 0} \int_{U_k} \exp\{-2c\rho(x,y)^{m/(m-1)}\alpha\} \\
&\qquad\qquad\qquad\qquad\qquad\qquad\qquad \times (1+\rho(x,y))^{s+d} d\mu(y) \\
&\le \frac{C'}{V(x,1)^2} \sum_{k\ge 0} e^{-2ck\alpha} V(x,(k+1)^{(m-1)/m}) \\
&\qquad\qquad\qquad\qquad\qquad\qquad\qquad \times (1+(k+1)^{(m-1)/m})^{s+d}.
\end{aligned}
$$

Using again (7.2), the last inequality gives

$$
J \le \frac{C''}{V(x,1)} (\sqrt{1+\zeta^2})^{s+2d+\frac{m}{m-1}},
$$

for some constant C''. Finally, by the Cauchy-Schwarz inequality

$$I \leq \frac{1}{2\pi}\sqrt{\frac{C''}{V(x,1)}}\left[\int_{\mathbb{R}}|\hat{g}(\zeta)|^2(\sqrt{1+\zeta^2})^{s+2d+\frac{m}{m-1}+2}d\zeta\right]^{1/2}$$

$$\times\left[\int_{\mathbb{R}}(\sqrt{1+\zeta^2})^{-2}d\zeta\right]^{1/2}.$$

This finishes the proof of the lemma. $\qquad\qquad\qquad\square$

LEMMA 7.17 *Assume again that the assumptions of Theorem 7.15 are satisfied and let $s > 0$. The following estimate holds for every $\varepsilon > 0$ and a.e. $x \in \Omega$:*

$$\int_{\Omega}|f(\sqrt[m]{A})(x,y)|^2(1+\rho(x,y))^s d\mu(y) \leq \frac{C_\varepsilon}{V(x,1)}\|f\|^2_{C^{s/2+\varepsilon}}$$

for all $f \in C^{s/2+\varepsilon}$ with $\mathrm{supp}(f) \subset [0,1]$. C_ε is a positive constant independent of f.

Proof. It follows from (7.38) that

$$\left[\int_{\Omega}|f(\sqrt[m]{A})(x,y)|^2 d\mu(y)\right]^{1/2} \leq \|g(A)\|_{\mathcal{L}(L^2)}\|p(1,x,.)\|_2$$

$$\leq \|g\|_\infty\|p(1,x,.)\|_2$$

$$\leq e\|f\|_\infty\|p(1,x,.)\|_2.$$

It follows from the Gaussian upper bound (7.3) and the fact that A is self-adjoint that

$$\|p(1,x,.)\|^2_2 = \int_{\Omega} p(1,x,y)p(1,y,x)d\mu(y) = p(2,x,x) \leq \frac{C}{V(x,1)}.$$

Hence,

$$\|f(\sqrt[m]{A})(x,.)\|_{L^2(\Omega,\mu)} \leq e\sqrt{\frac{C}{V(x,1)}}\|f\|_\infty. \qquad (7.41)$$

On the other hand, Lemma 7.16 gives

$$\|f(\sqrt[m]{A})(x,.)\|_{L^2(\Omega,(1+\rho(x,.))^s\mu)} \leq \sqrt{\frac{C}{V(x,1)}}\|f\|_{H^{\delta+s/2}}. \qquad (7.42)$$

The Stein-Weiss interpolation theorem,[3] applied to the operator defined by $f \to f(\sqrt[m]{A})(x,.)$, allows one to deduce from (7.41) and (7.42) that for

---

[3]See Bergh and Löfström [BeLö76], Theorem 5.5.3.

every $\theta \in (0,1)$

$$\|f(\sqrt[m]{A})(x,.)\|_{L^2(\Omega,(1+\rho(x,.))^{s\theta}\mu)} \le \frac{C(\theta)}{\sqrt{V(x,1)}}\|f\|_{[C^0,C^{\delta+s/2}]_{[\theta]}}.$$

Here $[C^0, C^{\delta+s/2}]_{[\theta]}$ denotes the classical space obtained by complex interpolation. In particular,[4] we have for all $\varepsilon > 0$ and $\theta \in (0,1)$

$$\|f(\sqrt[m]{A})(x,.)\|_{L^2(\Omega,(1+\rho(x,.))^{s\theta}\mu)} \le \frac{C(\theta,\varepsilon)}{\sqrt{V(x,1)}}\|f\|_{C^{s\theta/2+\delta\theta+\varepsilon}},$$

where $C(\theta,\varepsilon)$ is a positive constant which depends on θ and ε. Since $s > 0$ is arbitrary, one can replace s by $\frac{s}{\theta}$ in the previous inequality, and taking θ arbitrary small, we obtain the desired estimate. $\square$

We have proved weighted L^2-estimates for $f(\sqrt[m]{A})(x,.)$ when f has support in $[0,1]$. These estimates can be extended to functions with support in $[0,r]$ for all $r > 0$.

LEMMA 7.18 *Assume that the assumptions of Theorem 7.15 are satisfied and fix $r, s > 0$. The following estimate holds for every $\varepsilon > 0$ and a.e. $x \in \Omega$:*

$$\int_\Omega |f(\sqrt[m]{A})(x,y)|^2(1+r\rho(x,y))^s d\mu(y) \le \frac{C_\varepsilon}{V(x,\frac{1}{r})}\|\delta_r f\|^2_{C^{s/2+\varepsilon}}$$

for all $f \in C^{s/2+\varepsilon}$ with $\operatorname{supp}(f) \subset [0,r]$, where $\delta_r f(.) := f(r.)$ and C_ε is a constant independent of f and r.

Proof. Let $A' := r^{-m}A$ and $\rho' := r\rho$. Denote by $V'(x,R)$ the volume of the ball of center x and radius R for the metric ρ'. Note first that

$$V'(x,R) = V\left(x, \frac{R}{r}\right).$$

In particular, $V'(x,R)$ satisfies (7.2) with the same constants C and d.
The heat kernel $k(t,x,y)$ of A' satisfies

$$|k(t,x,y)| = |p(r^{-m}t,x,y)| \le \frac{C\exp(-c\frac{\rho'(x,y)^{m/(m-1)}}{t^{1/(m-1)}})}{\sqrt{V'(x,t^{1/m})V'(y,t^{1/m})}}.$$

Hence we can apply the previous lemma to A' and ρ' and conclude that

$$\int_\Omega |h(\sqrt[m]{A'})(x,y)|^2(1+\rho'(x,y))^s d\mu(y) \le \frac{C_\varepsilon}{V'(x,1)}\|h\|^2_{C^{s/2+\varepsilon}} \quad (7.43)$$

---

[4]See Bergh and Löfström [BeLö76] Theorem 6.4.5 and Triebel [Tri83], p. 113 and p. 38.

for all h with $\mathrm{supp}(h) \subset [0, 1]$. The constant C_ε depends only on ε and the constants involved in (7.2) and (7.3). Now, if $\mathrm{supp}(f) \subset [0, r]$, we apply the estimate (7.43) to $h = \delta_r f$ and obtain the lemma.   □

Remark. The previous lemma applied for $F(\lambda) := f(\lambda^m)$ gives

$$\int_\Omega |f(A)(x, y)|^2 (1 + r^{1/m} \rho(x, y))^s d\mu(y) \leq \frac{C_\varepsilon}{V(x, r^{-1/m})} \|\delta_{r^{1/m}} F\|^2_{C^{s/2+\varepsilon}}$$

for all $f \in C^{s/2+\varepsilon}$ with $\mathrm{supp}(f) \subset [0, r]$.

Proof of Theorem 7.15. First, we write $f(\lambda) = f(\lambda) - f(0) + f(0)$ and hence

$$f(\sqrt[m]{A}) = (f(.) - f(0))(\sqrt[m]{A}) + f(0)I.$$

Thus, replacing f by $f - f(0)$, we may assume in the sequel that $f(0) = 0$. Let $\varphi \in C_c^\infty(0, \infty)$ be a non-negative function satisfying (7.36). We have

$$f(\lambda) = \sum_{-\infty}^{+\infty} \varphi(2^{-n}\lambda) f(\lambda) = \sum_{-\infty}^{+\infty} f_n(\lambda) \text{ for all } \lambda \geq 0, \qquad (7.44)$$

where

$$f_n(\lambda) := \varphi(2^{-n}\lambda) f(\lambda).$$

Since

$$\left| \sum_{n=-N}^{N} f_n(\lambda) \right| \leq \|f\|_\infty \sum_{n=-N}^{N} \varphi(2^{-n}\lambda) \leq \|f\|_\infty \text{ for all } \lambda \geq 0, \ N \in \mathbb{N},$$

it follows from the classical properties of the functional calculus that the sequence $(\sum_{n=-N}^{N} f_n(\sqrt[m]{A}))_N$ converges strongly in $L^2(\Omega, \mu)$ to $f(\sqrt[m]{A})$ (see, e.g., Reed and Simon [ReSi80], Theorem VIII.5). Set

$$I_n := \int_\Omega |f_n(\sqrt[m]{A})(x, y)| d\mu(y).$$

By the Cauchy-Schwarz inequality, one has

$$I_n \leq \left[\int_\Omega |f_n(\sqrt[m]{A})(x, y)|^2 (1 + 2^n \rho(x, y))^{2s} d\mu(y) \right]^{1/2}$$

$$\times \left[\int_\Omega (1 + 2^n \rho(x, y))^{-2s} d\mu(y) \right]^{1/2}.$$

Lemma 7.18 applied for $f = f_n$ and $r = 2^n$ gives

$$\int_\Omega |f_n(\sqrt[m]{A})(x,y)|^2 (1 + 2^n \rho(x,y))^{2s} d\mu(y)$$

$$\leq \frac{C_\varepsilon}{V(x, 2^{-n})} \|\delta_{2^n} f_n\|^2_{C^{s+\varepsilon}}$$

$$= \frac{C_\varepsilon}{V(x, 2^{-n})} \|\varphi(.)f(2^n.)\|^2_{C^{s+\varepsilon}}.$$

Since one has for $s > \frac{d}{2}$ (see below)

$$\int_\Omega (1 + 2^n \rho(x,y))^{-2s} d\mu(y) \leq C'V(x, 2^{-n}), \qquad (7.45)$$

it follows that

$$I_n \leq C'_\varepsilon \|\varphi(.)f(2^n.)\|_{C^{s+\varepsilon}}.$$

Therefore,

$$\int_\Omega \left| \sum_{n=-N}^{N} f_n(\sqrt[m]{A})(x,y) \right| d\mu(y) \leq \sum_{n=-N}^{N} I_n$$

$$\leq C'_\varepsilon \sum_{n \in \mathbb{Z}} \|\varphi(.)f(2^n.)\|_{C^{s+\varepsilon}},$$

and this proves the theorem.

It remains to prove (7.45). More generally, we prove that there exists a positive constant C' such that for every $s > \frac{d}{2}$, $u > 0$, and $r > 0$

$$\int_{\rho(x,y) \geq u} (1 + r\rho(x,y))^{-2s} d\mu(y) \leq C'V\left(x, \frac{1}{r}\right) \min(1, (ur)^{d-2s}).$$

$$(7.46)$$

Assume that $ur \geq 1$. Then

$$\int_{\rho(x,y) \geq u} (1 + r\rho(x,y))^{-2s} d\mu(y)$$

$$\leq \sum_{k \geq 0} \int_{2^k u \leq \rho(x,y) \leq 2^{k+1}u} (r\rho(x,y))^{-2s} d\mu(y)$$

$$\leq (ur)^{-2s} \sum_{k \geq 0} (2^k)^{-2s} V(x, 2^{k+1}u)$$

$$\leq (ur)^{-2s} \sum_{k \geq 0} (2^k)^{-2s} C(2^{k+1}ur)^d V\left(x, \frac{1}{r}\right),$$

where we have used (7.2) since $2^{k+1}ur > 1$. This shows the desired inequality.

Assume now that $ur < 1$. We have

$$\int_{\rho(x,y)\geq u} (1 + r\rho(x,y))^{-2s}d\mu(y) \leq \int_{\rho(x,y)\geq\frac{1}{r}} (1 + r\rho(x,y))^{-2s}d\mu(y)$$

$$+ \int_{\rho(x,y)<\frac{1}{r}} 1d\mu(y). \qquad (7.47)$$

We apply the previous case (with $u = \frac{1}{r}$) and obtain

$$\int_{\rho(x,y)\geq\frac{1}{r}} (1 + r\rho(x,y))^{-2s}d\mu(y) \leq C'V\left(x, \frac{1}{r}\right).$$

The second term $\int_{\rho(x,y)<\frac{1}{r}} d\mu(y)$ in (7.47) is precisely $V(x, \frac{1}{r})$. This proves (7.46).   □

We apply now Theorem 7.15 to obtain L^p-bounds for wave equations. We shall consider for simplicity operators with heat kernel satisfying the Gaussian upper bound (7.3) with $m = 2$.

THEOREM 7.19 *Assume that (X, ρ, μ) satisfies the doubling property (7.2) and let Ω be an open subset of X. Assume that A is a non-negative self-adjoint operator on $L^2(\Omega, \mu)$ with heat kernel satisfying the Gaussian upper bound (7.3) with $m = 2$. For every $p \in [1, \infty)$ and $\alpha > d|1/2 - 1/p|$, the operator $(I + A)^{-\alpha/2}e^{it\sqrt{A}}$ extends from $L^2(\Omega, \mu) \cap L^p(\Omega, \mu)$ to a bounded operator on $L^p(\Omega, \mu)$. In addition, for every $\varepsilon > 0$,*

$$\|(I + A)^{-\alpha/2}e^{it\sqrt{A}}\|_{\mathcal{L}(L^p)} \leq C_\varepsilon(1 + |t|)^{d|1/2-1/p|+\varepsilon}.$$

Proof. Fix $t \in \mathbb{R}$ and set

$$f(\lambda) := (1 + \lambda^2)^{-\alpha/2}e^{it\lambda} \text{ for all } \lambda \geq 0.$$

Let $\psi \in C_c^\infty(\mathbb{R})$ be such that $\text{supp}(\psi) \subseteq [-1, 2]$, $\psi(\lambda) = 1$ on $[0, 1]$. We write

$$f = \psi f + (1 - \psi)f =: f_1 + f_2$$

with $\text{supp}(f_1) \subseteq [0, 2]$ and $\text{supp}(f_2) \subseteq [1, \infty)$.

We apply Lemma 7.18 to obtain

$$\int_\Omega |f_1(\sqrt{A})(x,y)|^2(1 + 2\rho(x,y))^{2s}d\mu(y) \leq \frac{C_\varepsilon}{V(x, 1/2)}\|f_1\|_{C^{s+\varepsilon}}^2$$

$$\leq \frac{C'_\varepsilon}{V(x, 1)}\|f_1\|_{C^{s+\varepsilon}}^2.$$

Therefore, by the Cauchy-Schwarz inequality and (7.46) we have for $s > \frac{d}{2}$

$$\int_\Omega |f_1(\sqrt{A})(x,y)| d\mu(y) \le \left[\int_\Omega |f_1(\sqrt{A})(x,y)|^2 (1 + 2\rho(x,y))^{2s} d\mu(y) \right]^{1/2}$$

$$\times \left[\int_\Omega (1 + 2\rho(x,y))^{-2s} d\mu(y) \right]^{1/2}$$

$$\le C' \|f_1\|_{C^{s+\varepsilon}}.$$

Using the expression $f(\lambda) := (1 + \lambda^2)^{-\alpha/2} e^{it\lambda}$, one obtains[5]

$$\|f_1\|_{C^s} \le C(1 + |t|^s). \tag{7.48}$$

Thus, $f_1(\sqrt{A}) \in \mathcal{L}(L^1(\Omega, \mu))$ and

$$\|f_1(\sqrt{A})\|_{\mathcal{L}(L^1)} \le C''_\varepsilon (1 + |t|^{\frac{d}{2}+\varepsilon}) \text{ for all } \varepsilon > 0, \tag{7.49}$$

where C'_ε is a positive constant depending on ε.

We consider now $f_2(\sqrt{A})$. Since $\text{supp}(f_2) \subseteq [1, \infty)$, the sum over $n \in \mathbb{Z}$ in (7.37) applied to f_2, is reduced to the sum over $n \ge 0$. Using the expression of f, one proves in a similar way as for (7.48) that

$$\|\varphi(.)f_2(2^n.)\|_{C^s} \le C(1 + |t|^s) 2^{-n(\alpha-s)}.$$

Therefore, for $\alpha > s > \frac{d}{2}$,

$$\sum_{n \ge 0} \|\varphi(.)f_2(2^n.)\|_{C^s} \le C'(1 + |t|^s).$$

We conclude from Theorem 7.15 that for $\alpha > \frac{d}{2}$

$$\|f_2(\sqrt{A})\|_{\mathcal{L}(L^1)} \le C'_\varepsilon (1 + |t|^{\frac{d}{2}+\varepsilon}) \text{ for all } \varepsilon > 0. \tag{7.50}$$

We deduce from (7.49) and (7.50) that for $\alpha > \frac{d}{2}$

$$\|(I + A)^{-\alpha/2} e^{it\sqrt{A}}\|_{\mathcal{L}(L^1)} \le C_\varepsilon (1 + |t|^{\frac{d}{2}+\varepsilon}) \text{ for all } \varepsilon > 0. \tag{7.51}$$

The theorem for $p \in (1, 2)$ follows by interpolation. Indeed, fix $\alpha > \frac{d}{2}$ and define

$$F(z) := (I + A)^{-z\alpha/2} e^{it\sqrt{A}}.$$

---

[5] A way to prove (7.48) is to consider first $\|f_1\|_{C^{[s]}}$ and $\|f_1\|_{C^{[s]+1}}$, where $[s]$ denotes the integer part of s. The estimate (7.48) for arbitrary $s > \frac{d}{2}$ follows from the inequality $\|f\|_{C^s} \le \|f\|_{C^{[s]}}^{1-s+[s]} \|f\|_{C^{[s]+1}}^{s-[s]}$, see, e.g., Triebel [Tri78], p. 202.

For $z = iv$, $\|F(iv)\|_{\mathcal{L}(L^2)} \leq 1$ and if $z = 1 + iv$, the previous proof gives (7.51) with α replaced by $\alpha + iv$ for all $v \in \mathbb{R}$. In addition, $z \to F(z)u$ is holomorphic with values in $L^2(\Omega, \mu)$ for all $u \in L^2(\Omega, \mu)$. By the Stein interpolation theorem, for every $p \in (1, 2)$, $F(\frac{2}{p} - 1) \in \mathcal{L}(L^p(\Omega, \mu))$ with norm bounded by $C_\varepsilon (1 + |t|)^{d(\frac{1}{p} - \frac{1}{2}) + \varepsilon}$. This proves the desired result for $p \in (1, 2)$. By duality, the same result holds for $p \in (2, \infty)$. $\qquad\square$

Consider the wave type equation

$$\begin{cases} \frac{d^2}{dt^2} u(t) + Au(t) &= 0, \ t \in \mathbb{R} \\ u(0) &= f, \\ \frac{du}{dt}(0) &= g. \end{cases}$$

Formally, the solution is given by

$$u(t) = \cos(t\sqrt{A})f + \sin(t\sqrt{A})A^{-1/2}g,$$

where

$$\cos(t\sqrt{A}) = \frac{e^{it\sqrt{A}} + e^{-it\sqrt{A}}}{2} \quad \text{and} \quad \sin(t\sqrt{A}) = \frac{e^{it\sqrt{A}} - e^{-it\sqrt{A}}}{2i}.$$

One obtains by Theorem 7.19 that

$$\begin{aligned} \|u(t)\|_p &= \|(I + A)^{-\alpha}(I + A)^\alpha u(t)\|_p \\ &\leq C_\varepsilon (1 + |t|)^{d|1/2 - 1/p| + \varepsilon} [\|(I + A)^\alpha f\|_p + \|(I + A)^\alpha A^{-1/2} g\|_p], \end{aligned}$$

for all $\alpha > d|1/2 - 1/p|$ and $\varepsilon > 0$.

As explained in Section 7.3 for Schrödinger equations, using[6]

$$e^{it\sqrt{A}} = (I + A_p)^{-(\beta - \alpha/2)}(I + A_p)^{-\alpha/2} e^{it\sqrt{A}}(I + A_p)^\beta$$

for $\beta > \frac{\alpha}{2} > \frac{d}{2}|\frac{1}{2} - \frac{1}{p}|$, one obtains

$$e^{it\sqrt{A}}(D(A_p^\beta)) \subseteq D(A_p^{\beta - \alpha/2}) \quad \text{for } \beta > \frac{\alpha}{2} > \frac{d}{2}\left|\frac{1}{2} - \frac{1}{p}\right|. \qquad (7.52)$$

From this one recovers regularity properties of the solution $u(t)$ from regularity properties of initial data f and g.

---

[6]Again, $-A_p$ denotes the corresponding generator of the semigroup $(e^{-tA})_{t \geq 0}$ on $L^p(\Omega, \mu)$.

7.4.2　The wave equation in $L^p(\mathbb{R}^d, dx)$

In the previous section we considered wave type equations in a abstract setting. Clearly, one has better regularity results for the solution $u(t) = \cos(t\sqrt{A})f + \sin(t\sqrt{A})A^{-1/2}g$ if the exponent α in Theorem 7.19 is smaller. In this section, we shall consider the classical wave equation in Euclidean space. We shall see that in this particular case, the exponent α can be improved.

Let A be $-\Delta := \frac{\partial^2}{\partial x_1^2} + \cdots + \frac{\partial^2}{\partial x_d^2}$ on $L^2(\mathbb{R}^d, dx)$. The operator $-\sqrt{-\Delta}$ is the generator of the Poisson semigroup, whose kernel is given by

$$k(z, x, y) = \Gamma\left(\frac{d+1}{2}\right)\pi^{-(d+1)/2}\frac{z}{(z^2 + |x-y|^2)^{\frac{d+1}{2}}} \tag{7.53}$$

for all $z \in \mathbb{C}^+, x, y \in \mathbb{R}^d$. This precise expression allows one to obtain a better result than Theorem 7.19. We have

THEOREM 7.20 *For every $p \in [1, \infty)$, the operator $(I + \sqrt{-\Delta})^{-\alpha}e^{it\sqrt{-\Delta}}$ is bounded on $L^p(\mathbb{R}^d, dx)$ for $\alpha > (d-1)\left|\frac{1}{p} - \frac{1}{2}\right|$, with norm estimate*

$$\|(I + \sqrt{-\Delta})^{-\alpha}e^{it\sqrt{-\Delta}}\|_{\mathcal{L}(L^p)} \le C_\alpha(1 + |t|)^\alpha \ for \ all \ t \in \mathbb{R}.$$

Proof. We prove that for $d \ge 2$

$$\|e^{-z\sqrt{-\Delta}}\|_{\mathcal{L}(L^p)} \le C\left(\frac{|z|}{\Re z}\right)^{\frac{d-1}{2}} \text{ for all } z \in \mathbb{C}^+ \tag{7.54}$$

and for $d = 1$

$$\|e^{-z\sqrt{-\Delta}}\|_{\mathcal{L}(L^p)} \le C_\varepsilon\left(\frac{|z|}{\Re z}\right)^\varepsilon \text{ for all } z \in \mathbb{C}^+, \ \varepsilon > 0. \tag{7.55}$$

Once (7.54) and (7.55) are proved, Theorem 7.20 follows from these two estimates by the same proof as that of Theorem 7.12.

We shall prove (7.54) and (7.55) for $p = 1$. The estimates for $p \in (1, 2)$ follow by interpolation, and for $p \in (2, \infty)$ by duality.

Consider first the case $d \ge 2$. We shall denote by Const all inessential constants.

Using the expression of the Poisson kernel (7.53), we have

$$\int_{\mathbb{R}^d} |k(z, x, y)|dx = \Gamma(\frac{d+1}{2})\pi^{-(d+1)/2}\int_{\mathbb{R}^d}\frac{|z|}{|z^2 + |x|^2|^{\frac{d+1}{2}}}dx$$

$$\le \text{Const}\int_0^\infty \frac{|z|r^{d-1}}{|z^2 + r^2|^{\frac{d+1}{2}}}dr.$$

Since $|z^2 + r^2|^2 = (|z|^2 - r^2)^2 + 4r^2(\Re z)^2$, we have

$$\int_{\mathbb{R}^d} |k(z, x, y)| dx \leq \text{Const} |z| \int_0^\infty \frac{r^{d-1}}{[(|z|^2 - r^2)^2 + 4r^2(\Re z)^2]^{\frac{d+1}{4}}} dr$$

$$= \text{Const} |z|^{d+1} \int_0^\infty \frac{t^{d-1} dt}{[|z|^4(1 - t^2)^2 + 4t^2(\Re z)^2|z|^2]^{\frac{d+1}{4}}}$$

$$= \text{Const} \left(\frac{|z|}{\Re z}\right)^{\frac{d+1}{2}} \int_0^\infty \frac{t^{d-1} dt}{[(\frac{|z|}{\Re z})^2(1 - t^2)^2 + 4t^2]^{\frac{d+1}{4}}}.$$

We split the last integral into the sum $\int_{|t-1| \leq \frac{1}{2}} + \int_{|t-1| > \frac{1}{2}}$. The integral over $|t - 1| > \frac{1}{2}$ is bounded by

$$\int_{|t-1| > \frac{1}{2}} \frac{t^{d-1}}{[(\frac{|z|}{\Re z})^2(1 - t^2)^2]^{\frac{d+1}{4}}} dt = C \left(\frac{\Re z}{|z|}\right)^{\frac{d+1}{2}}, \qquad (7.56)$$

where C is a positive constant. The integral over $|t - 1| \leq \frac{1}{2}$ is bounded by

$$\left(\frac{3}{2}\right)^{d-2} \int_{1/2}^{3/2} \frac{t}{[(1 - t^2)^2(\frac{|z|}{\Re z})^2 + 1]^{\frac{d+1}{4}}} dt.$$

Making the change of variables $u := (1 - t^2) \frac{|z|}{\Re z}$, we see that the last integral can be dominated by

$$\text{Const} \frac{\Re z}{|z|} \int_{\mathbb{R}} \frac{1}{(1 + u^2)^{\frac{d+1}{4}}} du.$$

Using this and (7.56) yields (7.54).

Assume now that $d = 1$. We follow the same proof as above and obtain

$$\int_{\mathbb{R}} |k(z, x, y)| dx \leq \text{Const} \frac{|z|}{\Re z} \int_0^\infty \frac{1}{|(1 - t^2)^2(\frac{|z|}{\Re z})^2 + 4t^2|^{1/2}} dt.$$

We split again the integral into $\int_{|t-1| \leq \frac{1}{2}} + \int_{|t-1| > \frac{1}{2}}$ and dominate the second term by

$$\int_{|t-1| > \frac{1}{2}} \frac{1}{|(1 - t^2)^2(\frac{|z|}{\Re z})^2|^{1/2}} dt = C' \frac{\Re z}{|z|}.$$

The first integral is dominated by

$$2 \int_{1/2}^{3/2} \frac{t}{[(1 - t^2)^2(\frac{|z|}{\Re z})^2 + 1]^{1/2}} dt.$$

Using again the change of variables $u := (1 - t^2)\frac{|z|}{\Re z}$ and the obvious inequality $1 + u^2 \geq \frac{1}{2}(1 + |u|)^2$, we can dominate the last integral (up to a constant) by

$$\frac{\Re z}{|z|} \int_0^{\frac{5}{4}\frac{|z|}{\Re z}} \frac{1}{1+u}du = \frac{\Re z}{|z|} \ln(1 + \frac{5}{4}\frac{|z|}{\Re z}).$$

This implies (7.55).   □

Using Theorem 7.20, we obtain as in (7.52) that $e^{it\sqrt{-\Delta}}h \in W^{\beta-\alpha,p}(\mathbb{R}^d)$ for all $h \in W^{\beta,p}(\mathbb{R}^d)$, $p \in (1,\infty)$ and $\beta > \alpha > (d-1)|1/2 - 1/p|$. This allows one to obtain regularity properties of the solution to the wave equation

$$\begin{cases} \frac{d^2}{dt^2}u(t) - \Delta u(t) & = 0, \;\; t \in \mathbb{R}, \\ u(0) & = f, \\ \frac{du}{dt}(0) & = g \end{cases}$$

from regularity properties of f and g.

7.5 SINGULAR INTEGRAL OPERATORS ON IRREGULAR DOMAINS

Denote again by (X, ρ, μ) a metric measured space and let T be a bounded linear operator on $L^2(X, \mu)$. We shall say that T is of weak type (1,1) if there exists a constant C such that

$$\mu(\{x, |Tu(x)| > \alpha\}) \leq \frac{C}{\alpha}\|u\|_1$$

for all $\alpha > 0$ and all $u \in L^2(X,\mu) \cap L^1(X,\mu)$. The smallest constant C for which this holds for all $\alpha > 0$ and $u \in L^2(X,\mu) \cap L^1(X,\mu)$ will be denoted $\|T\|_{L^1(X,\mu) \to L^{1,w}(X,\mu)}$.

By the Marcinkiewicz interpolation theorem, every bounded linear operator on $L^2(X,\mu)$ which is of weak type (1,1) extends from $L^2(X,\mu) \cap L^p(X,\mu)$ to a bounded operator on $L^p(X,\mu)$ for all $p \in (1,2]$.

The following theorem gives a criterion under which a bounded operator on $L^2(X,\mu)$ is of weak type (1,1). This criterion is given in terms of the associated kernel. We shall say that T has an associated kernel $k(x,y)$, if k is a measurable function on $X \times X$ and

$$Tu(x) = \int_X k(x,y)u(y)d\mu(y) \;\; \text{for a.e. } x \notin \text{supp}(u) \qquad (7.57)$$

for every $u \in L^2(X,\mu)$ with bounded support in X.

THEOREM 7.21 *Assume that (X, ρ, μ) satisfies the doubling property (7.1). Let T be a bounded linear operator on $L^2(X, \mu)$ which has an associated kernel $k(x, y)$ in the sense of (7.57). Let $(e^{-tA})_{t \geq 0}$ be a strongly continuous semigroup with a heat kernel $p(t, x, y)$ which satisfies the Gaussian upper bound (7.3) (for a.e. $x, y \in X$). Assume that the composite operators Te^{-tA} have associated kernels $k_t(x, y)$ (in the sense of (7.57)) and there exist constants W and $\delta > 0$ such that*

$$\int_{\rho(x,y) \geq \delta t^{1/m}} |k(x, y) - k_t(x, y)| d\mu(x) \leq W \qquad (7.58)$$

for a.e. $y \in X$ and all $t > 0$. Then the operator T is of weak type $(1, 1)$ and there exists a constant C, independent of T, such that

$$\|T\|_{L^1(X,\mu) \to L^{1,w}(X,\mu)} \leq C(W + \|T\|_{\mathcal{L}(L^2)}).$$

Proof. We shall use the Calderón-Zygmund decomposition for integrable functions; see Coifman and Weiss [CoWe71].

Let $f \in L^1(X, \mu) \cap L^2(X, \mu)$ and $\alpha > \|f\|_1(\mu(X))^{-1}$ (the latter term is 0 if $\mu(X) = \infty$). There exist a constant c, independent of f and α, and a decomposition

$$f = g + b = g + \sum_i b_i,$$

such that the following four properties hold:

(i) $|g(x)| \leq c\alpha$ for a.e. $x \in X$;

(ii) there exists a sequence of balls $Q_i := B(x_i, r_i)$ so that the support of each b_i is contained in Q_i and

$$\int_X |b_i(x)| d\mu(x) \leq c\alpha\mu(Q_i);$$

(iii) $\displaystyle\sum_i \mu(Q_i) \leq \frac{c}{\alpha}\|f\|_1;$

(iv) each point of X is contained in at most a finite number N of the balls Q_i.

Note that conditions (ii) and (iii) imply that $\|b\|_1 \leq c^2\|f\|_1$ and hence that $\|g\|_1 \leq (1 + c^2)\|f\|_1$. We also deduce from this and (i) that $g \in L^2(X, \mu)$.

Throughout this proof, we shall denote by $C_0, C_1, \ldots$ all inessential constants. We first observe that

$$\mu(\{x, |Tf(x)| > \alpha\}) \leq \mu(\{x, |Tg(x)| > \alpha/2\}) + \mu(\{x, |Tb(x)| > \alpha/2\}).$$

Using the facts that T is bounded on $L^2(X, \mu)$ and that $|g(x)| \leq c\alpha$, it follows that

$$
\begin{aligned}
\mu(\{x, |Tg(x)| > \alpha/2\}) &\leq 4\alpha^{-2}\|Tg\|_2^2 \\
&\leq 4\alpha^{-2}\|T\|_{\mathcal{L}(L^2)}^2\|g\|_2^2 \\
&\leq \frac{4c}{\alpha}\|T\|_{\mathcal{L}(L^2)}^2\|g\|_1 \\
&\leq 4(1+c^2)\frac{c}{\alpha}\|T\|_{\mathcal{L}(L^2)}^2\|f\|_1.
\end{aligned}
$$

Thus,

$$
\mu(\{x, |Tg(x)| > \alpha/2\}) \leq \frac{4c(1+c^2)\|T\|_{\mathcal{L}(L^2)}^2}{\alpha}\|f\|_1 = \frac{C_0\|T\|_{\mathcal{L}(L^2)}^2}{\alpha}\|f\|_1.
$$
(7.59)

Fix b_i and choose $t_i = r_i^m$ (where m is the constant appearing in the Gaussian bound (7.3)), where r_i is the radius of the ball $Q_i = B(x_i, r_i)$. We write

$$
Tb_i(x) = Te^{-t_i A}b_i(x) + (T - Te^{-t_i A})b_i(x).
$$

To analyze $Te^{-t_i A}b_i(x)$, we first estimate $e^{-t_i A}b_i$. Since (7.3) is equivalent to (7.5) and b_i has support contained in Q_i, we have

$$
\begin{aligned}
|e^{-t_i A}b_i(x)| &\leq C\int_X \frac{1}{V(x, t_i^{1/m})}\exp\left\{-c\frac{\rho(x,y)^{m/(m-1)}}{t_i^{1/(m-1)}}\right\}|b_i(y)|d\mu(y) \\
&\leq \frac{C'}{V(x, t_i^{1/m})}\exp\left\{-c_0\frac{\rho(x,x_i)^{m/(m-1)}}{t_i^{1/(m-1)}}\right\} \\
&\quad\times \int_{B(x_i, t_i^{1/m})}|b_i(y)|d\mu(y) \\
&\leq \frac{C'c}{V(x, t_i^{1/m})}\exp\left\{-c_0\frac{\rho(x,x_i)^{m/(m-1)}}{t_i^{1/(m-1)}}\right\}\alpha\mu(Q_i) \\
&\leq \frac{C''}{V(x, t_i^{1/m})}\alpha\int_X \exp\left\{-c'\frac{\rho(x,y)^{m/(m-1)}}{t_i^{1/(m-1)}}\right\}\chi_{Q_i}(y)d\mu(y).
\end{aligned}
$$

Note that we have used the property (ii) of the Calderón-Zygmund decomposition to obtain the third inequality. Here C', C'', c_0, and c' are positive constants.

We have then proved

$$
|e^{-t_i A}b_i(x)| \leq \frac{C''}{V(x, t_i^{1/m})}\alpha\int_X \exp\left\{-c'\frac{\rho(x,y)^{m/(m-1)}}{t_i^{1/(m-1)}}\right\}\chi_{Q_i}(y)d\mu(y).
$$

Let now $u \in L^2(X, \mu)$. Using this inequality, we have

$$|(u; e^{-t_i A} b_i)|$$

$$\leq \int_X |u(x)| |e^{-t_i A} b_i(x)| d\mu(x)$$

$$\leq C'' \alpha \int_X \left(\int_X \frac{1}{V(x, t_i^{1/m})} \exp \left\{ -c' \frac{\rho(x,y)^{m/(m-1)}}{t_i^{1/(m-1)}} \right\} |u(x)| d\mu(x) \right)$$
$$\times \chi_{Q_i}(y) d\mu(y)$$

$$\leq C_1 \alpha \int_X \left(\int_X \frac{1}{V(y, t_i^{1/m})} \exp \left\{ -c'' \frac{\rho(x,y)^{m/(m-1)}}{t_i^{1/(m-1)}} \right\} |u(x)| d\mu(x) \right)$$
$$\times \chi_{Q_i}(y) d\mu(y),$$

where we have used (7.4) to estimate $\frac{1}{V(x, t_i^{1/m})} \exp \left\{ -c' \frac{\rho(x,y)^{m/(m-1)}}{t_i^{1/(m-1)}} \right\}$ by

$\frac{1}{V(y, t_i^{1/m})} \exp \left\{ -c'' \frac{\rho(x,y)^{m/(m-1)}}{t_i^{1/(m-1)}} \right\}$, where $c'' \in (0, c')$ is a constant. On the
other hand,

$$\frac{1}{V(y, t_i^{1/m})} \int_X \exp \left\{ -c'' \frac{\rho(x,y)^{m/(m-1)}}{t_i^{1/(m-1)}} \right\} |u(x)| d\mu(x)$$

$$= \frac{1}{V(y, t_i^{1/m})} \sum_{k \geq 0} \int_{k t_i^{1/m} \leq \rho(x,y) \leq (k+1) t_i^{1/m}} \exp \left\{ -c'' \frac{\rho(x,y)^{m/(m-1)}}{t_i^{1/(m-1)}} \right\}$$
$$\times |u(x)| d\mu(x)$$

$$\leq \frac{1}{V(y, t_i^{1/m})} \left(\int_{B(y, t_i^{1/m})} |u(x)| d\mu(x) \right.$$

$$\left. + \sum_{k \geq 1} e^{-c'' k^{m/(m-1)}} \int_{B(y, (k+1) t_i^{1/m})} |u(x)| d\mu(x) \right)$$

$$\leq \left(1 + \sum_{k \geq 1} \frac{V(y, (k+1) t_i^{1/m})}{V(y, t_i^{1/m})} e^{-c'' k^{m/(m-1)}} \right) \mathcal{M}u(y)$$

$$\leq C \left(1 + \sum_{k \geq 1} (k+1)^d e^{-c'' k^{m/(m-1)}} \right) \mathcal{M}u(y),$$

where

$$\mathcal{M}u(y) := \sup_{r > 0} \frac{1}{V(y, r)} \int_{B(y,r)} |u(x)| d\mu(x),$$

is the Hardy-Littlewood maximal operator.

It follows from the previous estimates that there exists a positive constant C_1' such that

$$|(u; e^{-t_i A} b_i)| \le C_1' \alpha(\mathcal{M}u; \chi_{Q_i}).$$

Since this inequality holds for all $u \in L^2(X, \mu)$ and every i and since the Hardy-Littlewood maximal operator is bounded on $L^2(X, \mu)$ (see Christ [Chr91a]), it follows that there exists a constant C_2 such that

$$\left\| \sum_i e^{-t_i A} b_i \right\|_2 \le C_2 \alpha \left\| \sum_i \chi_{Q_i} \right\|_2. \tag{7.60}$$

We use properties (iii) and (iv) of the Calderón-Zygmund decomposition to obtain from (7.60)

$$\left\| \sum_i e^{-t_i A} b_i \right\|_2 \le C_2' \alpha \left(\sum_i \mu(Q_i) \right)^{1/2} \le c^{1/2} C_2' \alpha^{1/2} \|f\|_1^{1/2}.$$

Using again the fact that T is bounded on $L^2(X, \mu)$, we obtain

$$\mu\left(\left\{ x, \left| \sum_i T e^{-t_i A} b_i(x) \right| > \alpha/4 \right\}\right) \le 16\alpha^{-2} \left\| \sum_i T e^{-t_i A} b_i \right\|_2^2$$

$$\le 16\alpha^{-2} \|T\|_{\mathcal{L}(L^2)}^2 \left\| \sum_i e^{-t_i A} b_i \right\|_2^2$$

$$\le C_3 \alpha^{-1} \|T\|_{\mathcal{L}(L^2)}^2 \|f\|_1.$$

Thus,

$$\mu\left(\left\{ x, \left| \sum_i T e^{-t_i A} b_i(x) \right| > \alpha/4 \right\}\right) \le C_3 \alpha^{-1} \|T\|_{\mathcal{L}(L^2)}^2 \|f\|_1. \tag{7.61}$$

We estimate now the term $T \sum_i (I - e^{-t_i A}) b_i$. Denote by $B_i := B(x_i, (1 + \delta) t_i^{1/m})$ the ball with the same center as Q_i and radius $(1 + \delta) t_i^{1/m}$, where δ is the constant in (7.58). One has

$$\mu\left(\left\{ x, \left| \sum_i (T - T e^{-t_i A}) b_i(x) \right| > \alpha/4 \right\}\right)$$

$$\le \sum_i \mu(B_i) + \sum_i 4\alpha^{-1} \int_{X \setminus B_i} |(T - T e^{-t_i A}) b_i(x)| d\mu(x).$$

Using (7.1) and property (iii) of the Calderón-Zygmund decomposition, one finds a constant C_4 such that

$$\sum_i \mu(B_i) \le C_4 \sum_i \mu(Q_i) \le cC_4\alpha^{-1}\|f\|_1. \qquad (7.62)$$

Finally, we use (7.58) to control the last term. Since b_i is supported in $Q_i = B(x_i, t_i^{1/m})$, we have

$$\int_{X\backslash B_i} |(T - Te^{t_i A})b_i(x)|d\mu(x)$$
$$\le \int_{X\backslash B_i} \left| \int_X k(x,y) - k_{t_i}(x,y)b_i(y)d\mu(y) \right| d\mu(x)$$
$$\le \int_X |b_i(y)| \left[\int_{\rho(x,y)\ge \delta t_i^{1/m}} |k(x,y) - k_{t_i}(x,y)|d\mu(x) \right] d\mu(y)$$
$$\le W\|b_i\|_1.$$

Thus, using (ii) and (iii) of the Calderón-Zygmund decomposition one obtains

$$\sum_i \int_{X\backslash B_i} |(T - Te^{t_i A})b_i(x)|d\mu(x) \le W \sum_i \|b_i\|_1 \le Wc^2\|f\|_1. \qquad (7.63)$$

We obtain from (7.59), (7.61), (7.62) and (7.63) that

$$\|T\|_{L^1(X,\mu)\to L^{1,w}(X,\mu)} \le C(1 + W + \|T\|^2_{\mathcal{L}(L^2)}). \qquad (7.64)$$

Set

$$W(T) := \sup_{y\in X, t>0} \int_{\rho(x,y)\ge \delta t^{1/m}} |k(x,y) - k_t(x,y)|d\mu(x).$$

Observe that $W(\|T\|^{-1}_{\mathcal{L}(L^2)}T) = \|T\|^{-1}_{\mathcal{L}(L^2)}W(T)$ and apply (7.64) to the operator $\|T\|^{-1}_{\mathcal{L}(L^2)}T$ to obtain

$$\|T\|_{L^1(X,\mu)\to L^{1,w}(X,\mu)} \le C(2\|T\|_{\mathcal{L}(L^2)} + W(T)).$$

This proves the theorem.   □

Remark. The fact that $(e^{-tA})_{t\ge 0}$ is a semigroup has not been used in the proof. We could replace $(e^{-tA})_t$ in the previous theorem by any family of bounded operators $(A_t)_{t>0}$, such that each A_t is given by a kernel satisfying the Gaussian upper bound (7.3). Note also that one can weaken (7.3) to include situations where one has a polynomial upper bound, rather than a Gaussian one. For all this, see Duong and McIntosh [DuMc99a].

One advantage of the previous theorem is that it allows one to prove boundedness on $L^p(X, \mu)$ for operators given by singular integrals (in the sense of (7.57)) without assuming any regularity on their associated kernels. This suggests that the theorem holds for operators acting on $L^2(\Omega, \mu)$, where Ω is an arbitrary open subset of X, just by extending the kernels by 0 outside $\Omega \times \Omega$. This is indeed true and we formulate it in the next result.

Let (X, ρ, μ) be as above so that the doubling condition (7.1) holds. Let Ω be any open subset of X and $(e^{-tA})_{t \geq 0}$ be a strongly continuous semigroup on $L^2(\Omega, \mu)$ with kernel $p(t, x, y)$ satisfying the Gaussian upper bound (7.3). Again, $V(x, t^{1/m})$ in (7.3) denotes the volume of the ball of X (and not of Ω!) of center x and radius $t^{1/m}$.
We have

THEOREM 7.22 *Assume that (X, ρ, μ) satisfies the doubling property (7.1) and let Ω be any open subset of X. Let T be a bounded linear operator on $L^2(\Omega, \mu)$ with an associated kernel $k(x, y)$ (in the sense of (7.57), with Ω in place of X). Let $(e^{-tA})_{t \geq 0}$ be a strongly continuous semigroup with a heat kernel $p(t, x, y)$ which satisfies the Gaussian upper bound (7.3). Assume that the composite operators Te^{-tA} have associated kernels $k_t(x, y)$ (in the sense of (7.57)) and there exist constants W and $\delta > 0$ such that*

$$\int_{\rho(x,y) \geq \delta t^{1/m}} |k(x, y) - k_t(x, y)| d\mu(x) \leq W \qquad (7.65)$$

for a.e. $y \in \Omega$ and all $t > 0$. Then the operator T is of weak type $(1, 1)$ and there exists a constant C such that

$$\|T\|_{L^1(\Omega, \mu) \to L^{1,w}(\Omega, \mu)} \leq C(W + \|T\|_{\mathcal{L}(L^2(\Omega))}).$$

As we already mentioned, by the Marcinkiewicz interpolation theorem, the operator T in Theorem 7.22 extends from $L^2(\Omega, \mu) \cap L^p(\Omega, \mu)$ to a bounded operator on $L^p(X, \mu)$ for all $p \in (1, 2]$.

Proof. Define

$$\tilde{T}u(x) := \begin{cases} T(\chi_\Omega u)(x), & x \in \Omega, \\ 0, & x \notin \Omega. \end{cases}$$

It is clear that T is of weak type $(1, 1)$ on Ω if and only if $\tilde{T}$ is of weak type $(1, 1)$ on X. Now we apply Theorem 7.21 to prove that $\tilde{T}$ is of weak type $(1, 1)$ on X.

The operator $\tilde{T}$ has an associated kernel $\tilde{k}(x,y)$, given by

$$\tilde{k}(x,y) := \begin{cases} k(x,y) & \text{if } x,y \in \Omega, \\ 0 & \text{otherwise.} \end{cases}$$

We define in the same way, $\widetilde{e^{-tA}}$, $\tilde{p}(t,x,y), \widetilde{Te^{-tA}}$ and $\tilde{k}_t(x,y)$. Clearly, $\widetilde{Te^{-tA}} = \tilde{T}\widetilde{e^{-tA}}$. The assumptions of Theorem 7.22 imply that $\tilde{T}$ and $\widetilde{Te^{-tA}}$ satisfy the assumptions of Theorem 7.21 (except the fact that $\widetilde{e^{-tA}}$ is not necessarily a semigroup on $L^2(X,\mu)$ but, as mentioned in the remark following the proof of Theorem 7.21, the semigroup property is not needed in that theorem). We then obtain the weak type $(1,1)$ assertion for $\tilde{T}$ on X and this gives the same result for T on Ω. $\qquad\qquad\square$

7.6 SPECTRAL MULTIPLIERS

Theorem 7.15 gives a condition on f which allows one to extend $f(\sqrt[m]{A})$ from $L^2(\Omega,\mu) \cap L^1(\Omega,\mu)$ to a bounded operator on $L^1(\Omega,\mu)$. There are, however, interesting examples of functions f for which the condition there is not satisfied. An example is given by $f(\lambda) = \lambda^{is}$, where $s \in \mathbb{R}$ is fixed. The corresponding operator $f(A) = A^{is}$ is the imaginary power of A. Theorem 7.15 cannot be applied to prove boundedness on $L^p(\Omega,\mu)$ of A^{is}. This boundedness on $L^p(\Omega,\mu)$ for $1 < p < \infty$, will be achieved by applying Theorem 7.22.

THEOREM 7.23 *Assume that (X,ρ,μ) satisfies the doubling property (7.2). Let Ω be an open subset of X and assume that A is a non-negative self-adjoint operator on $L^2(\Omega,\mu)$ with a heat kernel $p(t,x,y)$ satisfying the Gaussian upper bound (7.3). Denote again by φ a non-negative C_c^∞ function satisfying (7.36). Let $f : [0,\infty) \to \mathbb{C}$ be a bounded function such that*

$$\sup_{t>0} \|\varphi(.)f(t.)\|_{C^s} < \infty \qquad (7.66)$$

for some $s > d/2$. Then $f(A)$ is of weak type $(1,1)$ and is bounded on $L^p(\Omega,\mu)$ for all $p \in (1,\infty)$. In addition,

$$\|f(A)\|_{L^1(\Omega,\mu)\to L^{1,w}(\Omega,\mu)} \leq C_s[\sup_{t>0} \|\varphi(.)f(t.)\|_{C^s} + |f(0)|] \qquad (7.67)$$

for some positive constant C_s, independent of f.

Results of this type are called spectral multipliers. Note that if $A = -\Delta$ is (minus) the Laplacian on $L^2(\mathbb{R}^d, dx)$, then the operator $T = f(-\Delta)$

satisfies

$$\widehat{Tu}(\xi) = f(|\xi|^2)\widehat{u}(\xi), \ \xi \in \mathbb{R}^d,$$

for all $u \in L^2(\Omega, dx)$, where $\widehat{\ }$ denotes the Fourier transform. Thus, $T = f(-\Delta)$ can be viewed as the multiplier by $f(|\xi|^2)$. For this particular operator, the previous result holds with the condition (7.66) replaced by

$$\sup_{t>0} \|\varphi(.)f(t.)\|_{W^{s,2}(\mathbb{R})} < \infty \tag{7.68}$$

for some $s > d/2$, where $W^{s,2}(\mathbb{R})$ denotes the classical Sobolev space. See Hörmander [Hör60].

Assume that f is of class $C^{[d/2]+1}$ where $[.]$ denotes the integer part. If f satisfies

$$\sup_{\lambda>0} |\lambda^k f^{(k)}(\lambda)| < \infty \text{ for } k = 0, 1, ..., [d/2]+1 \tag{7.69}$$

then f satisfies (7.66). Indeed, for every $k, j \leq [d/2]+1$, one has

$$|\varphi^{(j)}(\lambda)t^k f^{(k)}(t\lambda)| = |\varphi^{(j)}(\lambda)\lambda^{-k}||(t\lambda)^k f^{(k)}(t\lambda)|$$
$$\leq |\varphi^{(j)}(\lambda)\lambda^{-k}| \sup_{\lambda>0} |\lambda^k f^{(k)}(\lambda)|.$$

The term $|\varphi^{(j)}(\lambda)\lambda^{-k}|$ is uniformly bounded since φ has support contained in $[\frac{1}{4}, 1]$. This shows that (7.66) holds with $s = [d/2]$ and also with $s = [d/2]+1$. Now (7.66) for $d/2 < s \leq [d/2]+1$ follows from the interpolation inequality[7]

$$\|u\|_{C^s} \leq \|u\|_{C^{[s]}}^{1-s+[s]} \|u\|_{C^{[s]+1}}^{s-[s]}. \tag{7.70}$$

Proof of Theorem 7.23. As in the proof of Theorem 7.15, we write $f(\lambda) = f(\lambda) - f(0) + f(0)$ and hence

$$f(\sqrt[m]{A}) = (f(.) - f(0))(\sqrt[m]{A}) + f(0)I.$$

Replacing f by $f - f(0)$, we may assume in the sequel that $f(0) = 0$. Observe also that f satisfies (7.66) if and only if the function $\lambda \to f(\lambda^m)$ satisfies the same property. For this reason, we shall consider $f(\sqrt[m]{A})$ rather than $f(A)$.

We have again for all $\lambda \geq 0$

$$f(\lambda) = \sum_{n=-\infty}^{+\infty} \varphi(2^{-n}\lambda)f(\lambda) =: \sum_{n=-N}^{N} f_n(\lambda).$$

---

[7]See Triebel [Tri78], p. 202.

As mentioned in the proof of Theorem 7.15, this equality implies that the sequence $\sum_{n=-N}^{N} f_n(\sqrt[m]{A})$ converges strongly in $L^2(\Omega, \mu)$ to $f(\sqrt[m]{A})$. We shall prove that $\sum_{n=-N}^{N} f_n(\sqrt[m]{A})$ is of weak type $(1,1)$ with bound independent of N. This together with the strong convergence in $L^2(\Omega, \mu)$ implies the theorem.

Now, in order to prove a weak type $(1,1)$ bound for $\sum_{n=-N}^{N} f_n(\sqrt[m]{A})$, we apply Theorem 7.22. We prove that

$$
\sup_{u,x} \int_{\rho(x,y) \geq u} \left| \sum_{n=-N}^{N} \left[f_n(\sqrt[m]{A})(x,y) \right. \right.
$$

$$
\left. \left. - (e^{-u^m A} f_n(\sqrt[m]{A}))(x,y) \right] \right| d\mu(y) < \infty, \quad (7.71)
$$

where the supremum is taken for all $u > 0$ and $x \in \Omega$. Here we use again the notation $T(x,y)$ for the associated kernel of an operator T in the sense of (7.57).

Let $g_n(\lambda) := f_n(\lambda)(1 - e^{-(u\lambda)^m}) = f(\lambda)\varphi(2^{-n}\lambda)(1 - e^{-(u\lambda)^m})$. We have to estimate

$$
I_n := \int_{\rho(x,y) \geq u} |g_n(\sqrt[m]{A})(x,y)| d\mu(y).
$$

By the Cauchy-Schwarz inequality,

$$
I_n \leq \left[\int_\Omega |g_n(\sqrt[m]{A})(x,y)|^2 (1 + 2^n \rho(x,y))^{2s} d\mu(y) \right]^{1/2}
$$

$$
\times \left[\int_{\rho(x,y) \geq u} (1 + 2^n \rho(x,y))^{-2s} d\mu(y) \right]^{1/2}.
$$

We apply Lemma 7.18 with $f = g_n$ and $r = 2^n$ to obtain for all $s > \frac{d}{2}$

$$
\int_\Omega |g_n(\sqrt[m]{A})(x,y)|^2 (1 + 2^n \rho(x,y))^{2s} d\mu(y) \leq \frac{C_\varepsilon}{V(x, 2^{-n})} \|\delta_{2^n} g_n\|_{C^{s+\varepsilon}}^2.
$$
$$
(7.72)
$$

On the other hand, there exists a constant C_m such that[8]

$$
\|\delta_{2^n} g_n\|_{C^{s+\varepsilon}} \leq C_m \sup_{t>0} \|\varphi(.)f(t.)\|_{C^{s+\varepsilon}} \min(1, u2^n). \quad (7.73)
$$

Using (7.46), (7.72), and (7.73), it follows that

$$
I_n \leq C''_\varepsilon \min(1, u2^n) \min(1, (u2^n)^{\frac{d}{2}-s}) \sup_{t>0} \|\varphi(.)f(t.)\|_{C^{s+\varepsilon}}.
$$

---

[8]One can use again (7.70) in order to prove (7.73).

Hence for $s > \frac{d}{2}$

$$\sum_{-\infty}^{+\infty} I_n \leq C_\varepsilon'' \left(\sum_{u2^n \leq 1} u2^n + \sum_{u2^n > 1} (u2^n)^{\frac{d}{2}-s} \right) \sup_{t>0} \|\varphi(.)f(t.)\|_{C^{s+\varepsilon}}$$

$$= C_\varepsilon \sup_{t>0} \|\varphi(.)f(t.)\|_{C^{s+\varepsilon}},$$

where C_ε is a positive constant independent of u. This proves (7.71) and we conclude by Theorem 7.22 that $\sum_{n=-N}^{N} f_n(\sqrt[m]{A})$ is of weak type $(1,1)$ with

$$\left\| \sum_{n=-N}^{N} f_n(\sqrt[m]{A}) \right\|_{L^1(\Omega,\mu) \to L^{1,w}(\Omega,\mu)} \leq C_\varepsilon' \left[\|f\|_\infty + \sup_{t>0} \|\varphi(.)f(t.)\|_{C^{s+\varepsilon}} \right],$$

for all $N \geq 0$. This proves the theorem.   □

The following corollary gives L^p-estimates for imaginary powers of self-adjoint operators. More precisely,

COROLLARY 7.24 *Assume that Ω and A are as in Theorem 7.23. Then for each $u \in \mathbb{R}$, the operator A^{iu} is of weak type $(1,1)$ and extends from $L^p(\Omega,\mu) \cap L^2(\Omega,\mu)$ to a bounded operator on $L^p(\Omega,\mu)$ for all $p \in (1,\infty)$ with norm estimate*

$$\|A^{iu}\|_{\mathcal{L}(L^p(\Omega,\mu))} \leq C_\varepsilon (1+|u|)^{d|\frac{1}{2}-\frac{1}{p}|+\varepsilon} \text{ for all } u \in \mathbb{R}. \qquad (7.74)$$

This estimate holds for every $\varepsilon > 0$ (here C_ε is a constant independent of u).

Proof. We apply Theorem 7.23 with $f(\lambda) = \lambda^{iu}$. We use again the inequality (7.70) to obtain

$$\sup_{t>0} \|\varphi(.)f(t.)\|_{C^s} \leq C(1+|u|)^s$$

for $s > \frac{d}{2}$. It follows that A^{iu} is of weak type $(1,1)$ with

$$\|A^{iu}\|_{L^1(\Omega,\mu) \to L^{1,w}(\Omega,\mu)} \leq C_\varepsilon (1+|u|)^{\frac{d}{2}+\varepsilon} \qquad (7.75)$$

for all $\varepsilon > 0$. This and the fact that $\|A^{iu}\|_{\mathcal{L}(L^2(\Omega,\mu))} \leq 1$ imply by the Marcinkiewicz interpolation theorem that A^{iu} extends to a bounded operator on $L^p(\Omega,\mu)$ for $1 < p < \infty$, with norm bounded by $C_{p,\varepsilon}(1+|u|)^{\frac{d}{2}+\varepsilon}$. Choosing p arbitrary close to 1 and applying the Riesz-Thorin interpolation theorem, we obtain (7.74).   □

The next result concerns the situation where the volume has polynomial growth. We assume that $V(x, r)$ satisfies

$$C_1 V(r) \leq V(x, r) \leq C_2 V(r) \text{ for all } x \in X, r > 0, \qquad (7.76)$$

where C_1, C_2 are positive constants and $V(.)$ is a function such that

$$V(r) = O(r^d) \text{ as } r \to 0 \text{ and } V(r) = O(r^D) \text{ as } r \to \infty. \qquad (7.77)$$

Here d and D are positive constants. Of course, (7.1) holds in this situation and Theorem 7.23 applies with condition $s > \frac{\max(d,D)}{2}$ in (7.66). The next result asserts that for compactly supported functions, $s > D/2$ is actually enough. We have

THEOREM 7.25 *Let Ω, A be as in Theorem 7.23 and assume that (7.76) and (7.77) hold. Set $f := f_0 + f_\infty$ with f_0 supported in $[0, 2]$ and f_∞ supported in $[1, \infty)$. Assume that*

$$\sup_{t>0} \|\varphi(.) f_0(t.)\|_{C^s} < \infty \text{ for some } s > D/2 \qquad (7.78)$$

and

$$\sup_{t>0} \|\varphi(.) f_\infty(t.)\|_{C^s} < \infty \text{ for some } s > \frac{\max(d, D)}{2}. \qquad (7.79)$$

Then $f(A)$ is of weak type $(1,1)$ and extends to a bounded operator on $L^p(\Omega, \mu)$ for all $p \in (1, \infty)$.

Proof. By Theorem 7.23 and (7.79), $f_\infty(A)$ is of weak type $(1, 1)$. We only have to show that if (7.78) is satisfied, then $f_0(A)$ is of weak type $(1, 1)$.

Let φ be as in the proof of Theorem 7.23. Since the support of f_0 is contained in $[0, 2]$ we have

$$f_0(\lambda) = \sum_{-\infty}^{+\infty} \varphi(2^{-n}\lambda) f_0(\lambda) = \sum_{-\infty}^{3} \varphi(2^{-n}\lambda) f_0(\lambda).$$

Therefore the proof of Theorem 7.23 applies once we prove

$$\int_{\rho(x,y) \geq u} (1 + 2^n \rho(x, y))^{-2s} d\mu(y) \leq C' v(x, 2^{-n}) \min(1, (u2^n)^{D-2s})$$

for all $u > 0$ and $n \leq 0$.

We show that for all $r \in (0, 1]$ and $s > \frac{D}{2}$

$$\int_{\rho(x,y) \geq u} (1 + r\rho(x, y))^{-2s} d\mu(y) \leq C' V\left(x, \frac{1}{r}\right) \min(1, (ur)^{D-2s}).$$

$$(7.80)$$

The proof is similar to that of (7.46). If $ur \geq 1$, then

$$\int_{\rho(x,y)\geq u} (1+r\rho(x,y))^{-2s}d\mu(y)$$

$$\leq \sum_{k\geq 0} \int_{2^k u \leq \rho(x,y) \leq 2^{k+1}u} (r\rho(x,y))^{-2s}d\mu(y)$$

$$\leq (ru)^{-2s} \sum_{k\geq 0} (2^k)^{-2s} V(x, 2^{k+1}u).$$

Since $ur \geq 1$ and $r < 1$, we have

$$V(x, 2^{k+1}u) = V\left(x, 2^{k+1}ur\frac{1}{r}\right) = O\left(\left(2^{k+1}ur\frac{1}{r}\right)^D\right).$$

Thus,

$$V(x, 2^{k+1}u) \leq C(2^k)^D (ur)^D V\left(x, \frac{1}{r}\right)$$

which gives the desired inequality. If $ur < 1$, we write again (7.47) to obtain (7.80). □

7.7 RIESZ TRANSFORMS ASSOCIATED WITH UNIFORMLY ELLIPTIC OPERATORS

Let Ω be an arbitrary open subset of $\mathbb{R}^d$. We denote by dx the Lebesgue measure and by $(.;.)$ the scalar product of $L^2(\Omega, dx)$. Consider the uniformly elliptic operator

$$A = -\sum_{k,j=1}^d D_j(a_{kj}D_k) + a_0$$

acting on $L^2(\Omega, dx)$ and subject to Dirichlet boundary conditions (see Chapter 4). A is the operator associated with the form

$$\mathfrak{a}(u, v) = \sum_{k,j=1}^d \int_\Omega a_{kj}(x)D_k u \overline{D_j v}dx + \int_\Omega a_0 u\overline{v}dx,$$

with domain $D(\mathfrak{a}) = H_0^1(\Omega)$. We assume that

$$a_0, a_{jk} = a_{kj} \in L^\infty(\Omega, \mu, \mathbb{R}), \sum_{k,j=1}^d a_{kj}\xi_k\overline{\xi_j} \geq \eta|\xi|^2 \text{ for all } \xi \in \mathbb{C}^d,$$

$$(7.81)$$

where $\eta > 0$ is a constant. Under this assumption $\mathfrak{a}$ is a closed symmetric form (see Chapter 4). Its associated operator A is then self-adjoint. If in addition, a_0 is non-negative, then A is non-negative (or accretive) in the sense that

$$(Au; u) \geq 0 \text{ for all } u \in D(A).$$

Denote by $A^{1/2}$ the square root of A. It is a classical fact that (see the next chapter)

$$\mathfrak{a}(u, v) = (A^{1/2}u; A^{1/2}v) \text{ for all } u, v \in D(\mathfrak{a}) = H_0^1(\Omega).$$

Let us denote the gradient by ∇. It follows from this equality and assumption (7.81) that

$$\eta \||\nabla u|\|_2^2 \leq \sum_{k,j=1}^{d} \int_{\Omega} a_{kj}(x) D_k u \overline{D_j u} dx$$
$$\leq \mathfrak{a}(u, u)$$
$$= \|A^{1/2}u\|_2^2.$$

In particular, the Riesz transform operator $\nabla A^{-1/2}$ extends to a bounded operator on $(L^2(\Omega, dx))^d$. We shall denote by $\nabla A^{-1/2}$ its bounded extension. The next result shows that $\nabla A^{-1/2}$ is also bounded on $(L^p(\Omega, dx))^d$ for all $p \in (1, 2]$.

THEOREM 7.26 *Assume that (7.81) holds and that a_0 is non-negative on Ω. For each $k \in \{1, \ldots, d\}$, $D_k A^{-1/2}$ is of weak type $(1, 1)$ on Ω. In particular, $D_k A^{-1/2}$ extends to a bounded operator on $L^p(\Omega, dx)$ for all $p \in (1, 2]$.*

Proof. We have seen above that $D_k A^{-1/2}$ is bounded on $L^2(\Omega, dx)$. By Marcinkiewicz interpolation theorem, it suffices to prove that $D_k A^{-1/2}$ is of weak type $(1, 1)$. This will be achieved by applying Theorem 7.22. By Theorem 6.10 the heat kernel $p(t, x, y)$[9] satisfies a Gaussian upper bound, and therefore it is enough to prove that for some positive constant W,

$$\int_{|x-y| \geq \sqrt{t}} |k_t(x, y)| dx \leq W, \ t > 0, y \in \Omega, \tag{7.82}$$

where $k_t(x, y)$ denotes the associated kernel of $D_k A^{-1/2}(I - e^{-tA})$.

The operator $A^{-1/2}$ is given by the formula

$$A^{-1/2}u = \frac{1}{2\sqrt{\pi}} \int_0^{\infty} e^{-sA} u \frac{ds}{\sqrt{s}}.$$

---

[9]This is $p_{H_0^1}(t, x, y)$ in Theorem 6.10.

Thus,

$$D_k A^{-1/2}(I - e^{-tA})$$

$$= \frac{1}{2\sqrt{\pi}} \int_0^\infty D_k e^{-sA} \frac{ds}{\sqrt{s}} - \frac{1}{2\sqrt{\pi}} \int_0^\infty D_k e^{-(s+t)A} \frac{ds}{\sqrt{s}}$$

$$= \frac{1}{2\sqrt{\pi}} \int_0^\infty D_k e^{-sA} \left(\frac{1}{\sqrt{s}} - \chi_{\{s>t\}} \frac{1}{\sqrt{s-t}} \right) ds.$$

The kernel $k_t(x, y)$ is then given by

$$k_t(x, y) = \frac{1}{2\sqrt{\pi}} \int_0^\infty D_{k,x} p(s, x, y) \left(\frac{1}{\sqrt{s}} - \chi_{\{s>t\}} \frac{1}{\sqrt{s-t}} \right) ds, \quad (7.83)$$

where $D_{k,x} p(s, x, y)$ is the partial derivative with respect to the x variable, and this makes sense because of (6.48) and the fact that $e^{-tA}(L^2(\Omega, dx)) \subseteq H_0^1(\Omega)$ for all $t > 0$.

Using the weighted gradient estimate of Theorem 6.19 (assertion 2)) and the Cauchy-Schwarz inequality, one obtains

$$\int_{|x-y| \geq \sqrt{t}} |D_{k,x} p(s, x, y)| dx$$

$$\leq \left[\int_\Omega |D_{k,x} p(s, x, y)|^2 e^{2\beta |x-y|^2 / s} dx \right]^{1/2} \left[\int_{|x-y| \geq \sqrt{t}} e^{-2\beta |x-y|^2 / s} dx \right]^{1/2}$$

$$\leq C s^{-d/4 - 1/2} \left[\int_{|x-y| \geq \sqrt{t}} e^{-2\beta |x-y|^2 / s} dx \right]^{1/2}.$$

Here $\beta > 0$ and $C > 0$ are constants. On the other hand,

$$\int_{|x-y| \geq \sqrt{t}} e^{-2\beta |x-y|^2 / s} dx \leq e^{-\beta t / s} \int_{\mathbb{R}^d} e^{-\beta |x-y|^2 / s} dx$$

$$\leq C e^{-\beta t / s} s^{d/2}.$$

Hence, for all $t > 0$ and $s > 0$

$$\int_{|x-y| \geq \sqrt{t}} |D_{k,x} p(s, x, y)| dx \leq C' e^{-\frac{\beta t}{2s}} s^{-1/2}, \quad (7.84)$$

where C' is a positive constant. Using this and (7.83) it follows that

$$
\int_{|x-y|\geq\sqrt{t}} |k_t(x,y)|dx
$$

$$
\leq \frac{C'}{2\sqrt{\pi}} \int_0^\infty |\frac{1}{\sqrt{s}} - \chi_{\{s>t\}}\frac{1}{\sqrt{s-t}}|e^{-\frac{\beta t}{2s}}s^{-1/2}ds
$$

$$
= C''\left[\int_0^t e^{-\frac{\beta t}{2s}}\frac{ds}{s} + \int_t^\infty \left(\frac{1}{\sqrt{s-t}} - \frac{1}{\sqrt{s}}\right)e^{-\frac{\beta t}{2s}}s^{-1/2}ds \right]
$$

$$
= C''[I_1 + I_2].
$$

The first term $I_1 := \int_0^t e^{-\frac{\beta t}{2s}}\frac{ds}{s}$ satisfies

$$
I_1 = \int_0^1 e^{-\frac{\beta}{2u}}\frac{du}{u} < \infty.
$$

The second one satisfies

$$
I_2 \leq \int_t^\infty \left(\frac{1}{\sqrt{s-t}} - \frac{1}{\sqrt{s}}\right)s^{-1/2}ds
$$

$$
= \int_0^\infty \left(\frac{1}{\sqrt{u}} - \frac{1}{\sqrt{u+t}}\right)\frac{du}{\sqrt{u+t}}
$$

$$
= \int_0^\infty \left(\frac{1}{\sqrt{ut^{-1}(ut^{-1}+1)}} - \frac{1}{ut^{-1}+1}\right)\frac{du}{t}
$$

$$
= \int_0^\infty \left(\frac{1}{\sqrt{v(v+1)}} - \frac{1}{v+1}\right)dv.
$$

The last term is finite and it is independent of t. Thus we have proved (7.82). $\square$

The above theorem holds also for uniformly elliptic operators A that are subject to Neumann or mixed boundary conditions. In this case, one assumes, for example, that Ω has the extension property (more precisely, one assumes that $H^1(\Omega)$ satisfies the Sobolev inequality (6.19), which indeed holds if Ω has the extension property). The result is now reformulated as follows: for every $\varepsilon > 0$, the Riesz transform $D_k(\varepsilon I + A)^{-1/2}$ is of weak type $(1,1)$ and extends to a bounded operator on $L^p(\Omega, dx)$ for $1 < p \leq 2$. The proof is exactly the same as the previous one. The Gaussian upper bound of the heat kernel that we need in the proof is satisfied if Ω has the extension property, see Theorem 6.10.

Let A be as in Theorem 7.26. Recall that the semigroup $(e^{-tA})_{t\geq 0}$ is sub-Markovian (see Chapter 4) and let us denote by $-A_p$ the generator of the corresponding semigroup on $L^p(\Omega, dx), 1 \leq p < \infty$. As a consequence of Theorem 7.26, for every $p \in (1, 2]$,

$$\|\nabla u\|_p \leq C\|A_p^{1/2}u\|_p \text{ for all } u \in D(A_p^{1/2}), \quad (7.85)$$

where $C > 0$ is a constant.

If A is subject to Neumann or mixed boundary conditions and Ω has the extension property, then (7.85) holds with A_p replaced by $\varepsilon I + A_p$ for any $\varepsilon > 0$.

In order to prove (7.85), let $u \in D(A_p^{1/2})$ and write for $t > 0$

$$\nabla e^{-tA}u = \nabla A_2^{-1/2}A_2^{1/2}e^{-tA}u.$$

The term on the right-hand side makes sense, since $e^{-tA}(L^p(\Omega, \mu)) \subseteq L^2(\Omega, dx)$ and $e^{-tA}(L^2(\Omega, dx)) \subseteq D(A_2^{1/2})$ for all $t > 0$. By Theorem 7.26

$$\|\nabla e^{-tA}u\|_p \leq C\|A_2^{1/2}e^{-tA}u\|_p$$
$$= C\|e^{-tA}A_p^{1/2}u\|_p$$
$$\leq C\|A_p^{1/2}u\|_p.$$

We have proved that

$$\|\nabla e^{-tA}u\|_p \leq C\|A_p^{1/2}u\|_p \text{ for all } t > 0. \quad (7.86)$$

This inequality implies that for every $t > 0$, $\|e^{-tA}u\|_{W^{1,p}(\Omega)} \leq C'$ for some positive constant C'. Since $e^{-tA}u \to u$ in $L^p(\Omega, dx)$ as $t \to 0$, it follows that $u \in W^{1,p}(\Omega)$ and one obtains from (7.86) that $\|\nabla u\|_p \leq C\|A_p^{1/2}u\|_p$. When A is subject to Neumann or mixed boundary conditions, the arguments are the same. We only have to replace A_p by $\varepsilon I + A_p$.

Remark.   1) Since $D(A_p) \subseteq D(A_p^{1/2})$ it follows from (7.85) that $D(A_p) \subseteq W^{1,p}(\Omega)$ for all $p \in (1, 2]$.

2) The results in this section hold also for non-symmetric operators $A := -\sum_{k,j=1}^{d} D_j(a_{kj}D_k) + \sum_{k=1}^{d} b_k D_k - D_k(c_k \cdot) + a_0$ with complex-valued coefficients a_{kj}, b_k, c_k, a_0, and acting on $L^2(\mathbb{R}^d, dx)$. We assume that the heat kernel of A has a Gaussian upper bound. One has now to replace A by $A + \lambda I$ for some positive constant λ in the statments. What is needed in order to apply the previous proof to the nonsymmetric case is the boundedness of $D_k(A + \lambda I)^{-1/2}$ on $L^2(\mathbb{R}^d, dx)$, which is indeed true as we shall see in the next chapter.

3) The boundedness assumption of a_0 was not used in the proof of Theorem 7.26. This theorem holds with $0 \leq a_0 \in L^1_{\text{loc}}(\Omega)$. The operator A is

now defined as the operator associated with the form

$$\mathfrak{a}(u, v) = \int_{\Omega} a_{kj} D_k u \overline{D_j v} dx + \int_{\Omega} a_0 u \overline{v} dx,$$

$$D(\mathfrak{a}) = \left\{ u \in H_0^1(\Omega), \int_{\Omega} a_0 |u|^2 dx < \infty \right\}.$$

Similarly, we can replace $H_0^1(\Omega)$ by $H^1(\Omega)$ if the latter space satisfies (6.19) (which is the case if Ω has the extension property).

7.8 GAUSSIAN LOWER BOUNDS

This section is devoted to Gaussian lower bounds. The aim is to show how such lower bounds can be obtained from Gaussian upper bounds and Hölder continuity of the heat kernel.

We shall assume that the heat kernel $p(t, x, y)$ satisfies

$$|p(t, x, y) - p(t, x', y)| \leq \frac{Ct^{-\eta/m}}{V(y, t^{1/m})} \rho(x, x')^{\eta} \text{ for all } x, x', y \in X, \quad (7.87)$$

where η and C are positive constants and m is as in (7.3). In the case where $C' r^d \leq V(x, r) \leq C r^d$ for all $x \in X$ and $r > 0$, Theorem 6.6 gives a criterion for the validity of (7.87).

The first step needed in order to obtain a lower bound is given in the following result.

PROPOSITION 7.27 *Assume that the doubling property (7.1) holds and assume that the heat kernel $p(t, x, y)$ satisfies (7.87). If $p(t, x, y)$ satisfies the on-diagonal lower bound:*

$$p(t, x, x) \geq \frac{c}{V(x, t^{1/m})} \text{ for all } t > 0, x \in X \quad (7.88)$$

for some constant $c > 0$, then there exist constants $c' > 0$ and $\beta > 0$ such that

$$p(t, x, y) \geq \frac{c'}{V(y, t^{1/m})} \text{ for all } t, x, y \text{ such that } \rho(x, y) \leq \beta t^{1/m}. \quad (7.89)$$

Proof. It follows from (7.87) and (7.88) that

$$p(t, x, y) \geq p(t, y, y) - \frac{Ct^{-\eta/m}}{V(y, t^{1/m})} \rho(x, y)^\eta$$

$$\geq \frac{c}{V(y, t^{1/m})} - \frac{Ct^{-\eta/m}}{V(y, t^{1/m})} \rho(x, y)^\eta$$

$$\geq \frac{c_0}{V(y, t^{1/m})} [1 - t^{-\eta/m} \rho(x, y)^\eta]$$

$$\geq \frac{c_0}{2V(y, t^{1/m})}$$

for $\rho(x, y)^\eta \leq \frac{1}{2} t^{\eta/m}$, which shows (7.89). $\qquad \square$

Now we show that the on-diagonal lower bound (7.88) can be deduced from a Gaussian upper bound and the conservation property.

PROPOSITION 7.28 *Assume that X satisfies the doubling property (7.1) and let A be a self-adjoint operator on $L^2(X, \mu)$ with a heat kernel $p(t, x, y)$ satisfying the Gaussian upper bound (7.3) for $x, y \in X$. Assume also that $\int_X p(t, x, y)d\mu(y) = 1$ for all $t > 0$ and a.e. $x \in X$.[10] Then there exists a constant $c > 0$ such that for all $t > 0$ and a.e. $x \in X$*

$$p(t, x, x) \geq \frac{c}{V(x, t^{1/m})}.$$

Proof. Fix a positive constant α. We have

$$\int_{X \setminus B(x, \alpha t^{1/m})} p(t, x, y)d\mu(y)$$

$$\leq \int_{X \setminus B(x, \alpha t^{1/m})} \frac{C}{V(x, t^{1/m})} \exp\left\{-c\left(\frac{\rho^m(x, y)}{t}\right)^{\frac{1}{m-1}}\right\} d\mu(y)$$

$$\leq Ce^{-\frac{c}{2}\alpha^{m/m-1}} \int_X \frac{1}{V(x, t^{1/m})} \exp\left\{-\frac{c}{2}\left(\frac{\rho^m(x, y)}{t}\right)^{\frac{1}{m-1}}\right\} d\mu(y)$$

$$\leq C'e^{-\frac{c}{2}\alpha^{m/m-1}}.$$

For the fact that $\int_X \frac{1}{V(x, t^{1/m})} \exp\{-\frac{c}{2}(\frac{\rho^m(x,y)}{t})^{\frac{1}{m-1}}\}d\mu(y)$ is bounded by a constant, see the proof of Proposition 7.1.

Now for α large enough, the term $C'e^{-\frac{c}{2}\alpha^{m/m-1}}$ is smaller than $1/2$. Thus,

$$\int_{B(x, \alpha t^{1/m})} p(t, x, y)d\mu(y) = 1 - \int_{X \setminus B(x, \alpha t^{1/m})} p(t, x, y)d\mu(y)$$

$$\geq 1/2$$

---

[10]This is the conservation (or the Markov) property $e^{-tA}1 = 1$.

for α large enough. It follows that

$$
\begin{aligned}
p(2t, x, x) &= \int_X p(t, x, y)p(t, y, x)d\mu(y) \\
&= \int_X |p(t, x, y)|^2 d\mu(y) \\
&\geq \int_{B(x, \alpha t^{1/m})} |p(t, x, y)|^2 d\mu(y) \\
&\geq \frac{1}{V(x, \alpha t^{1/m})} \left(\int_{B(x, \alpha t^{1/m})} p(t, x, y)d\mu(y) \right)^2 \\
&\geq \frac{1}{4V(x, \alpha t^{1/m})}.
\end{aligned}
$$

Using (7.1) (or (7.2)), the last term is estimated from below by $\frac{C}{V(x,(2t)^{1/m})}$ for some positive constant C. □

In order to deduce a Gaussian lower bound from the Gaussian upper bound we shall assume that X has the chain condition: there exists a constant C such that for every $x, y \in X$ and every positive $n \in \mathbb{N}$, there exists a sequence of points $x_i, 0 \leq i \leq n$, in X such that $x_0 = x$, $x_n = y$ and

$$
\rho(x_i, x_{i+1}) \leq C \frac{\rho(x, y)}{n} \text{ for all } i = 0, ..., n-1.
$$

The sequence $x_i, 1 \leq i \leq n$, is referred to as a chain connecting x and y.

THEOREM 7.29 *Assume that X satisfies the doubling property (7.1) and the chain condition. Let A be a self-adjoint operator on $L^2(X, \mu)$ with heat kernel $p(t, x, y)$ satisfying the Gaussian upper bound (7.3) and the Hölder estimate (7.87). Assume that $p(t, x, y) \geq 0$ for all $t > 0, x, y \in X$ and satisfies the on-diagonal lower bound (7.88). Then there exist constants $C_0, c_0 > 0$ such that*

$$
p(t, x, y) \geq \frac{C_0}{\sqrt{V(x, t^{1/m})V(y, t^{1/m})}} \exp \left\{ -c_0 \left(\frac{\rho^m(x, y)}{t} \right)^{\frac{1}{m-1}} \right\},
$$
(7.90)

for all $x, y \in X$ and $t > 0$.

Proof. Fix a positive $n \in \mathbb{N}$ and $x, y \in X$. Denote by $x_i, 0 \leq i \leq n$, a chain connecting x and y. Let $r := \frac{\rho(x, y)}{n}$ and β be as in Proposition 7.27.

By the semigroup property and positivity assumption of $p(t, x, y)$, we have

$$p(t, x, y)$$

$$= \int_X \cdots \int_X p\left(\frac{t}{n}, x, z_1\right) p\left(\frac{t}{n}, z_1, z_2\right) \cdots p\left(\frac{t}{n}, z_{n-1}, y\right)$$
$$d\mu(z_1)...d\mu(z_{n-1})$$

$$\geq \int_{B(x_1,r)} \cdots \int_{B(x_{n-1},r)} p\left(\frac{t}{n}, x, z_1\right) p\left(\frac{t}{n}, z_1, z_2\right) \cdots p\left(\frac{t}{n}, z_{n-1}, y\right)$$
$$d\mu(z_1)...d\mu(z_{n-1}).$$

Let $z_0 = x$ and $z_n = y$. Clearly,

$$\rho(z_i, z_{i+1}) \leq \rho(x_i, x_{i+1}) + 2r \leq (C+2)\frac{\rho(x, y)}{n}, \ i = 0, .., n-1.$$

If $(C + 2)\rho(x, y) \leq \beta t^{1/m}$, then $\rho(x, y) \leq \beta t^{1/m}$. In this case (7.90) follows from (7.88) and Proposition 7.27. We may then assume that $(C + 2)\rho(x, y) > \beta t^{1/m}$. We choose $n \geq 2$ to be the smallest integer such that

$$(C+2)\frac{\rho(x, y)}{n^{\frac{m-1}{m}}} \leq \beta t^{1/m}.$$

This gives

$$\rho(z_i, z_{i+1}) \leq \beta \left(\frac{t}{n}\right)^{1/m}, \ i = 0, ..., n-1.$$

Applying now Proposition 7.27, we obtain

$$p(t, x, y)$$

$$\geq \int_{B(x_1,r)} \cdots \int_{B(x_{n-1},r)} \frac{c^n d\mu(z_1)...d\mu(z_{n-1})}{V(x, (\frac{t}{n})^{1/m})V(z_1, (\frac{t}{n})^{1/m})...V(z_{n-1}, (\frac{t}{n})^{1/m})}.$$

By the doubling property (7.2), we have

$$V(z_i, (t/n)^{1/m}) \leq C\left(1 + (t/n)^{1/m}\frac{1}{r}\right)^d V(x_i, r), \ i = 0, ..., n-1.$$

Using the definition of n one obtains $(t/n)^{1/m}\frac{1}{r} \leq C'$ for some constant C'. Thus,

$$V(z_i, (t/n)^{1/m}) \leq CC'^d V(x_i, r).$$

Inserting this in the previous lower estimate of $p(t, x, y)$ yields

$$p(t, x, y) \geq C''\frac{c^n}{V(x, (t/n)^{1/m})} \geq C''\frac{e^{-c'n}}{V(x, (t/n)^{1/m})} \tag{7.91}$$

for some positive constants C'' and c'. By definition of n, one has

$$n - 1 \leq \left(\frac{C+2}{\beta} \right)^{m/(m-1)} \frac{\rho(x,y)^{m/(m-1)}}{t^{1/(m-1)}}.$$

This and (7.91) give the Gaussian lower bound:

$$p(t,x,y) \geq \frac{C_0}{V(x,t^{1/m})} \exp \left\{ - c_0 \left(\frac{\rho^m(x,y)}{t} \right)^{\frac{1}{m-1}} \right\},$$

which is equivalent to (7.90) (see (7.5)). □

By Proposition 7.28, one can replace the assumption (7.88) in Theorem 7.29 by the conservation property.

Notes

There are other consequences of Gaussian upper bounds of heat kernels. We mention briefly the following further applications.

It was shown by Duong and Robinson [DuRo96] that if (X, μ, ρ) satisfies the doubling condition (7.1) and A has an H^∞ functional calculus on $L^2(X, \mu)$ and the heat kernel of A has a Gaussian upper bound, then A has an H^∞ functional calculus on $L^p(X, \mu)$ for $1 < p < \infty$. A similar result also holds for operators acting on $L^2(\Omega, \mu)$ with Ω any open subset of X. This was proved by Duong and McIntosh [DuMc99a].

Gaussian upper bounds were also used to prove maximal $L^p - L^q$ a priori estimates for the solution to the evolution equation $\frac{du}{dt}(t) = Au(t) + f(t)$, $u(0) = 0$. Such results have been proved by Hieber and Prüss [HiPr97] and Coulhon and Duong [CoDu00].

Section 7.1. Gaussian upper bounds for complex time play a central role in our investigation of L^p spectral theory. The extension of the Gaussian upper bound from $t > 0$ to complex t was first proved by Davies [Dav89] (Theorem 3.4.8) in the case where $V(x, t^{1/m})$ is replaced by $t^{d/m}$ (actually, $m = 2$ in [Dav89]). Theorems 7.2 and 7.3 are proved in Carron, Coulhon, and Ouhabaz [CCO02]. A weaker version of these results appears in Duong and Robinson [DuRo96]. Theorem 7.4 is taken from [CCO02], and its proof uses an idea from [DuRo96].

Section 7.2. It was first proved by Helffer and Sjöstrand [HeSj89] that for self-adjoint operators on Hilbert spaces, the formula (7.26) coincides with the usual functional calculus defined by spectral measures. Subsequently, Davies [Dav95a] and [Dav95d] extends the Helffer-Sjöstrand calculus to the Banach space setting. Jensen and Nakamura [JeNa94] have also used the Helffer-Sjöstrand calculus to study L^p-mapping properties of Schrördinger operators. Theorem 7.8 is due to Davies [Dav95a]. See also [Dav95d].

Our approach to prove L^p spectral independence is due to Davies [Dav95b]. Theorem 7.10 was proved by Davies [Dav95b] in the case where the volume $V(x, r)$ has polynomial growth. See also [Dav95c] for an abstract result. A different approach was used by Hempel and Voigt [HeVo86] to prove p-independence of the spectrum for Schrödinger operators. Arendt [Are94] has proved that for nonsymmetric uniformly elliptic operators, the component of the resolvent set which contains the left half-plane is the same in all L^p-spaces (therefore, for self-adjoint operators the spectrum is p-independent). His approach is similar to that of Hempel and Voigt [HeVo86]. Kunstmann [Kun99] extends this result and proves that the spectrum for nonsymmetric elliptic operators whose heat kernel has a Gaussian upper bound is p-independent. For operators with unbounded singular drift terms, p-independance of the spectrum for p is some interval containing 2 was proved by Liskevich, Sobol, and Vogt [LSV02]. Theorem 7.10 can be applied to the Laplace-Beltrami operator on a Riemannian manifold with non-negative Ricci curvature. In the setting of manifolds, a better result was proved by Shubin [Shu92] and Sturm [Stu93] who only assume that the volume grows uniformly subexponentialy.

Section 7.3. For the Laplace operator Δ acting on $L^p(\mathbb{R}^d, dx)$, Lanconelli [Lan68] has proved that $e^{it\Delta}$ sends the Sobolev space $W^{\alpha, p}$ into $L^p(\mathbb{R}^d, dx)$ where $\alpha > 2d|\frac{1}{2} - \frac{1}{p}|$. Equivalently, this means that $(I - \Delta)^{-\alpha/2} e^{it\Delta}$ acts boundedly on $L^p(\mathbb{R}^d, dx)$ for $\alpha > 2d|\frac{1}{2} - \frac{1}{p}|$. Riesz means associated with the Schrödinger group $(e^{it\Delta})_{t \in \mathbb{R}}$ have been studied by Sjöstrand [Sjö70]. He proved that for integer $k > d|\frac{1}{2} - \frac{1}{p}|$, the operator $t^{-k} \int_0^t (t - s)^{k-1} e^{is\Delta} ds$ acts as a bounded operator on $L^p(\mathbb{R}^d, dx)$, and its norm can be bounded uniformly in $t > 0$. He also proved that this is not the case if $k < |\frac{1}{2} - \frac{1}{p}|$. Extensions of Lanconelli's and Sjöstrand's previously mentioned results were given by Boyadzhiev and de Laubenfels [BodeL93] and by El Mennaoui [ElM92]. The results are formulated in [BodeL93] in the language of C-semigroups and in [ElM92] in that of integrated semigroups. Related results can also be found in Arendt et al. [ABHN01] (Theorem 8.3.9) and El Mennaoui and Keyantuo [ElKe96]. On Lie groups with polynomial growth and manifolds with non-negative Ricci curvature, similar results have been obtained by Alexopoulos [Ale94a]. Less sharp results in this setting have been announced by Lohoué [Loh92]. In the abstract setting of operators on metric spaces, Theorems 7.11 and 7.12 are proved by Carron, Coulhon, and Ouhabaz [CCO02]. The proof of Theorem 7.11 uses an idea from El Mennaoui [ElM92]. A more recent result of Blunck [Blu03a] shows the boundedness of Riesz means on L^p for p close to 2 for operators whose semigroups satisfy a weighted norm estimate.

Section 7.4. This section is mainly taken from Ouhabaz [Ouh03b]. The proofs use the same techniques as in Duong, Ouhabaz, and Sikora [DOS02]. A related result to Theorem 7.15 is proved in an unpublished paper by Hebisch [Heb95]. Theorem 7.19 was proved by Alexopoulos [Ale94a] in the setting of Lie groups and Riemannian manifolds with non-negative Ricci curvature. Related results for wave equations with a potential can be found in Marshall, Strauss, and Wainger

[MSW80], Beals and Strauss [BeSt93] and Zhong [Zha95]. For the classical wave equation, the sharper result in Theorem 7.20 is due to Peral [Per80] and Miyachi [Miy80]. Their results even include the limit case $\alpha = (d-1)|\frac{1}{p} - \frac{1}{2}|$. Müller and Stein [MüSt99] have proved a version of Theorem 7.20 for a sub-Laplacian on the Heisenberg group. Some of the arguments used in the simpler proof of Theorem 7.20 given here are taken from Galé and Pyltik [GaPy97].

Section 7.5. Theorems 7.21 and 7.22 are due to Duong and McIntosh [DuMc99a]. The proof of Theorems 7.21 uses a similar reasoning as in Hebisch [Heb90] and Duong and Robinson [DuRo96]. These theorems have been applied successfully in [DuMc99a] to prove the existence of an H^∞ functional calculus in L^p for operators acting on arbitrary domains, in Coulhon and Duong [CoDu99] to study Riesz transforms on Riemannian manifolds and in Duong, Ouhabaz, and Sikora [DOS02] to prove spectral multiplier results. Recently, Blunck and Kunstmann [BlKu03b] have extended Theorems 7.21 and 7.22 to the case of operators without associated kernel. They have replaced the Gaussian upper bound in the assumptions by a (more general) weighted norm estimate of the semigroup.

Section 7.6. Theorem 7.23 is a particular case of a more general result proved in Duong, Ouhabaz, and Sikora [DOS02]. As mentioned in (7.68), for the Euclidean Laplacian, Theorem 7.23 holds with $W^{s,2}$ instead of C^s in the condition (7.66). This is a well-known result of Hörmander [Hör60]. Some conditions under which one can replace in (7.66) C^s by a Sobolev space $W^{s,p}$ are given in [DOS02]. Note that if no additional assumptions are made, Theorem 7.23 cannot hold with $W^{s,2}$ in place of C^s in (7.66). This can be seen by considering the harmonic oscillator $-\Delta + x^2$. Thangavelu [Tha89] has proved that, for the harmonic oscillator $A :=$ $-\Delta + x^2$ in one dimension, the corresponding Bochner-Riesz means $\sigma_R^\alpha(A)$, where $\sigma_R^\alpha(\lambda) = ((1 - \frac{\lambda}{R})^+)^\alpha$, are uniformly bounded on all $L^p(\mathbb{R}, dx)$ only if $\alpha > 1/6$. If Theorem 7.23 holds for the harmonic oscillator with $W^{s,2}$ in place of C^s, then the uniform boundedness of Bochner-Riesz means must hold for all $\alpha > 0$.

For extensions of Hörmander's multiplier theorem to sub-Laplacians on some Lie groups, see Christ [Chr91b], Mauceri and Meda [MaMe90], and the references therein. Related results to Theorem 7.23 (but not on arbitrary domains) have been obtained by Hebisch [Heb95] and Alexopoulos [Ale99].

Section 7.7. There are several works on Riesz transforms associated with the Laplace-Beltrami operator on manifolds; see Strichartz [Str83], Bakry [Bak87], and the references therein. On Lie groups of polynomial growth, Saloff-Coste [SaC90] has used Gaussian upper bounds and regularity (with respect to the space variable) of the heat kernel to prove L^p-boundedness of Riesz transforms associated with sub-Laplacians. Coulhon and Duong [CoDu99] have proved that on a Riemannian manifold or even on an arbitrary domain, the Gaussian upper bound for the heat kernel suffices to prove boundedness on L^p for $1 < p \leq 2$ of Riesz transforms $\nabla(-\Delta)^{-1/2}$. Their proof was then used by Duong and McIntosh [DuMc99b] to prove Theorem 7.26. Similar results for divergence form operators on $\mathbb{R}^d$ can be found in Auscher and Tchamitchian [AuTc98]. See also Blunck and Kunstmann [BlKu03a] and Hofmann and Martell [HoMa03]. In [AuTc98] one also finds results on the reverse inequality $\|A^{1/2}u\|_p \leq C\|\nabla u\|_p$. For Riesz transforms on L^p with $p > 2$, the situation is very different from the case $p < 2$. An example of a uniformly elliptic operator for which the corresponding Riesz transform is not bounded on L^p for any $p > 2 + \varepsilon$ (where $\varepsilon > 0$ is a constant) was given by Kenig and can be found in [AuTc98]. The recent works by Auscher [Aus03] and Auscher et al. [ACDH03] investigate on manifolds the Riesz transforms on L^p for $p > 2$.

Section 7.8. The results in this section have been proved by several authors in different situations. We refer the reader to Coulhon [Cou03], Davies [Dav89], Grigor'yan, Hu and Lau [GHL03] and Saloff-Coste [SaC02]. Our presentation follows mainly [Cou03]. Let us also mention that on a complete Riemannian manifold that satisfies the doubling condition (7.1), relationships between the validity of Gaussian upper and lower bounds, Harnack inequalities and L^2-Poincaré inequalities on balls have been established by Grigor'yan [Gri91] and Saloff-Coste [SaC92].

Chapter Eight

A REVIEW OF THE KATO SQUARE ROOT
PROBLEM

T. Kato asked in the early 1960's the question whether for a divergence form uniformly elliptic operator A, the domain of the square root $A^{1/2}$ coincides with the domain of the sesquilinear form of A. The question was solved in few particular cases but the general case was open for several decades, and was known as the Kato square root problem. The problem was solved in the case of dimension one by Coifman, McIntosh, and Meyer [CMM82]. The case of arbitrary dimension has been solved only recently by Hofmann, Lacey, and McIntosh [HLM02] for second-order divergence form operators whose heat kernel has a Gaussian upper bound,[1] and by Auscher, Hofmann, Lacey, McIntosh, and Tchamitchian [AHLMT02] in the general case of second-order operators with complex-valued coefficients. In these articles, the authors proved that for a uniformly elliptic operator A in divergence form, $D(A^{1/2}) = H^1(\mathbb{R}^d)$, and $\|A^{1/2}u\|_2$ and $\||\nabla u|\|_2$ are equivalent.

In this chapter, we shall review the Kato square root problem. We first discuss the problem in the abstract setting of operators associated with sesquilinear forms. In this general case, it has a negative answer. We then return to the special case of uniformly elliptic operators and discuss some consequences of the main theorem. We shall not give detailed proofs but we provide the reader with references.

8.1 THE PROBLEM IN THE ABSTRACT SETTING

In this section we shall consider the square root problem in an abstract setting. We shall follow mainly the papers of Kato [Kat61], [Kat62], Lions [Lio62], and McIntosh [McI72], [McI90].

Let H be a Hilbert space with a scalar product $(.;.)$. Let $\mathfrak{a}$ be a densely defined, accretive, continuous, and closed form on H. Denote by A its associated operator. The square root $A^{1/2}$ is the unique m-accretive operator such that $(A^{1/2})^2 = A^{1/2}A^{1/2} = A$. Following Kato [Kat61], [Kat80], the

---

[1]This is the case if the coefficients are real-valued; see Chapter 6.

square root satisfies:

$$A^{1/2}u = \frac{2}{\pi} \int_0^\infty A(I + t^2 A)^{-1} u \, dt, \tag{8.1}$$

for $u \in D(A)$. Note that the term on the right-hand side makes sense. Indeed, we have[2]

$$\|(I + t^2 A)^{-1} Au\| \le \|Au\|$$

and

$$\|(I + t^2 A)^{-1} Au\| = t^{-2} \|u - (I + t^2 A)^{-1} u\|$$
$$\le 2t^{-2} \|u\|.$$

Note also that if A is sectorial (equivalently if $\mathfrak{a}$ is sectorial), then for every $u \in D(A)$, we have

$$A^{1/2}u = \frac{2}{\sqrt{\pi}} \int_0^\infty A e^{-t^2 A} u \, dt. \tag{8.2}$$

Again the term in the right-hand side makes sense since $\|Ae^{-t^2 A}u\| \le \|Au\|$ and $\|Ae^{-t^2 A}u\| \le ct^{-2}\|u\|$ for some positive constant c and all $t > 0$. The first inequality follows from the fact that the semigroup $(e^{-tA})_{t \ge 0}$ is contractive on H (see Proposition 1.51). The second one follows from the Cauchy formula and the fact that the semigroup generated by $-A$ is bounded holomorphic on H (see (1.19) and Theorem 1.54).

For self-adjoint operators, one always has $D(A^{1/2}) = D(\mathfrak{a})$.

THEOREM 8.1 *Assume that the form $\mathfrak{a}$ is symmetric, accretive, densely defined, and closed. Then $D(A^{1/2}) = D(\mathfrak{a})$ and*

$$\mathfrak{a}(u, v) = (A^{1/2}u; A^{1/2}v) \ for \ all \ u, v \in D(\mathfrak{a}).$$

The first step in proving this theorem is to prove that $D(A)$ is a core of $A^{1/2}$ (this does not use the symmetry assumption and holds in a more general setting, see [Kat61]). Once this is proved, one defines the form

$$\mathfrak{b}(u, v) := (A^{1/2}u; A^{1/2}v), \ u, v \in D(\mathfrak{b}) := D(A^{1/2}).$$

Clearly, $\mathfrak{a}(u, v) = \mathfrak{b}(u, v)$ for $u, v \in D(A)$. The domain $D(A)$ is a core of $\mathfrak{b}$ (since it is a core of $A^{1/2}$) and it is also a core of $\mathfrak{a}$ (see Lemma 1.25). This implies that the two forms $\mathfrak{a}$ and $\mathfrak{b}$ coincide.

---

[2]See Proposition 1.22 in Chapter 1.

More generally, one can define fractional powers A^α of any m-accretive operator A, for $\alpha \in (0,1)$. For $u \in D(A)$, $A^\alpha u$ is defined by:

$$A^\alpha u = \frac{\sin \pi \alpha}{\pi} \int_0^\infty \lambda^{\alpha-1} A(\lambda I + A)^{-1} u \, d\lambda. \qquad (8.3)$$

The operator A^α is m-accretive and one has $A^\alpha A^\beta = A^{\alpha+\beta}$ and $A^{\alpha*} = A^{*\alpha}$. See Kato [Kat61] or [Kat80]. It was proved by Kato [Kat61] that for $0 < \alpha < \frac{1}{2}$, one has $D(A^\alpha) = D(A^{*\alpha})$ but that A^α and $A^{*\alpha}$ do not necessarily have the same domain if $\frac{1}{2} < \alpha \leq 1$. The case $\alpha = \frac{1}{2}$ was left open.

The latter question when $\alpha = \frac{1}{2}$ was solved in the negative for general m-accretive operators by Lions [Lio62]. He proved that $D(A^\alpha)$ form a complex interpolation family. That is $D(A^\alpha) = [H, D(A)]_{[\alpha]}$, where the latter denotes the classical complex interpolation space. He then considered the operator $A := \frac{d}{dx}$ on $L^2([0,\infty), dx)$ with Dirichlet boundary condition at 0. The adjoint A^* is given by $-\frac{d}{dx}$ with domain $D(A^*) = H^1([0,\infty))$. Therefore,

$$D(A^\alpha) = [L^2([0,\infty)), H_0^1([0,\infty))]_{[\alpha]} = H_0^\alpha([0,\infty)),$$

and similarly

$$D(A^{*\alpha}) = [L^2([0,\infty)), H^1([0,\infty))]_{[\alpha]} = H^\alpha([0,\infty)).$$

Thus, since $H_0^\alpha([0,\infty)) \neq H^\alpha([0,\infty))$ for $\alpha \geq \frac{1}{2}$, one has $D(A^\alpha) \neq D(A^{*\alpha})$ for $\alpha \geq \frac{1}{2}$.

The above operator A considered by Lions for which he proved that $D(A^{1/2}) \neq D(A^{*1/2})$ is not the operator associated with a sesquilinear form. Special properties hold for operators associated with forms as the following theorem shows.

THEOREM 8.2 *Let $\mathfrak{a}$ be a densely defined, accretive, continuous, and closed sesquilinear form on H. Denote by A its associated operator. If two of the domains $D(A^{1/2}), D(A^{*1/2})$ and $D(\mathfrak{a})$ are equal, then all three are equal, and the form $\mathfrak{a}$ has the representation:*

$$\mathfrak{a}(u,v) = (A^{1/2}u; A^{*1/2}v) \text{ for all } u,v \in D(\mathfrak{a}) = D(A^{1/2}) = D(A^{*1/2}). \qquad (8.4)$$

This theorem was proved in Lions [Lio62] and Kato [Kat62]. Note that in these two articles, the form $\mathfrak{a}$ is supposed to be sectorial. Now if $\mathfrak{a}$ is as in the theorem, then $\mathfrak{a} + \varepsilon(.;.)$ is sectorial for $\varepsilon \in (0,1]$ (see the comments following Theorem 1.52) and by Lemma A2 in Kato [Kat61], one has

$D((A+\varepsilon I)^{1/2}) = D(A^{1/2})$ and $(A+\varepsilon I)^{1/2}u$ converges to $A^{1/2}u$ as $\varepsilon \to 0$. Thus, letting $\varepsilon \to 0$ one obtains the previous theorem from its analog for sectorial forms.

Kato's question on $D(A^{1/2}) = D(A^{*1/2})$ for operators associated with sesquilinear nonsymmetric forms was solved by McIntosh [McI72]. The question was again solved in the negative. McIntosh's example of a sectorial operator A for which $D(A^{1/2}) \neq D(A^{*1/2})$ is based on the following idea: For every integer $n \geq 2$, there exist bounded self-adjoint operators S_n and T_n on a finite dimensional Hilbert space H_n such that $0 \leq S_n \leq 1^3$, S_n is invertible and

$$\|S_n T_n - T_n S_n\|_{\mathcal{L}(H_n)} = 2 - \frac{1}{n} \text{ and } \|S_n T_n + T_n S_n\|_{\mathcal{L}(H_n)} \leq \frac{1}{n}.$$

Now one defines on H_n the operator $A_n := (S_n^{-1} + iT_n)^2$ and shows that A_n satisfies the properties:
1) A_n is sectorial with $|\Im(A_n u; u)| \leq \alpha |\Re(A_n u; u)|$, where $\alpha > 0$ is a constant independent of n.
2) A_n is invertible.
3) For each n, there exists an element $v \in H_n$ which does not satisfy the inequalities

$$(n-1)^{-1/2}\|A_n^{*1/2}v\| \leq \|A_n^{1/2}v\| \leq (n-1)^{1/2}\|A_n^{*1/2}v\|.$$

Finally, define $A = \bigoplus A_n$ on $H = \bigoplus H_n$. One obtains from the previous properties of A_n that A is sectorial, m-accretive with both $A^{1/2}$ and $A^{*1/2}$ invertible and for every $\beta > 0$ there exists $v \in D(A^{1/2}) \cap D(A^{*1/2})$ which does not satisfy the inequalities

$$\beta^{-1}\|A^{*1/2}v\| \leq \|A^{1/2}v\| \leq \beta\|A^{*1/2}v\|.$$

This together with the remark following the proof of Theorem 1.50 shows that the domains $D(A^{1/2})$ and $D(A^{*1/2})$ must be different.

The following example shows also how the case $\alpha = \frac{1}{2}$ is critical. Consider $d = 1$ and $A = -\frac{d}{dx}(a\frac{d}{dx})$. We assume that a is a real-valued bounded function such that $a \geq 1$. In this case, A is self-adjoint with domain

$$D(A) = \{u \in H^1(\mathbb{R}), au' \in H^1(\mathbb{R})\}.$$

---

[3]In the sense that $0 \leq (S_n u; u) \leq \|u\|^2$ for all $u \in H_n$.

Now by interpolation, $D(A^\alpha) = [L^2, D(A)]_{[\alpha]}$ and one obtains

$$D(A^\alpha) = D((A^{1/2})^{2\alpha}) = [L^2, H^1]_{[2\alpha]} = H^{2\alpha}(\mathbb{R}) \text{ for } 0 < \alpha < \frac{1}{2},$$

$$D(A^{1/2}) = H^1(\mathbb{R}),$$

$$D(A^\alpha) = \{f \in H^1(\mathbb{R}), af' \in H^{2\alpha-1}(\mathbb{R})\} \text{ for } \frac{1}{2} < \alpha < 1.$$

A proof of the latter equality can be found in David, Journé and Semmes [DJS85]. See also Auscher and Tchamitchian [AuTc92].

Summarizing, we have seen that $D(A^\alpha) = D(A^{*\alpha})$ for $\alpha \in (0, \frac{1}{2})$ and every m-accretive operator. The equality $D(A^{1/2}) = D(A^{*1/2})$ may be false even when A is a sectorial operator. McIntosh's example which disproves the equality $D(A^{1/2}) = D(A^{*1/2})$ (or equivalently, the equality $D(A^{1/2}) = D(\mathfrak{a})$, by Theorem 8.2) is not a differential operator. Therefore, the problem, as refined by McIntosh, became whether the last equality holds for divergence form uniformly elliptic operators. It is this special case which is now known as the Kato square root problem.
Kato was motivated by applications to wave equations as well as to nonautonomous evolution equations

$$\frac{du(t)}{dt} + A(t)u(t) + f(t) = 0, \ 0 \le t \le T.$$

Another motivation for the Kato square root problem lies in the interactions with other problems from harmonic analysis, e.g., L^2-boundedness of the Cauchy integral along Lipschitz curves ([CMM82]), quadratic estimates, $T(b)$ type theorems, ...

8.2 THE KATO SQUARE ROOT PROBLEM FOR ELLIPTIC OPERATORS

Consider on $L^2(\mathbb{R}^d, dx, \mathbb{C})$ ($d \ge 1$) the sesquilinear form

$$\mathfrak{a}(u, v) = \sum_{k,j=1}^{d} \int_{\mathbb{R}^d} a_{kj} D_k u \overline{D_j v} dx, \ \ D(\mathfrak{a}) = H^1(\mathbb{R}^d).$$

We assume as usual that the uniform ellipticity condition (U.Ell) holds. That is

$$a_{kj} \in L^\infty(\mathbb{R}^d, \mathbb{C}), 1 \le k, j \le d, \ \Re \sum_{k,j=1}^{d} a_{kj}(x)\xi_k \overline{\xi_j} \ge \eta|\xi|^2 \text{ for all } \xi \in \mathbb{C}^d,$$

$$(8.5)$$

where $\eta > 0$ is a constant and the inequality $\Re \sum_{k,j=1}^d a_{kj}(x)\xi_k\overline{\xi_j} \geq \eta|\xi|^2$ holds for a.e. $x \in \mathbb{R}^d$.

The form $\mathfrak{a}$ is densely defined, accretive, continuous and closed. It is even sectorial. Its associated operator is the uniformly elliptic operator in divergence form $A = -\sum_{k,j=1}^d D_j(a_{kj}D_k)$.

As we have mentioned in the last section, the Kato square root problem for the elliptic operator A is the problem of identifying $D(A^{1/2})$ with $H^1(\mathbb{R}^d)$, the domain of its sesquilinear form. In contrast to the case of operators in an abstract setting, the answer in this case is yes.

THEOREM 8.3 *Under the uniform ellipticity condition (U.Ell), $D(A^{1/2}) = H^1(\mathbb{R}^d)$ and there exists a constant C such that*

$$C^{-1}\|A^{1/2}u\|_2 \leq \||\nabla u|\|_2 \leq C\|A^{1/2}u\|_2 \text{ for all } u \in H^1(\mathbb{R}^d). \quad (8.6)$$

This theorem was proved by Coifman, McIntosh, and Meyer [CMM82] in the case $d = 1$, by Hofmann, Lacey, and McIntosh [HLM02] for arbitrary dimension under the restriction of a pointwise decay of the heat kernel of A, and by Auscher, Hofmann, Lacey, McIntosh, and Tchamitchian [AHLMT02] in the general case stated here.

Other previously known results for arbitrary dimension have been proved by Coifman, Deng and Meyer [CDM83], Fabes, Jerison, and Kenig [FJK85]. In these two articles, the authors have independently proved the validity of the Kato square root problem for operators A with matrix $\mathcal{E} := (a_{kj})_{1 \leq k,j \leq d}$ satisfying $\|\mathcal{E} - I_d\|_\infty \leq \varepsilon(d)$ for some small ε depending on the dimension d (I_d is the identity matrix). See also Journé [Jou91]. McIntosh [McI85] has proved Theorem 8.3 in the case where the coefficients a_{kj} are $W^{s,2}$ multipliers[4] for some $s > 0$. The previously mentioned result from [CDM83] and [FJK85] was extended by Auscher, Hofmann, Lewis, and Tchamitchian [AHLT01] in the sense that I_d can be replaced by any symmetric matrix with real-valued coefficients.

The proof of Theorem 8.3 is very much based on techniques from harmonic analysis. These techniques have not been developed in the present book, and therefore in order to give an idea of the proof we shall restrict ourself to describing the main steps without going into the details. We shall be very concise; the reader is referred to [HLM02] and [AHLMT02] for a detailed proof and for more information.

First, it is enough to prove

$$\|A^{1/2}f\|_2 \leq C\||\nabla f|\|_2 \text{ for all } f \in C_c^\infty(\mathbb{R}^d). \quad (8.7)$$

---

[4]That is, $a_{kj}u \in W^{s,2}$ for all $u \in W^{s,2}$.

Indeed, by a density argument this will hold for all $f \in H^1(\mathbb{R}^d)$ and gives $H^1(\mathbb{R}^d) \subseteq D(A^{1/2})$. Using (8.7) for the adjoint A^* gives also $H^1(\mathbb{R}^d) \subseteq D(A^{*1/2})$. Now if $f \in D(A)$, we write

$$\eta |||\nabla f|||_2^2 \le \Re\mathfrak{a}(f, f)$$
$$= \Re(A^{1/2}f; A^{*1/2}f)$$
$$\le \|A^{1/2}f\|_2 \|A^{*1/2}f\|_2$$
$$\le C \|A^{1/2}f\|_2 |||\nabla f|||_2.$$

Therefore,

$$|||\nabla f|||_2 \le C' \|A^{1/2}f\|_2. \tag{8.8}$$

This and the fact that $D(A)$ is a core of $D(A^{1/2})$ imply that $D(A^{1/2}) \subseteq H^1(\mathbb{R}^d)$. Hence $D(A^{1/2}) = H^1(\mathbb{R}^d)$ and assertion (8.6) of the theorem follows from (8.7) and (8.8).

Using the representation (8.1) in a slightly different form and standard orthogonality argument of Littlewood-Paley theory, one shows that (8.7) follows from the quadratic estimate[5]

$$\int_0^\infty \|(I + t^2 A)^{-1} tAf\|_2^2 \frac{dt}{t} \le C \int_{\mathbb{R}^d} |\nabla f(x)|^2 dx. \tag{8.9}$$

Define for a $\mathbb{C}^d$-valued function $F = (f_1, \ldots, f_d)$,

$$\theta_t(F)(x) := -(I + t^2 A)^{-1} t \operatorname{div}((a_{kj}(x))_{k,j}.F(x))$$
$$:= -\sum_{k,j=1}^d (I + t^2 A)^{-1} t D_k(a_{kj}(x) f_j(x))$$

and $\gamma_t(x)$ the $\mathbb{C}^d$-vector with components $-\sum_k (I + t^2 A)^{-1} t D_k a_{kj}(x), 1 \le j \le d$,

$$\gamma_t(x) := \left(-\sum_k (I + t^2 A)^{-1} t D_k a_{kj}(x) \right)_{1 \le j \le d}.$$

Thus, $\theta(\nabla f) = -(I + t^2 A)^{-1} tAf$ and (8.9) can be rewritten as

$$\int_0^\infty \|\theta_t(\nabla f)\|_2^2 \frac{dt}{t} \le C \int_{\mathbb{R}^d} |\nabla f(x)|^2 dx. \tag{8.10}$$

---

[5]Note that one may assume that the coefficients a_{kj} are smooth, and hence Af in (8.9) makes sense. The constants involved in the estimates must however be independent of the smoothness of the coefficients.

Denote by P_t the operator of convolution with $t^{-d}p(\frac{x}{t})$ where p is a smooth function supported in the unit ball of $\mathbb{R}^d$ and such that $\int_{\mathbb{R}^d} p(x)dx = 1$. Using the Gaffney type estimates:[6]

$$\int_{F_0} |(I+t^2 A)^{-1}f|^2 dx + \int_{F_0} |t\nabla(I+t^2 A)^{-1}f|^2 dx \le Ce^{-cd_0/t}\int_{E_0} |f|^2 dx$$

for all f with $\text{supp}(f) \subseteq E_0$,

$$\int_{F_0} |t(I+t^2 A)^{-1}divF|^2 dx \le Ce^{-cd_0/t}\int_{E_0} |F|^2 dx$$

for all F $\mathbb{C}^d$-valued function with $\text{supp}(F) \subseteq E_0$, one obtains

$$\int_0^\infty \|\gamma_t(x) \cdot (P_t\nabla f) - \theta_t P_t \nabla f\|_2^2 \frac{dt}{t} \le C\|\nabla f\|_2^2. \qquad (8.11)$$

Here and in what follows $u \cdot v := u_1 v_1 + \cdots + u_d v_d$ for $u = (u_1, ..., u_d)$ and $v = (v_1, ..., v_d) \in \mathbb{C}^d$. Using (8.11) and the fact that P_t commutes with partial derivatives one proves

$$\int_0^\infty \|\gamma_t(x) \cdot (P_t^2 \nabla f) - \theta_t \nabla f\|_2^2 \frac{dt}{t} \le C\|\nabla f\|_2^2. \qquad (8.12)$$

Therefore, using the Carleson inequality and (8.12), the desired estimate (8.10) follows from the Carleson estimate

$$\sup_Q \frac{1}{|Q|} \int_Q \int_0^{l(Q)} |\gamma_t(x)|^2 \frac{dxdt}{t} < \infty, \qquad (8.13)$$

where the supremum is taken over all cubes Q of $\mathbb{R}^d$, $l(Q)$ denotes the sidelength of Q, and $|Q|$ is its measure.

The proof of Theorem 8.3 is now reduced to proving that $|\gamma_t(x)|^2 \frac{dxdt}{t}$ is a Carleson measure (which means (8.13)). This is the most technical part of the proof. It relies on ideas of the type "$T(b)$ theorem" for Carleson measures. It is proved by Auscher and Tchamitchian [AuTc98] that if for each cube Q we have a function $F_Q : 5Q \to \mathbb{C}$ with the following properties:

i) $\int_{5Q} |\nabla F_Q|^2 dx \le C|Q|$,

ii) $AF_Q \in L^2(5Q)$ and $\int_{5Q} |AF_Q|^2 dx \le C\frac{|Q|}{l(Q)^2}$,

---

[6]Here E_0, F_0 are any closed sets of $\mathbb{R}^d$ and $d_0 := \text{dist}(E_0, F_0)$ the distance between E_0 and F_0. Gaffney type estimates are off diagonal estimates in an average sense. They hold for general divergence form uniformly elliptic operators with complex-valued coefficients. These type of estimates have been proved by Gaffney for Laplace-Beltrami operators on Riemannian manifolds.

iii) $\displaystyle\sup_Q \int_0^{l(Q)} \int_Q |\gamma_t(x)|^2 \frac{dxdt}{t}$

$\displaystyle\leq C \sup_Q \int_0^{l(Q)} \int_Q |\gamma_t(x) \cdot P_t \nabla F_Q(x)|^2 \frac{dxdt}{t}$

(where C is a positive constant), then $|\gamma_t(x)|^2 \frac{dxdt}{t}$ is a Carleson measure. One can also replace F_Q by a finite number (which does not depend on the cube) of functions $F_{n,Q}$. It is this type of techniques together with stopping time arguments which have been used in [HLM02] and [AHLMT02] to prove the desired Carleson estimate and then obtain Theorem 8.3.

8.3 SOME CONSEQUENCES

We start with the observation that the result of Theorem 8.3 carries over from operators of the type $-\sum_{k,j=1}^d D_j(a_{kj}D_k)$ to operators with lower order terms,

$$A := -\sum_{k,j=1}^d D_j(a_{kj}D_k) + \sum_{k=1}^d b_k D_k - D_k(c_k\cdot) + a_0.$$

The operator A is the operator associated with the form

$$\mathfrak{a}(u,v) = \sum_{k,j=1}^d \int_{\mathbb{R}^d} a_{kj} D_k u \overline{D_j v} + \sum_{k=1}^d \int_{\mathbb{R}^d} b_k \overline{v} D_k u + c_k u \overline{D_k v} + a_0 u \overline{v} dx$$

(8.14)

with domain $D(\mathfrak{a}) = H^1(\mathbb{R}^d)$. We assume that (8.5) holds and that the coefficients $b_k, c_k, a_0 \in L^\infty(\mathbb{R}^d, dx, \mathbb{C})$.

By adding a positive constant λ to a_0 if necessary, we obtain an accretive form. Its associated operator $A + \lambda I$ is accretive and we have

COROLLARY 8.4 *The following equalities hold for some constant λ:*

$$D((A + \lambda I)^{1/2}) = D((A^* + \lambda I)^{1/2}) = H^1(\mathbb{R}^d). \qquad (8.15)$$

Therefore, the norms $\|(A+\lambda I)^{1/2}u\|_2 + \|u\|_2$ and $\|u\|_{H^1(\mathbb{R}^d)}$ are equivalent.

Proof. We follow the same idea as in the proof of Proposition 1 in [AuTc92]. Denote by $B = -\sum_{k,j=1}^d D_j(a_{kj}D_k)$ the principal part of A, by $\mathfrak{a}_1$ the sesquilinear form obtained from $\mathfrak{a}$ by taking $c_k = 0$ for each k. Denote by B_1 the operator associated with $\mathfrak{a}_1$. It follows from the definition of the operator associated with a form that $D(B) = D(B_1)$. From Theorem 8.3 and a previously mentioned result of Lions, it follows that

$$D((B_1 + \lambda I)^{1/2}) = [L^2, D(B_1)]_{[1/2]} = [L^2, D(B)]_{[1/2]} = H^1(\mathbb{R}^d).$$
(8.16)

This and Theorem 8.2 give

$$H^1(\mathbb{R}^d) = D((B_1 + \lambda I)^{*1/2}).$$

Since B_1^* is the operator associated with the adjoint form $\mathfrak{a}_1^*$ and

$$\mathfrak{a}^*(u, v) = \mathfrak{a}_1^*(u, v) + \sum_{k=1}^{d} \int_{\mathbb{R}^d} \overline{c_k v} D_k u \, dx$$

we obtain as in (8.16) that $D((A^* + \lambda I)^{1/2}) = H^1(\mathbb{R}^d)$. We conclude again by Theorem 8.2.   □

Denote again by

$$A := -\sum_{k,j=1}^{d} D_j(a_{kj} D_k) + \sum_{k=1}^{d} b_k D_k - D_k(c_k \cdot) + a_0$$

the uniformly elliptic operator associated with the form $\mathfrak{a}$ defined in (8.14). The following theorem shows that the Riesz transforms associated with $A + \lambda I$ are bounded on $L^p(\mathbb{R}^d, dx)$ for $1 < p \leq 2$, provided the heat kernel of A satisfies a Gaussian upper bound.

THEOREM 8.5 *Assume that the heat kernel $p(t, x, y)$ of A has a Gaussian upper bound[7]:*

$$|p(t, x, y)| \leq Ct^{-d/2} \exp\left[-c\frac{|x - y|^2}{t}\right] e^{wt},$$

where $C, c,$ and w are positive constants. Then there exists a positive constant λ such that for each $k \in \{1, ..., d\}$ the Riesz transform $D_k(A + \lambda I)^{-1/2}$ is a weak type (1,1) and extends from $L^2(\mathbb{R}^d, dx) \cap L^p(\mathbb{R}^d, dx)$ to a bounded operator on $L^p(\mathbb{R}^d, dx)$ for each $p \in (1, 2]$.

The proof is the same as that of Theorem 7.26. One of the main ingredient needed there is the boundedness of the Riesz transform $D_k(A + \lambda I)^{-1/2}$ on $L^2(\mathbb{R}^d, dx)$, which now holds since $D((A + \lambda I)^{1/2}) = H^1(\mathbb{R}^d)$ by the previous corollary.

A consequence of Theorem 8.5 is that $D(A_p^{1/2}) \subseteq W^{1,p}(\mathbb{R}^d)$ for $1 < p \leq 2$, where $-A_p$ is the generator on $L^p(\mathbb{R}^d, dx)$ of the semigroup induced by $(e^{-tA})_{t \geq 0}$ (see the comments following the proof of Theorem 7.26).

---

[7]By Theorem 6.10, a Gaussian upper bound holds if the coefficients a_{kj} are real-valued.

As mentioned in the notes of Chapter 7, if the heat kernel is not supposed to satisfy a Gaussian upper bound, one obtains the boundedness of $D_k(A + \lambda I)^{-1/2}$ on $L^p(\mathbb{R}^d, dx)$ for $p \leq 2$ and close enough to 2, see Blunck and Kunstmann [BlKu03a] and Hofmann and Martell [HoMa03].

The reverse inequality $\|A^{1/2}u\| \leq c\|\nabla u\|_p$ can be found in Auscher et al. [AHLMT02], Auscher and Tchamitchian [AuTc98] and Auscher [Aus03].

Let now $A := -\sum_{k,j=1}^{d} D_j(a_{kj}D_k)$. If the heat kernel of A satisfies a Gaussian upper bound

$$|p(t, x, y)| \leq Ct^{-d/2} \exp\left[-c\frac{|x - y|^2}{t}\right] \text{ for all } t > 0$$

then the semigroup $(e^{-tA^{1/2}})_{t \geq 0}$ generated by $-A^{1/2}$ extends to a uniformly bounded semigroup on $L^p(\mathbb{R}^d, dx), 1 \leq p < \infty$. Consider now the Neumann problem

$$\begin{cases} \frac{\partial^2 u(t)}{\partial t^2} = Au(t), \ t \geq 0, \\ \frac{\partial u(t)}{\partial t}\Big|_{t=0} = f, \end{cases}$$

The formal solution is $u(t) = -e^{-tA^{1/2}}A^{-1/2}f$. As mentioned in Auscher and Tchamitchian [AuTc98], by Theorem 8.5 one immediately has for each $p \in (1, 2]$

$$\||\nabla u(t)|\|_p \leq c\|A^{1/2}u(t)\|_p = c\|e^{-tA^{1/2}}f\|_p \leq c'\|f\|_p.$$

Thus

$$\sup_{t>0} \left(\|\frac{\partial u(t)}{\partial t}\|_p + \||\nabla u(t)|\|_p\right) \leq C\|f\|_p.$$

If the operator A has lower order terms, this estimate holds provided one replaces A by $A + \lambda I$ for some positive constant λ.

Bibliography

[Ada75] R.A. Adams, *Sobolev Spaces*, Pure and Applied Mathematics 65. Academic Press, 1975.

[ARS95] S. Albeverio, M. Röckner, and W. Stannat, A remark on coercive forms and associated semigroups, in *Partial Differential Operators and Mathematical Physics (Holzhau, 1994)*, Oper. Theory Adv. Appl. 78, Birkhäuser, 1995. 1-8.

[Ale94a] G. Alexopoulos, Oscillating multipliers on Lie groups and Riemannian manifolds, *Tohôku Math. J.* 46 (4) (1994) 457-468.

[Ale94b] G. Alexopoulos, Spectral multipliers on Lie groups of polynomial growth, *Proc. Amer. Math. Soc.* 120 (3) (1994) 973-979.

[Ale99] G. Alexopoulos, L^p bounds for spectral multipliers from Gaussian estimates of the heat kernel, to appear in *Proceedings of the ICMS Instructional Conference on Lie Groups and Partial Differential Equations, Edinburgh 1999*.

[Ama83] H. Amann, Dual semigroups and second order linear elliptic boundary value problems, *Israël J. Math.* 45 (2-3) (1983) 225-254.

[Ama95] H. Amann, *Linear and Quasilinear Parabolic Problems. Vol. I, Abstract Linear Theory*. Monographs in Mathematics 89, Birkhäuser 1995.

[AmEs96] H. Amann and J. Escher, Strongly continuous dual semigroups, *Ann. Mat. Pura Appl.* 171 (4) (1996) 41-62.

[Anc75] A. Ancona, Sur les espaces de Dirichlet: principes, fonctions de Green, *J. Math. Pures et Appl.* 54 (9) (1975) 75-124.

[Are84] W. Arendt, Kato's inequality: a characterization of generators of positive semigroups, *Proc. Royal Irish Acad.* 84 (1984) 155-174.

[Are87] W. Arendt, Vector valued Laplace transform and Cauchy problems, *Israël J. Math.* 59 (3) (1987) 327-352.

[ArBé92] W. Arendt and Ph. Bénilan, Inégalité de Kato et semi-groupes sous-markoviens, *Rev. Mat. Univ. Complut. Madrid* 5 (2-3) (1992) 279-308.

[Are94] W. Arendt, Gaussian estimates and interpolation of the spectrum in L^p, *Diff. Int. Eq.* 7 (5-6) (1994) 1153-1168.

[AEH97] W. Arendt, O. El Mennaoui, M. Hieber, Boundary values of holomorphic semigroups, *Proc. Amer. Mat. Soc.* 125 (3) (1997) 635-647.

[ArEl97] W. Arendt and A.F.M. ter Elst, Gaussian estimates for second order elliptic operators with boundary conditions, *J. Op. Theory* 38 (1) (1997) 87-130.

[Are01] W. Arendt, Different domains induce different heat semigroups on $C_0(\Omega)$, in *Evolution Equations and Their Applications in Physical and Life Sciences (Bad Herrenalb, 1998)*, Lecture Notes in Pure and Appl. Math. 215, Dekker, 2001. 1-14.

[ABHN01] W. Arendt, C.J.K. Batty, M. Hieber, and F. Neubrander, *Vector-Valued Laplace Transforms and Cauchy Problems*, Monographs in Mathematics 96, Birkhäuser, 2001.

[ArWa03] W. Arendt and M. Warma, Dirichlet and Neumann boundary conditions: what is in between? *J. Evol. Eq.* 3 (1) (2003) 119-135.

[Aro68] D.G. Aronson, Nonnegative solutions of linear parabolic equations, *Ann. Scuola Norm. Sup. Pisa* 22 (3) (1968) 607-694.

[AuTc92] P. Auscher and Ph. Tchamitchian, Conjecture de Kato sur les ouverts de $\mathbb{R}$, *Rev. Mat. Iberoam.* 8 (2) (1992) 149-199.

[ACT96] P. Auscher, Th. Coulhon, and Ph. Tchamitchian, Absence du principe du maximum pour certaines équations paraboliques complexes, *Coll. Math.* 71 (1) (1996) 87-95.

[AMcT98] P. Auscher, A. McIntosh, and Ph. Tchamitchian, Heat kernel of complex elliptic operator and applications, *J. Funct. Anal.* 152 (1) (1998) 22-73

[AuTc98] P. Auscher and Ph. Tchamitchian, *Square Root Problem for Divergence Operators and Related Topics*, Astérisque 249, Soc. Math. France, 1998.

[ABBO00] P. Auscher, L. Barthélemy, Ph. Bénilan, and E.M. Ouhabaz, Absence de la L^∞-contractivité pour les semi-groupes associés aux opérateurs elliptiques complexes sous forme divergence, *Potential Anal.* 12 (2) (2000) 169-189.

[AuTc01] P. Auscher and Ph. Tchamitchian, Gaussian estimates for second order elliptic divergence operators on Lipschitz and C^1 domains, in *Evolution Equations and Their Applications in Physical and Life Sciences (Bad Herrenalb, 1998)*, Lecture Notes in Pure and Appl. Math. 215, Dekker, 2001. 15-32.

[AHLT01] P. Auscher, S. Hofmann, J. Lewis, and Ph. Tchamitchian, Extrapolation of Carleson measures and the analyticity of Kato's square-root operators, *Acta Math.* 187 (2001) 161-190.

[AHLMT02] P. Auscher, S. Hofmann, M. Lacey, A. McIntosh, and Ph. Tchamitchian, The solution of the Kato square root problem for second order elliptic operators on $\mathbb{R}^n$, *Ann. of Math.* 156 (2) (2002) 633-654.

[ACDH03] P. Auscher, Th. Coulhon, X.T. Duong, and S. Hofmann, Riesz transforms on manifolds and heat kernel regularity. In preparation.

[Aus03] P. Auscher, On necessary and sufficient conditions for L^p-estimates of Riesz transforms associated to elliptic operators on $\mathbb{R}^n$ and related estimates. Preprint 2003.

[Bak87] D. Bakry, Etude des transformations de Riesz dans les variétés riemanniennes à courbure de Ricci minorée, in *Séminaire de Probabilités XXI,* Lecture Notes in Math. 1247. Springer, 1987. 137-172.

[Bak89] D. Bakry, Sur l'interpolation complexe des semi-groupes de diffusion, in *Séminaire de Probabilités XXIII,* Lecture Notes in Math. 1372. Springer, 1989. 1-20.

[BaEm85] M. Balabane and H. Emamirad, L^p estimates for Schrödinger evolution equations, *Trans. Amer. Math. Soc.* 292 (1) (1985) 357-373.

[Bar98] M. Barlow, Diffusions on fractals, in *Lectures on Probability Theory and Statistics*, Ecole d'été de Probabilités de Saint Flour, XXV 1995, Lecture Notes in Math. 1690, Springer, 1998. 1-121.

[Bar96] L. Barthélemy, Invariance d'un ensemble convexe fermé par un semi-groupe associé à une forme non-linéaire, *Abst. Appl. Anal.* 1 (3) (1996) 237-262.

[BeSt93] M. Beals and W. Strauss, L^p estimates for the wave equation with a potential, *Comm. P. D. E.* 18 (7-8) (1993) 1365-1397.

[Bén78] Ph. Bénilan, Opérateurs accrétifs et semi-groupes dans les espaces $L^p(1 \leq p \leq +\infty)$, in *Functionnal Analysis and Numerical Analysis, Japan-France Seminar,* H. Fujita (ed), Japan Society for the Advancement of Science, 1978.

[BCP90] Ph. Bénilan, M. Crandall, and A. Pazy, *Nonlinear Evolution Equations in Banach Spaces,* (unpublished manuscript) Besançon, 1990.

[BéCr91] Ph. Bénilan and M. Crandall, Completely accretive operators, in *Semigroup Theory and Evolution Equations,* Ph. Clement, B. de Pagter, and E. Mitidieri (Eds). Lecture Notes Pure Appl. Math. 135, Marcel Dekker, 1991. 41-75.

[Bér86] P.H. Bérard, *Spectral Geometry: Direct and Inverse Problems,* Lecture Notes in Math. 1207, Springer, 1986.

[BeLö76] J. Bergh and J. Löfström, *Interpolation Spaces,* Springer-Verlag, 1976.

[BeDe58] A. Beurling and J. Deny, Espaces de Dirichlet, *Acta Math.* 99 (1958) 203-224.

[BeDe59] A. Beurling and J. Deny, Dirichlet spaces, *Proc. Nat. Acad. Sci. U.S.A.* 45 (1959) 208-215.

[BlKu03a] S. Blunck and P.C. Kunstmann, Weak type (p, p) estimates for Riesz transforms, *Math. Z.* 247 (1) (2004) 137-148.

[BlKu03b] S. Blunck and P.C. Kunstmann, Calderon-Zygmund theory for non-integral operators and the H^∞ functional calculus, *Rev. Mat. Iberoam.* 19 (3) (2003) 919-942.

[Blu03a] S. Blunck, Generalized Gaussian estimates and Riesz means of Schrödinger groups. Preprint, 2003.

[Blu03b] S. Blunck, A Hörmander-type spectral multiplier theorem for operators without heat kernel, *Ann. Scuola Norm. Sup. Pisa Cl. Sci.* 2 (5) (2003) 449-459.

[BoHi91] N. Bouleau and F. Hirsch, *Dirichlet Forms and Analysis on Wiener Space,* De Gruyter Stud. Math. 14. De Gruyter, 1991.

[BodeL93] K. Boyadzhiev and R. deLaubenfels, Boundary values of holomorphic semigroups, *Proc. Amer. Math. Soc.* 118 (1) (1993) 113-118.

[Bre92] H. Brezis, *Analyse Fonctionnelle*, 3rd edition, Masson, 1992.

[BrPa70] H. Brezis and A. Pazy, Semigroups of non-linear contractions on
convex sets, *J. Funct. Anal.* 6 (1970) 237-281.

[CKS87] E. Carlen, S. Kusuoka, and D.W. Stroock, Upper bounds for symmetric Markov transition functions, *Ann. Inst. H. Poincaré, Probab. et
Statist.* 23, suppl. au no 2 (1987) 245-287.

[CCO02] G. Carron, Th. Coulhon, and E.M. Ouhabaz, Gaussian estimates
and L^p-boundedness of Riesz means, *J. Evol. Eq.* 2 (3) (2002) 299-317.

[ChKr64] R.V. Chacon and U. Krengel, Linear modulus of a linear operator, *Proc. Amer. Math. Soc.* 15 (1964) 553-559.

[Cha84] I. Chavel, *Eigenvalues in Riemannian Geometry*, Academic Press
1984.

[Chr91a] M. Christ, *Lectures on Singular Integral Operators*, CBMS, 1991.

[Chr91b] M. Christ, L^p bounds for spectral multipliers on nilpotent groups,
Trans. Amer. Math. Soc. 328 (1) (1991) 73-81.

[CHADP87] Ph. Clément, H.J.A.M. Heijmans, S. Angenent, C.J. van
Duijn, and B. van de Pagter, *One-parameter Semigroups*, North-
Holland, 1987.

[CoWe71] R.R. Coifman and G. Weiss, *Analyse Harmonique non-
Commutative sur Certains Espaces Homogènes*, Lecture Notes in Math.
242, Springer-Verlag 1971.

[CMM82] R. Coifman, A. McIntosh, and Y. Meyer, L'intégrale de Cauchy
définit un opérateur borné sur L^2 pour les courbes lipschitziennes, *Ann.
of Math.* 116 (2) (1982) 361-387.

[CDM83] R. Coifman, D. Deng, and Y. Meyer, Domaine de la racine
carrée de certains opérateurs différentiels accrétifs, *Ann. Institut Fourier
(Grenoble)* 33 (2) (1983) 123-134.

[Cou91] Th. Coulhon, Dimensions of continuous and discrete semigroups,
in *Semigroup Theory and Evolution Equations,* Ph. Clement, B. de
Pagter, and E. Mitidieri (eds.). Lecture Notes Pure Appl. Math. 135,
Marcel Dekker, 1991. 93-99.

[Cou92] Th. Coulhon, Inégalités de Gagliardo-Nirenberg pour les semi-groupes d'opérateurs et applications, *Potential Anal.* 1 (4) (1992) 343-353.

[Cou93] Th. Coulhon, Itération de Moser et estimation gaussienne du noyau de la chaleur, *J. Operator Theory* 29 (1) (1993) 157-165.

[Cou96] Th. Coulhon, Ultracontractivity and Nash type inequalities, *J. Func. Anal.* 141 (2) (1996) 510-539.

[Cou03] Th. Coulhon, Off-diagonal heat kernel lower bounds without Poincaré, *J. London Math. Soc.* 2nd Ser. 68 (3) (2003) 795-816.

[CoDu99] Th. Coulhon and X.T. Duong, Riesz transforms for $1 \leq p \leq 2$, *Trans. Amer. Math. Soc.* 351 (3) (1999) 1151-1169.

[CoDu00] Th. Coulhon and X.T. Duong, Maximal regularity and kernel bounds: observations on a theorem by Hieber and Prüss, *Adv. Diff. Eq.* 5 (1-3) (2000) 343-368.

[CFKS87] H.L. Cycon, R.G. Froese, W. Kirsch, and B. Simon, *Schrödinger Operators*, Texts and Monographs in Physics, Springer-Verlag, 1987.

[Dan00] D. Daners, heat kernel estimates for operators with boundary conditions, *Math. Nachr.* 217 (2000) 13-41.

[DaLi88] R. Dautray and J.L. Lions, *Analyse Mathématiques et Calcul Numérique*, Masson, 1988.

[DJS85] G. David, J.L. Journé, and S. Semmes, Opérateurs de Calderón-Zygmund, fonctions para-accrétives et interpolation, *Rev. Mat. Iberoam.* 1 (4) (1985) 1-56.

[Dav80] E.B. Davies, *One-parameter Semigroups*, Academic Press, 1980.

[Dav85] E.B. Davies, L^1 properties of second order elliptic operators, *Bull. London Math. Soc.* 17 (5) (1985) 417-436.

[Dav87] E.B. Davies, Explicit constants for Gaussian upper bounds on heat kernels, *Amer. J. Math.* 109 (2) (1987) 319-333.

[Dav89] E.B. Davies, *Heat Kernels and Spectral Theory*, Cambridge Tracts in Math. 92, Cambridge Univ. Press, 1989.

[Dav95a] E.B. Davies, The functional calculus, *J. London Math. Soc.* 2nd Ser. 52 (1995) 166-176.

[Dav95b] E.B. Davies, L^p spectral independence and L^1 analyticity, *J. London Math. Soc.* 2nd Ser. 52 (1995) 177-184.

[Dav95c] E.B. Davies, Uniformly elliptic operators with measurable coefficients, *J. Funct. Anal.* 132 (1) (1995) 141-169.

[Dav95d] E.B. Davies, *Spectral Theory and Differential Operators*, Cambridge Studies in Advanced Math. 42, Cambridge Univ. Press, 1995.

[Dav97a] E.B. Davies, Limits on L^p regularity of self-adjoint elliptic operators, *J. Diff. Eq.* 135 (1) (1997) 83-102.

[Dav97b] E.B. Davies, L^p spectral theory of higher-order elliptic differential operators, *Bull. London Math. Soc.* 29 (5) (1997) 513-546.

[DaPa89] E.B. Davies and M.M.H. Pang, Sharp heat kernel bounds for some Laplace operators, *Quart. J. Math.* 2nd Ser. 40 (159) (1989) 281-290.

[deL93] R. deLaubenfels, *Existence Families, Functional Calculi and Evolution Equations*, Lecture Notes in Math. 1570, Springer, 1993.

[Dev78] A. Devinatz, Self-adjointness of second order degenerate-elliptic operators, *Indiana Univ. Math. J.* 27 (2) (1978) 255-266.

[DuMc99a] X.T. Duong and A. McIntosh, Singular integral operators with non-smooth kernels on irregular domains, *Rev. Math. Iberoam.* 15 (2) (1999) 233-265.

[DuMc99b] X.T. Duong and A. McIntosh, The L^p-boundedness of Riesz transforms associated with divergence form operators, *Proc. of the Center for Mathematics and its Applications,* Australian National University, Canberra, 1999.

[DuOu99] X.T. Duong and E.M. Ouhabaz, Complex multiplicative perturbations of elliptic operators: heat kernel bounds and holomorphic functional calculus, *Diff. Integ. Eq.* 12 (3) (1999) 395-418.

[DOS02] X.T. Duong, E.M. Ouhabaz and A. Sikora, Plancherel estimates and sharp spectral multipliers, *J. Funct. Anal.* 196 (2) (2002) 443-485.

[DuRo96] X.T. Duong and D.W. Robinson, Semigroup kernels, Poisson bounds and holomorphic functional calculus, *J. Funct. Anal.* 142 (1) (1996) 89-128.

[ElM92] O. El Mennaoui, *Trace des Semi-groupes Holomorphes Singuliers à l'Origine et Comportement Asymptotique*, Thesis, Université de Franche-Comté, Besançon 1992.

[ElKe96] O. El Mennaoui and V. Keyantuo, On the Schrödinger equation in L^p spaces, *Math. Ann.* 304 (2) (1996) 293-302.

[EnNa99] K. Engel and R. Nagel, *One-parameter Semigroups for Linear Evolution Equations*, Graduate Texts in Math. 194, Springer, 1999.

[Epp89] J.B. Epperson, The hypercontractive approach to exactly bounding an operator with complex Gaussian kernel, *J. Funct. Anal.* 87 (1) (1989) 1-30.

[FJK85] E.B. Fabes, D. Jerison, and C. Kenig, Multilinear square functions and partial differential equations, *Amer. J. Math.* 107 (6) (1985) 1325-1368.

[FKS82] E.B. Fabes, C. Kenig, and R. Serapioni, The local regularity of solutions of degenerate elliptic equations, *Comm. Partial Differential Equations* 7 (1) (1982) 77-116.

[FaSt86] E.B. Fabes and D.W. Stroock, A new proof of Moser's parabolic Harnack inequality via the old ideas of Nash, *Arch. Rat. Mech. Anal.* 96 (4) (1986) 327-338.

[Fat83] H. O. Fattorini, *The Cauchy Problem*, Addison-Wesley, 1983.

[Fic56] G. Fichera, Sulle equazioni differentiali lineari ellipttico-paraboliche del secondo ordine, *Atti Accad. Naz. Lincei Mem. Cl.Sci. Fis. Mat. Nat. Sez. I* 5 (8) (1956) 1-30.

[FSSC98] B. Franchi, R. Serapioni, and F. Serra Cassano, Irregular solutions of linear degenerate elliptic equations, *Potential Anal.* 9 (3) (1998) 201-216.

[FrSe87] B. Franchi and R. Serapioni, Pointwise estimates for a class of strongly degenerate elliptic operators: a geometrical approach, *Ann. Scuola Norm. Sup. Pisa Cl. Sci.* 14 (4) (1987) 527-568.

[Fuk80] M. Fukushima, *Dirichlet Forms and Markov Processes*, North-Holland, 1980.

[FOT94] M. Fukushima, Y. Oshima, and M. Takeda, *Dirichlet Forms and Markov Processes*, De Gruyter Stud. Math. 19, De Gruyter 1994.

[GaPy97] J.E. Galé and T. Pytlik, Functional calculus for infinitesimal generators of holomorphic semigroups, *J. Funct. Anal.* 150 (2) (1997) 307-355.

[GiTr77] D. Gilbarg and N.S. Trudinger, *Elliptic Partial Differential Equations of Second Order*, Grund. Math. Wiss. 224, Springer, 1977.

[Gol85] J.A. Goldstein, *Semigroups of Linear Operators and Applications*, Oxford Univ. Press, 1985.

[Gri91] A. Grigor'yan, The heat equation on non-compact Riemannian manifolds, *Matem. Sbornik* 182 (1) (1991) 55-87. English translation: *Math. USSR Sb.* 72 (1) (1992) 47-77.

[Gri95] A. Grigor'yan, Upper bounds of derivatives of the heat kernel on an arbitrary complete manifold, *J. Funct. Anal.* 127 (2) (1995) 363-389.

[Gri97] A. Grigor'yan, Gaussian upper bounds for the heat kernel on arbitrary manifolds, *J. Diff. Geom.* 45 (1) (1997) 33-52.

[GHL03] A. Grigor'yan, J. Hu, and K.S. Lau, Heat kernels on metric-measure spaces and an application to semi-linear elliptic equations, *Trans. Amer. Math. Soc.* 355 (5) (2003) 2065-2095.

[Heb90] W. Hebisch, A multiplier theorem for Shrödinger operators, *Coll. Math.* 60/61 (2) (1990) 659-664.

[Heb95] W. Hebisch, Functional calculus for slowly decaying kernels. Preprint 1995.

[HeSj89] B. Helffer and J. Sjöstrand, Equation de Schrödinger avec champ magnétique et équation de Harper, in *Schrödinger Operators*, H. Holden and A. Jensen (eds.), Lecture Notes in Physics 345, Springer-Verlag 1989. 118-197.

[HeVo86] R. Hempel and J. Voigt, The spectrum of a Schrödinger operator in $L^p(\mathbb{R}^N)$ is p-independent, *Comm. Math. Phys.* 104 (2) (1986) 243-250.

[HSU77] H. Hess, R. Schrader, and D.A. Uhlenbrock, Domination of semigroups and generalization of Kato's inequality, *Duke Math. J.* 44 (4) (1977) 893-904.

[HSU80] H. Hess, R. Schrader, and D.A. Uhlenbrock, Kato's inequality and the spectral distribution of Laplacians on compact Riemannian manifolds, *J. Diff. Geom.* 15 (1) (1980) 27-37.

[Hie96] M. Hieber, Gaussian estimates and holomorphy of semigroups on L^p spaces. *J. London Math. Soc.* 2nd Ser. 54 (1) (1996), 148-160.

[HiPr97] M. Hieber and J. Prüss, Heat kernels and maximal $L^p - L^q$ estimates for parabolic evolution equations, *Comm. P.D.E.* 22 (9-10) (1997) 1647-1669.

[HiPh57] E. Hille and R.S. Phillips, *Functional Analysis and Semigroups*, Coll. Publ. 31, American Math. Society, 1957.

[HLM02] S. Hofmann, M. Lacey, and A. McIntosh, The solution of the Kato problem for divergence form elliptic operators with Gaussian heat kernel bounds, *Ann. of Math.* 2nd Ser. 156 (2) (2002) 623-631.

[HoMa03] S. Hofmann and J.M. Martell, L^p bounds for Riesz transforms and square roots associated to second order elliptic operators, *Pub. Mat.* 47 (2003) 497-515.

[Hör60] L. Hörmander, Estimates for translation invariant operators in L^p spaces, *Acta Math.* 104 (1960) 93-140.

[Jac01] N. Jacob: Generators of feller semigroups as generators of L^p-sub-Markovian semigroups, in *Evolution Equations and Their Applications in Physical and Life Sciences (Bad Herrenalb, 1998)*, Lecture Notes in Pure and Appl. Math. 215, Dekker, 2001. 15-32,

[JeNa94] A. Jensen and S. Nakamura, Mapping properties of functions of Schrödinger operators between L^p spaces and Besov spaces, in *Spectral and Scattering Theory and Applications*, K. Yajima (ed.), Adv. Stud. Pure Math. 23, Birkhaüser, 1994. 187-209.

[Jou91] J.L. Journé, Remarks on Kato's square-root problem, *Publ. Mat.* 35 (1991) 299-321.

[Kar01] S. Karrmann, Gaussian estimates for second order operators with unbounded coefficients, *J. Math. Anal. Appl.* 258 (1) (2001) 320-348.

[Kat55] T. Kato, Quadratic forms in Hilbert space and asymptotic perturbation series. Technical report No. 7, Univ. of Calif. 1955.

[Kat61] T. Kato, Fractional powers of dissipative operators, *J. Math. Soc. Japan* 13 (1961) 246-274.

[Kat62] T. Kato, Fractional powers of dissipative operators II, *J. Math. Soc. Japan* 14 (1962) 242-248.

[Kat73] T. Kato, Schrödinger operators with singular potentials, *Israël J. Math.* 13 (1973) 135-148.

[Kat78] T. Kato, Trotter's product formula for an arbitrary pair of self-adjoint contraction semigroups, in *Topics in Functional Analysis,* Adv. in Math. Suppl. Studies 3, Academic Press, 1978, 185-195.

[Kat80] T. Kato, *Perturbation Theory for Linear Operators*, 2nd edition, Springer-Verlag, 1980.

[Kip74] C. Kipnis, Majoration des semi-groupes de contractions de L^1 et applications, *Ann. Inst. H. Poincaré* B 10 (1974) 369-384.

[KrMa94] G.I. Kresin and V.G. Maz'ya, Criteria for validity of maximum modulus principle for solutions of linear parabolic systems, *Arkiv. Mat.* 32 (1) (1994) 121-155.

[Kub75] Y. Kubokawa, Ergodic theorems for contraction semigroups, *J. Math. Soc. Japan* 27 (1975) 184-193.

[Kun99] P.C. Kunstmann, Heat kernel estimates and L^p spectral independence of elliptic operators, *Bull. London Math. Soc.* 31 (3) (1999) 345-353.

[Kun02] P.C. Kunstmann, L^p-spectral properties of the Neumann Laplacian on horns, comets and stars, *Math. Z.* 242 (1) (2002) 183-201.

[Lan68] E. Lanconelli, Valutazioni in $L^p(\mathbb{R}^n)$ della soluzione del problema di Cauchy per l'equazione di Schrödinger, *Boll. Un. Mat. Ital.* 4 (1968) 591-607.

[Lan99] M. Langer, L^p-contractivity of semigroups generated by parabolic matrix differential operators, in *The Maz'ya Anniversary Collection*, Op. Theory: Adv. and App. 109, Birkhäuser, 1999, 307- 330.

[LaMi54] P. Lax and A. Milgram, Parabolic equations, in *Contributions to the Theory of Partial Differential Equations,* L. Bers, S. Bochner, and F. John (Eds.), Ann. Math. Studies 33, Princeton University Press, 1954, 167-190.

[LiYa86] P. Li and S.T. Yau, On the parabolic kernel of the Schrödinger operator, *Acta Math.* 156 (3-4) (1986) 153-201.

[Lio61] J.L. Lions, *Équations Différentielles Opérationnelles et Problèmes aux Limites*, Grund. Math. Wiss. III, Springer 1961.

[Lio62] J.L. Lions, Espaces d'interpolation et domaine de puissances fractionnaires d'opérateurs, *J. Math. Soc. Japan* 14 (1962) 233-241.

[LiSe93] V. A. Liskevich and Yu.A. Semenov, Some inequalities for submarkovian generators and their applications to perturbation theory, *Proc. Amer. Math. Soc.* 119 (4) (1993) 1171-1177.

[LiPe95] V.A. Liskevich and M.A. Perelmuter, Analyticity of submarkovian semigroups, *Proc. Amer. Math. Soc.* 123 (4) (1995) 1097-1104.

[LiSe96] V. A. Liskevich and Yu.A. Semenov, Some problems on Markov semigroups, in *Schrödinger Operators, Markov Semigroups, Wavelet Analysis, Operator Algebras*, M. Demuth et al. (eds.), Math. Top. 11, Akademie Verlag 1996. 163-217,

[Lis96] V. A. Liskevich, On C_0-semigroups generated by elliptic second order differential expressions on L^p-spaces, *Diff. Integ. Eqs.* 9 (4) (1996) 811-826.

[LSV02] V. Liskevich, Z. Sobol, and H. Vogt, On the L_p-theory of C_0-semigroups associated with second-order elliptic operators, II, *J. Funct. Anal.* 193 (1) (2002) 55-76.

[Loh92] N. Lohoué, Estimations des sommes de Riesz d'opérateurs de Schrödinger sur les variétés riemanniennes et les groupes de Lie, *C.R.A.S. Paris* 315 (1992) 13-18.

[LuPh61] G. Lumer and R.S. Phillips, Dissipative operators in a Banach space, *Pacific J. Math.* 11 (1961) 679-698.

[Lun95] A. Lunardi, *Analytic Semigroups and Optimal Regularity in Parabolic Equations*, Progr. Nonlin. Diff. Eq. Appl. 16, Birkhäuser, 1995.

[MaRö92] Z.M. Ma and M. Röckner, *Introduction to (Non-Symmetric) Dirichlet Forms*, Universitext, Springer-Verlag 1992.

[MVV01] A. Manavi, H. Vogt, and J. Voigt, Domination of semigroups associated with sectorial forms, Preprint 2001.

[MSW80] B. Marshall, W. Strauss, and S. Wainger, $L^p - L^q$ estimates for the Klein-Gordon equation, *J. Math. Pures et App.* 59 (1980) 417-440.

[MaMe90] G. Mauceri and S. Meda, Vector-valued multipliers on stratified groups, *Rev. Mat. Iberoam.* 6 (3-4) (1990) 141-154.

[Maz85] V. G. Maz'ja, *Sobolev Spaces*, Springer, 1985.

[MNP85] V. G. Maz'ja, S.A. Nazarov, and B.A. Plamenevskii, Absence of De Giorgi type theorems for strongly elliptic equations with complex coefficients, *J. Math. Sov.* 28 (1985) 726-739.

[McI72] A. McIntosh, On the comparability of $A^{1/2}$ and $A^{*1/2}$, *Proc. Amer. Math. Soc.* 32 (1972) 430-434.

[McI85] A. McIntosh, Square roots of elliptic operators, *J. Funct. Anal.* 61 (3) (1985) 307-327.

[McI90] A. McIntosh, The square root problem for elliptic operators: a survey, in *Functional-Analytic Methods For Partial Differential Equations (Tokyo, 1989),* H. Fujita, T. Ikebe, and S.T. Kuroda (eds.), Lecture Notes in Math. 1450, Springer, 1990, 122-140.

[MOR03] M. Melgaard, E.M. Ouhabaz, and G. Rozenblum, Lieb-Thirring inequality for the Aharonov-Bohm Hamiltonian and eigenvalue estimates, Mittag-Leffler Institut, Report No. 4, 2002/2003, to appear in *Ann. Inst. H. Poincaré.*

[MPRS02] G. Metafune, J. Prüss, A. Rhandi, and R. Schnaubelt, L^p-regularity for elliptic operators with unbounded coefficients, Preprint 2002, Martin-Luther Universität, Halle-Wittenberg.

[Miy80] A. Miyachi, On some estimates for the wave equation in L^p and H^p, *J. Fac. Sci. Tokyo Sci. IA* 27 (2) (1980) 331-354.

[MüSt99] D. Müller and E.M. Stein, L^p-estimates for the wave equation on the Heisenberg group, *Rev. Mat. Iberoam.* 15 (2) (1999) 297-334.

[Nag86] R. Nagel (ed.), *One Parameter Semigroups of Positive Operators*, Lecture Notes in Math. 1184, Springer, 1986.

[NaVo96] R. Nagel and J. Voigt, On inequalities for symmetric submarkovian operators, *Arch. der Math.* 67 (4) (1996) 308-311.

[Nel64] E. Nelson, Interaction of non-relativistic particles with a quantized scalar field, *J. Math. Phys.* 5 (1964) 1190-1197.

[NoSt91] J.R. Norris and D.W. Stroock, Estimates on the fundamental solution to heat flows with uniformly elliptic coefficients, *Proc. London Math. Soc.* 3rd Ser. 62 (2) (1991) 373-402.

[Oka80] N. Okazawa, Singular perturbations of m-accretive operators, *J. Math. Soc. Japan* 32 (1) (1980) 19-44.

[Oka91] N. Okazawa, Sectorialness of second order elliptic operators in divergence form, *Proc. Amer. Math. Soc.* 113 (3) (1991) 701-706.

[Ole67] O.A. Oleinik, Linear equations of second order with nonnegative characteristic form, *Amer. Math. Soc. Translations,* Ser. 2, Amer. Math. Soc. 65, 1967. 167-199.

[Osh92] Y. Oshima, On conservativeness and recurrence criteria for Markov processes, *Potential Anal.* 1 (2) (1992) 115-131.

[Ouh92a] E.M. Ouhabaz, *Propriétés d'Ordre et de Contractivité pour les Semi-Groupes et Applications aux Opérateurs Elliptiques*, PhD Thesis, Université de Franche-Comté, Besançon 1992.

[Ouh92b] E.M. Ouhabaz, L^∞-contractivity of semigroups generated by sectorial forms, *London Math. Soc.* 2nd Ser. 46 (3) (1992) 529-542.

[Ouh93] E.M. Ouhabaz, Semi-groupes sous-markoviens engendrés par des opérateurs matriciels, *Math. Ann.* 296 (4) (1993) 667-676.

[Ouh95] E.M. Ouhabaz, Gaussian estimates and holomorphy of semigroups, *Proc. Amer. Math. Soc.* 123 (5) (1995) 1465-1474.

[Ouh96] E.M. Ouhabaz, Invariance of closed convex sets and domination criteria for semigroups, *Potential Anal.* 5 (6) (1996) 611-625.

[Ouh99] E.M. Ouhabaz, L^p contraction semigroups for vector-valued functions, *Positivity* 3 (1) (1999) 83-93.

[Ouh98] E.M. Ouhabaz, Heat kernels of multiplicative perturbations: Hölder estimates and Gaussian lower bounds, *Indiana Univ. Math. J.* 47 (4) (1998) 1481-1495.

[Ouh01] E.M. Ouhabaz, The spectral bound and principal eigenvalues of Schrödinger operators on Riemannian manifolds, *Duke Math. J.* 110 (1) (2001) 1-35.

[Ouh02] E.M. Ouhabaz, Gaussian upper bounds for heat kernels of second-order elliptic operators with complex coefficients on arbitrary domains, *J. Op. Theory* 51 (2) (2004) 335-360.

[Ouh03a] E.M. Ouhabaz, Sharp Gaussian bounds for heat kernels of second-order divergence form operators. Preprint, 2003.

[Ouh03b] E.M. Ouhabaz, Gaussian upper bounds and L^p estimates for wave equations. Preprint 2003.

[Paz83] A. Pazy, *Semigroups of Linear Operators and Applications to Partial Differential Equations*, Applied Math. Sciences 44, Springer, 1983.

[Per80] J. C. Peral, L^p estimates for the wave equation, *J. Funct. Anal.* 36 (1) (1980) 114-145.

[Phi59] R.S. Phillips, Dissipative operators and hyperbolic systems of partial differential equations, *Trans. Amer. Math. Soc.* 90 (1959) 193-254.

[PrEi84] F.O. Porper and S.D. Eidel'man, Two sided estimates of fundamental solutions of second order parabolic equations and some applications, *Russian Math. Surveys* 39 (3) (1984) 119-178.

[Prü93] J. Prüss, *Evolutionary Integral Equations and Applications*, Monographs in Mathematics 87, Birkhäuser 1993.

[ReSi75] M. Reed and B. Simon, *Methods of Modern Mathematical Physics*, Vol. II. Academic Press, 1975.

[ReSi79] M. Reed and B. Simon, *Methods of Modern Mathematical Physics*, Vol. IV. Academic Press, 1979.

[ReSi80] M. Reed and B. Simon, *Methods of Modern Mathematical Physics*, Vol. I. Academic Press, 1980.

[Rob91] D.W. Robinson, *Elliptic Operators on Lie Groups*, Oxford Univ. Press, 1991.

[Rob96] D.W. Robinson, Basic semigroup theory, *Proceedings of the Center For Mathematics and Its Applications,* 34, Australian National University, 1996.

[RoSo97] G. Rozenblum and M. Solomyak, The Cwikel-Lieb-Rozenblum estimator for generators of positive semigroups and semigroups dominated by positive semigroups, *Algebra i Analiz* 9 (6) (1997) 214-236 (Russian). Translation in *St. Petersbug Math. J.* 9 (6) (1998) 1195-1211.

[Roz00] G. Rozenblum, Domination of semigroups and estimates for eigenvalues, *Algebra i Analiz* 12 (5) (2000) 158-177 (Russian). Translation in *St. Petersbug Math. J.* 12 (5) (2001) 831-845.

[SaC90] L. Saloff-Coste, Analyse sur les groupes de Lie à croissance polynomiale, *Arkiv. Mat.* 28 (2) (1990) 315-331.

[SaC92] L. Saloff-Coste, A Note on Poincaré, Sobolev, and Harnack inequalities, *Duke Math. J., I.M.R.N.* 2 (1992) 27-38.

[SaC95] L. Saloff-Coste, Parabolic Harnack inequality for divergence form second order differential operators, *Potential Anal.* 4 (4) (1995) 429-467.

[SaC02] L. Saloff-Coste, *Aspects of Sobolev-Type Inequalities,* London Math. Soc. Lecture Note Series 289, Cambridge Univ. Press, 2002.

[Sch74] H.H. Schaefer, *Banach Lattices and Positive Operators*, Grund. Math. Wiss. 215, Springer-Verlag, 1974.

[Shi97] I. Shigekawa, L^p contraction semigroups for vector valued functions, *J. Funct. Anal.* 147 (1) (1997) 69-108.

[Shu92] M. Shubin, Spectral theory of elliptic operators on non-compact manifolds, *Astérisque* 207 (1992) 37-108.

[Sik96] A. Sikora, Sharp pointwise estimates on heat kernels, *Quart. J. Math.* 2nd Ser. 47 (187) (1996) 371-382.

[Sim78] B. Simon, A canonical decomposition for quadratic forms with applications to monotone convergence theorems. *J. Funct. Anal.* 28 (3) (1978) 377-385.

[Sim79] B. Simon, Kato's inequality and the comparison of semigroups, *J. Funct. Anal.* 32 (1) (1979) 97-101.

[Sim82] B. Simon, Schrödinger semigroups, *Bull. Amer. Math. Soc.* 7 (3) (1982) 447-526.

[Sjö70] S. Sjöstrand, On the Riesz means of the solutions of the Schrödinger equation, *Ann. Scuola Norm. Sup. Pisa* 24 (1970) 331-348.

[SoVo02] Z. Sobol and H. Vogt, On the L^p-theory of C_0-semigroups associated with second-order elliptic operators, I, *J. Func. Anal.* 193 (1) (2002) 24-54.

[Sto93] P. Stollmann, Closed ideals in Dirichlet spaces, *Potential Anal.* 2 (3) (1993) 263-268.

[StVo96] P. Stollmann and J. Voigt, Perturbation of Dirichlet forms by measures, *Potential Anal.* 5 (2) (1996) 109-138.

[Str83] R.S. Strichartz, Analysis of the Laplacian on the complete Riemannian manifold, *J. Funct. Anal.* 52 (1) (1983) 48-79.

[Str86] R.S. Strichartz, L^p contractive projections and the heat semigroups for differential forms, *J. Funct. Anal.* 65 (3) (1986) 348-357.

[Str88] D.W. Stroock, Diffusion semigroups corresponding to uniformly elliptic divergence form operators, in *Séminaire de Probabilités XXII*, Lecture Notes in Math. 1321, Springer, 1988. 316-347.

[StVa79] D.W. Stroock and S.R.S. Varadhan, *Multidimentional Diffusion Processes*, Grund. Math. Wiss. 233, Springer 1979.

[Stu93] K.T. Sturm, On the L^p spectrum of uniformly elliptic operators on Riemannian manifolds, *J. Funct. Anal.* 118 (2) (1993) 442-453.

[TFR00] K. Taira, A. Favini and S. Romanelli, Feller semigroups generated by degenerate elliptic operators, *Semigroup Forum* 60 (2) (2000) 296-309.

[Tan79] H. Tanabe, *Equations of Evolution*, Pitman, 1979.

[Tha89] S. Thangavalu, Summability of Hermite expansions. I, *Trans. Amer. Math. Soc.* 314 (1) (1989) 119-142.

[Tho98] S. Thomaschewski, *Positivität und Majorisierung von Variationell Definierten Halbgruppen*, Diplomarbeit, Universität Ulm 1998.

[Tri78] H. Triebel, *Interpolation Theory, Function Spaces, Differential Operators*. North-Holland Math. Library, North-Holland, 1978.

[Tri83] H. Triebel, *Theory of Function Spaces*. Monographs in Mathematics 84, Birkhäuser 1983.

[Var85] N.Th. Varopoulos, Hardy-Littlewood theory for semigroups, *J. Funct. Anal.* 63 (2) (1985) 240-260.

[Var88] N.Th. Varopoulos, Analysis on Lie groups, *J. Funct. Anal.* 76 (2) (1988) 346-410.

[VSC92] N.Th. Varopoulos, L. Saloff-Coste, and Th. Coulhon, *Analysis and Geometry on Groups*, Cambridge Tracts in Math. 100, Cambridge Univ. Press 1992.

[Voi92] J. Voigt, One-parameter semigroups acting simultaneously on different L^p-spaces, *Bull. Soc. Roy. Sci. Liège* 61 (6) (1992) 465-470.

[Wei79] F.B. Weissler, Two-point inequalities, the hermite semigroup, and the Gauss-Weierstrass semigroup, *J. Funct. Anal.* 32 (1) (1979) 102-121.

[Won81] B. Wong-Dzung, *L^p-Theory of Degenerate-Elliptic and Parabolic Operators of Second Order*, PhD Thesis, Univ. California 1981.

[Won83] B. Wong-Dzung, L^p-Theory of degenerate-elliptic and parabolic operators of second order, *Proc. Royal Soc. Edinburgh* 95 A (1983) 95-113.

[Yosh71] A. Yoshikawa, Fractional powers of operators, interpolation theory and imbedding theorems, *J. Fac. Sci. Univ. Tokyo Sect. IA Math.* 18 (1971) 335–362.

[Yos65] K. Yosida, *Functional Analysis*, Grund. Math. Wiss. 123, Springer-Verlag 1965.

[Zha95] J. Zhang, The $L^p - L^q$ estimates for the wave equation with a nonnegative potential, *Comm. P. D. E.* 20 (1-2) (1995) 315-334.

Index